D1218809

ADVANCES IN THIN LAYER CHROMATOGRAPHY

ADVANCES IN THIN LAYER CHROMATOGRAPHY

CLINICAL AND ENVIRONMENTAL APPLICATIONS

Edited by

JOSEPH C. TOUCHSTONE
Department of Obstetrics and Gynecology
School of Medicine
University of Pennsylvania

A Wiley-Interscience Publication

JOHN WILEY & SONS
New York • Chichester • Brisbane • Toronto • Singapore

Library of Congress Cataloging in Publication Data

Main entry under title:

Advances in thin layer chromatography.

"Proceedings of the Second Biennial Symposium on Thin Layer Chromatography held in Philadelphia, Penna., December 1980"—Pref.
"A Wiley-Interscience publication."
Includes index.
1. Thin layer chromatography—Congresses.
2. Biological chemistry—Technique—Congresses.
3. Chemistry, Clinical—Technique—Congresses.
I. Touchstone, Joseph C. II. Symposium on Thin Layer Chromatography (2nd : 1980 : Philadelphia, Pa.)

QP519.9.T55A38 543′.08956 81-23146
ISBN 0-471-09936-8 AACR2

Printed in the United States of America

10 9 8 7 6 5 4 3 2 1

Contributors

Mariann Anglin
Wilmington Medical Center
Wilmington, Delaware

Richard Apotheker
Kontes
Vineland, New Jersey

Bart W. Bartelsman
New England Nuclear
Boston, Massachusetts

Thomas E. Beesley
Whatman
Clifton, New Jersey

William J. Begue
Eli Lilly
Indianapolis, Indiana

Joel Bitman
U. S. Department of Agriculture
Beltsville, Maryland

W. Emmett Braselton
Michigan State University
East Lansing, Michigan

John J. Coupal
Veteran's Hospital
Lexington, Kentucky

Frank H. DeLand
Veteran's Hospital
Lexington, Kentucky

Tibor Devenyi
Hungarian Academy
Budapest, Hungary

Deborah P. DiZio
ICI Americas
Wilmington, Delaware

Eugenia Dolar
Wilmington Medical Center
Wilmington, Delaware

Marlaine Domoto
Iowa State University
Ames, Iowa

Alvin L. Donaho
Eli Lilly
Indianapolis, Indiana

Robert A. Elverson
Crozer-Chester Medical Center
Chester, Pennsylvania

Robert M. Eppley
U. S. Food and Drug Administration
Washington, D. C.

Herman R. Felton
Analtech, Inc.
Newark, Delaware

Heinz Filthuth
Berthold
Wildbad, West Germany

Dennis L. Fost
New England Nuclear
Boston, Massachusetts

Melvin E. Getz
U. S. Department of Agriculture
Beltsville, Maryland

Martin Gurkin
MCB
Gibbstown, New Jersey

Paul R. Handy
Eli Lilly
Indianapolis, Indiana

Gerald J. Hansen
University of Pennsylvania
Philadelphia, Pennsylvania

Roger C. Hatch
University of Georgia
Athens, Georgia

Dennis P. H. Hsieh
University of California
Davis, California

Walter Hyde
Iowa State University
Ames, Iowa

Haleem J. Issaq
Frederick Cancer Research Center
Frederick, Maryland

Anthony F. Heald
ICI Americas
Wilmington, Delaware

Anant V. Jain
University of Georgia
Athens, Georgia

John D. Johnson
Helena Laboratories
Beaumont, Texas

Euishin E. Kim
Veteran's Hospital
Lexington, Kentucky

Kwan Y. Lee
Proctor and Gamble
Cincinatti, Ohio

Sidney S. Levin
University of Pennsylvania
Philadelphia, Pennsylvania

Robert E. Levitt
University of Pennsylvania
Philadelphia, Pennsylvania

Marvin L. Lewbart
Crozer-Chester Medical Center
Chester, Pennsylvania

James J. Leyden
University of Pennsylvania
Philadelphia, Pennsylvania

Joseph O. Malbica
ICI Americas
Wilmington, Delaware

David A. Maltby
Michigan State University
East Lansing, Michigan

Kenneth J. McGinley
University of Pennsylvania
Philadelphia, Pennsylvania

Rhonda Moore
Iowa State University
Ames, Iowa

Shari L. Ohringer
New England Nuclear
Boston, Massachusetts

Richard Pfeiffer
Iowa State University
Ames, Iowa

Edwin M. Richardson
Wilmington Medical Center
Wilmington, Delaware

Dexter Rogers
Consultant
Mays Landing, New Jersey

Thomas B. Roos
Dartmouth College
Hanover, New Hampshire

Michael R. Ruggieri
University of Pennsylvania
Philadelphia, Pennsylvania

Linda M. St. Onge
E. I. Du Pont
Wilmington, Delaware

Fred D. Sancillio
Applied Analytical Ind.
Wilmington, North Carolina

Robert J. Schock
Michigan State University
East Lansing, Michigan

Peter M. Scott
Health and Welfare
Ottawa, Canada

Constance S. Seckel
Children's Hospital
Ohio State University
Columbus, Ohio

Joseph Sherma
Lafayette College
Easton, Pennsylvania

Seth Shulman
Bioscan
Washington, D. C.

Howard R. Sloan
Children's Hospital
Ohio State University
Columbus, Ohio

Roger D. Soloway
University of Pennsylvania
Philadelphia, Pennsylvania

Karen E. Soroka
U. S. Department of Agriculture
Beltsville, Maryland

Michael E. Stack
U. S. Food and Drug
Administration
Washington, D. C.

H. Michael Stahr
Iowa State University
Ames, Iowa

Gary E. Stolzenberg
U. S. Department of Agriculture
Fargo, Nevada

Michael H. Thomas
U. S. Department of Agriculture
Beltsville, Maryland

Joel J. Thrasher
U. S. Food and Drug Administration
Washington, D. C.

Steven R. Tonsager
Michigan State University
East Lansing, Michigan

Joseph C. Touchstone
University of Pennsylvania
Philadelphia, Pennsylvania

Ramsay Ventalchalam
University of Illinois
Urbana, Illinois

Ernest H. Wake
California American Water Company
National City, California

Nick C. Wan
University of California
Davis, California

Roger A. Ward
California American Water Company
National City, California

Tom B. Watkins
University of Delaware
Newark, Delaware

Harris H. Wisneski
U. S. Food and Drug Administration
Washington, D. C.

David L. Wood
U. S. Department of Agriculture
Beltsville, Maryland

Carolyn M. Zelop
Lafayette College
Easton, Pennsylvania

Albert Zlatkis
University of Houston
Houston, Texas

Preface

This volume represents proceedings of the Second Biennial Symposium on Thin Layer Chromatography held in Philadelphia, Pennsylvania in December 1980, which has become an international event. The success of the symposium was evident in the response of both the attendees and the exhibitors of the equipment. The quality and content of the program were greatly enhanced by reports of the newer advances, which included high-performance and reverse-phase thin layer chromatography. The general acceptance of the techniques was further shown by the excellent papers in both the clinical and, particularly, environmental areas.

The advances in the field include automation of the technique. With further developments in this area, it is conceivable that the procedure will become more widely recognized since, of the three chromatographies, thin layer chromatography (TLC) may, in fact, be more versatile because more samples can be processed at the same time. Furthermore, there is more leverage in the number of detection capabilities inherent in TLC. TLC, in fact, can be a very sensitive quantitative method when the fluorescent derivatives are considered. The time factor using high-performance TLC can be diminished to as little as 5 minutes. All these factors together tend to make the methodology one that should be seriously considered.

The manufacturers have continually upgraded the materials available, and high-performance as well as reverse-phase TLC are now becoming accepted techniques.

The aim of this book is also to present the latest results in both the clinical and environmental areas as presented at the symposium. The chapters were edited in order to have conformity in nomenclature and format. No attempt was made to change the style or conclusions of the authors as is done in the refereeing of professional journals. We feel that this volume will serve as a platform for those who become frustrated in editorial processes.

In conclusion, we are greatly appreciative of the time given by the participants in the symposium, both in their presentations as well as the preparation of the manuscripts. This cooperation has made it possible to publish the proceedings in the shortest possible time. The efforts of our secretaries, typists, and all those who gave of their time made the undertaking a pleasant success and not a drudgery. We are also indebted to the exhibitors who contributed to the success of the meeting.

The editors are indebted to their home bases for continued support without which this project would not have succeeded. Drs. Herman Felton, Haleem Issaq, and H. Michael Stahr assisted in the editorial process.

Joseph C. Touchstone

Philadelphia, Pennsylvania

Contents

ADVANCES IN THIN LAYER CHROMATOGRAPHY

CHAPTER 1

Chemically Bonded Reverse-Phase TLC

Thomas E. Beesley

INTRODUCTION

During the past two years, thin layer chromatography (TLC) has shown significant momentum in three major areas. The first of these was the immediate response to the availability of the "high-performance" line of silica gel plates prepared with silica in the 2 to 7 µm range. Results from these high-resolution separations brought about the second major trend that was renewed interest for *in situ* quantitation by densitometry. Following closely behind these developments, and of major interest in this paper, was the availability of chemically bonded reverse-phase materials for TLC.

The history of reverse-phase separations on TLC began in earnest in the early 1960s. Separations then were accomplished on silica layers impregnated with a variety of heavy oils, nonpolar solvents, or silicones. Some excellent results were obtained, but the technique suffered from both the elutability of the impregnating material at moderate to high solvent polarities, poor retention of moderately polar analytes, incomplete coverage, poor reproducibility, and long development times.

Chemical bonding onto the surface of silica allows one not only the ability to vary chain length to alter selectivity, but to enhance retention by chain spacing. Understanding of these principles allows for the creation of an optimized structure for the greatest possible resolution capabilities. Much of this work has been investigated by high-performance liquid chromatography (HPLC), and HPLC has been the benefactor in that reverse-phase TLC can now be used to screen solvent systems rapidly and at low cost. The visual impact of TLC quickly identifies the suitability of a mobile phase for HPLC.

The complex surface of silica (Figure 1), if not viewed correctly, can be reacted on in a variety of ways to give a host

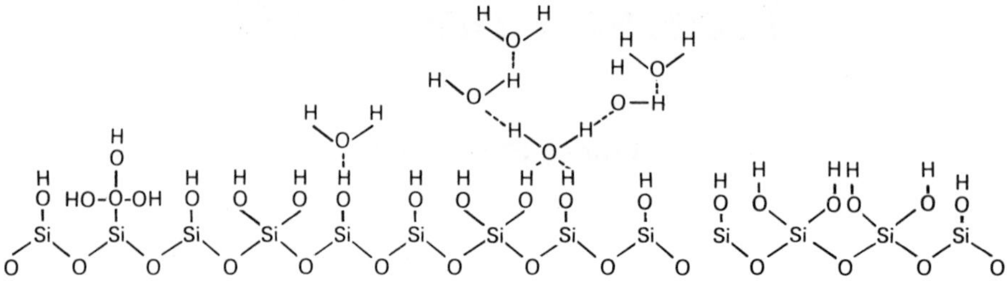

Figure 1. Some possible structures of the surface of silica gel (R. P. W. Scott, Contemporary Liquid Chromatography, Techniques of Chemistry, Wiley, New York, 11, 209 (1976).

of products, all with different selectivities. The results are obvious -- carbon chain length is not the only criterion for retention. Two very important features in preparing reverse-phase TLC have been found. First, it is not desirable to load as much C_{18} as is chemically possible onto the surface of the silica. Second, the remaining chromatographically available surface hydroxyls must be blocked or capped. These are not diametrically opposed positions.

The first point has been demonstrated by Scott and Simpson (1), who showed that C_{18} loads in excess of 12% result in chain-chain interactions that cause them to collapse on one another in mobile phases where the water composition is in excess of 40% (Figure 2).

It is then desirable when one wishes to have the broadest possible application for a product to use a lower carbon chain. As stated by Brinkman and DeVries (2), with mobile phases containing more than 30% water, RP-coated HPTLC plates (18 to 20%) cannot be used conveniently because of nonwettability of the stationary phase. KC_{18} plates from Whatman can be used irrespective of the proportion of water, provided 3% NaCl is added to the mobile-phase solvent mixture on the preferred 3% ammonium acetat

Residual hydroxyl groups remaining on the surface after only a 12% carbon load must be silanated with smaller carbons since they will be of the more reactive variety and will exhibit strong

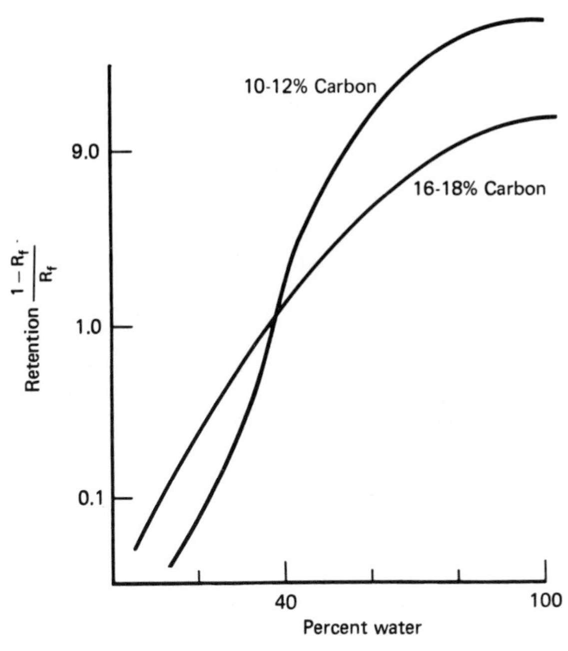

Figure 2. Retention versus carbon load.

retention of polar materials leading to broad tailing bands (Figure 3, plate A).

Once such a material is prepared, it has a disadvantage as far as TLC is concerned in that it will be quite hydrophobic, and therefore little capillary action will take place in aqueous solvents. This can be quantitatively overcome with the proper amphoteric binder that will work with mobile phases of up to 100% water. Brinkman and DeVries (2), demonstrated separation of aromatic oxy compounds, amino phenols, and aromatic acids in mobile phases containing 70 to 80% aqueous sodium chloride on Whatman KC_{18} layers. The presence of 0.05M (3.0%) sodium chloride is used primarily as a "mass ion" effect to salt out the binder. There is a slight increase in retention when salt is used, but in correlating to HPLC, where the salt is not required, the effect is hardly noticeable.

In designing the surface structure for reverse-phase TLC, a chromatographic model had to be constructed by which we could monitor not only the amount of carbon added but also the uniformity of its distribution. The polyaromatic hydrocarbon chrysene

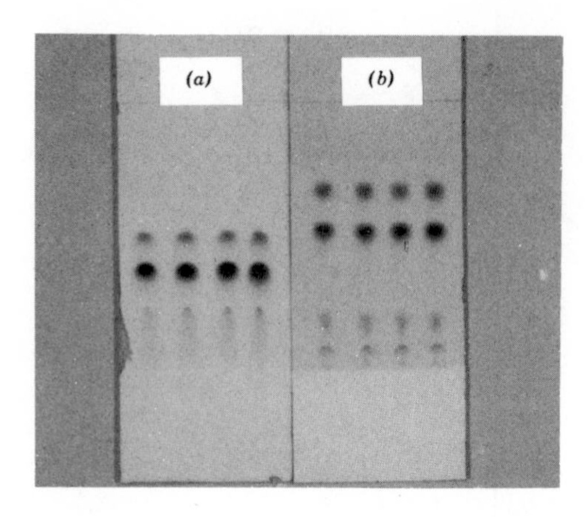

Figure 3. Effect of degree of silanation on band shape on RPTLC.

offered us such a model. A standard curve of percent carbon and R_f gave a straight-line relationship, and 12.5% carbon load was chosen as an ideal target, demonstrating the maximum resolution of polyaromatic hydrocarbons (PAHs) chosen (Figure 4). For more than 24 accepted experiments, the standard deviation was 3.38%.

SAMPLE APPLICATION

The results of applying this method to some commercially available reverse-phase plates can be seen in Table 1. Sample application presents some unique problems since water or high percentages of water tend to diffuse the applied spot. Pure organic solvents, while penetrating the layer rapidly, have the tendency to spread because of their eluting power on reversed phase. There are two possible answers to this dilemma. Use of preadsorbent RP-TLC handles all solutions the same, water or organic, and therefore is a method of choice. The second method is to use low spotting volumes of methanol, or preferably ethanol. As on conventional layers, the spotting volume on RP-TLC has an effect on the

TABLE I. SUMMARY OF COMMERCIALLY AVAILABLE REVERSE-PHASE PRODUCTS AND THEIR RESPONSE TO PAH SEPARATION[a]

	Chrysene	Anthracene	Biphenyl	Development Time (min)
E Merck RP-8	0.36	0.45	--	45
E Merck RP-18	0.14	0.28	0.39	33
Analtech RPS	0.42	0.52	0.73	86
Fluka Dodecyl	0.39	0.39	0.42	90
Whatman KC_{18}	0.23	0.33	0.40	34

[a]Solvent system: methanol/water; 90/10. Detection: 254 nm UV quench.

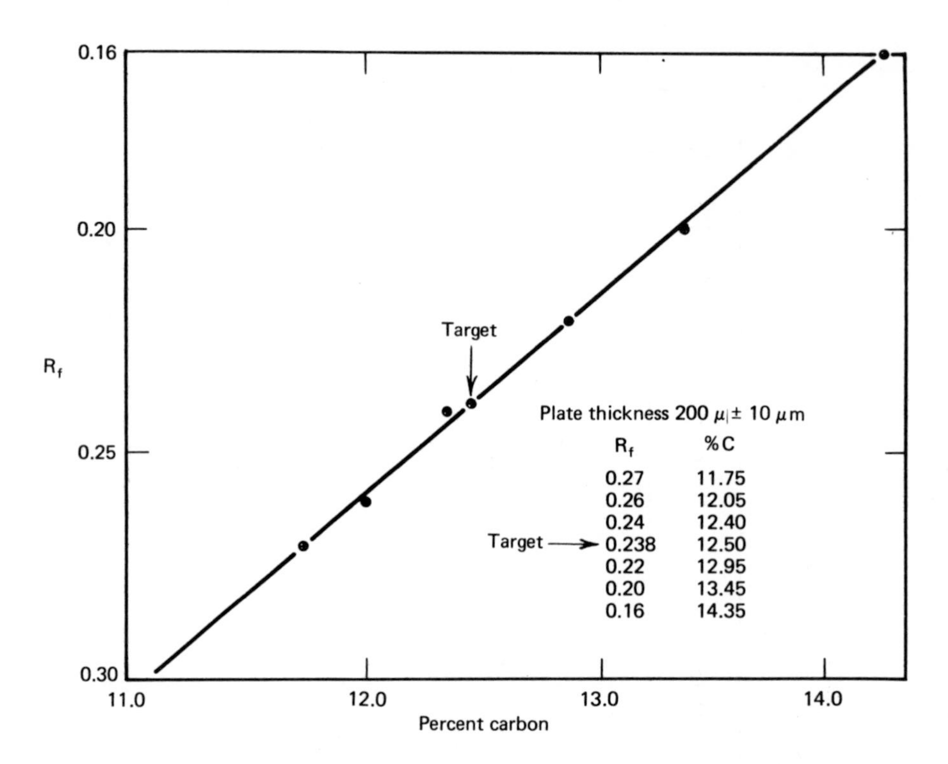

Figure 4. Precent carbon by chrysene method.

ideal here (Figure 5).

As with all TLC, sample load has a profound effect. High efficiencies (> 1000) can be obtained if sample size is kept below 500 ng. At 500 ng to 2 μg, N drops to 500.

SOLVENT SYSTEMS

Solvent compositions for reverse-phase TLC are uniquely simple. The use of methanol/water or acetonitrile/water accounts for a large portion of the work done on these layers. The water may be buffered to further enhance the separation (Table II).

Of a more profound nature is the ability to effect separations by solvent selectivity or to use the hydrocarbon surface to adsorb materials that will interact with the analyte in a unique

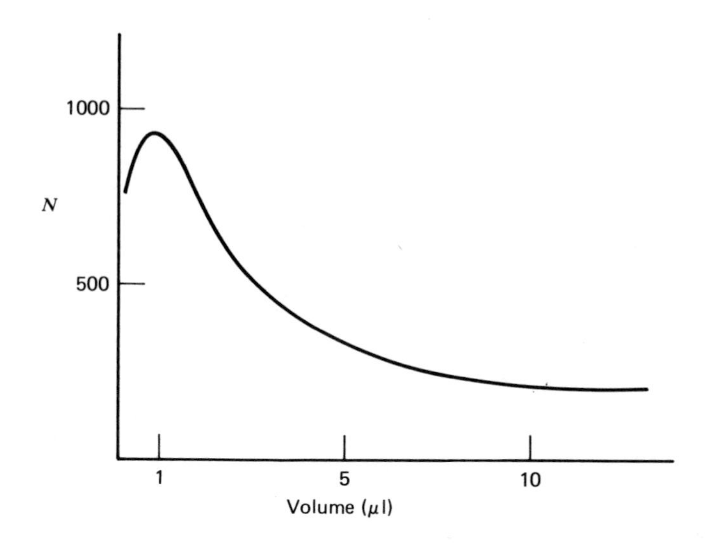

Figure 5. Efficiency/volume on $KC_{18}F$.

way. In the first case, a specific group of solvents that includes THF, $MeCl_2$, DMSO, and DMF, when added in low concentrations of up to 5% to the normal alcohol/water systems, cause dramatic changes in selectivity. The mechanism appears to be the association of these electron donors to one of the analytes preferentially. A knowledge of the chemistry of the interested species may help in the choice. The results to date have been somewhat remarkable. The second method of adsorbed ions is not widely used as yet, but the ability of the C_{18} chain to adsorb strongly molecules containing a minimum of four carbons has been demonstrated by Scott and Simpson (2) (Figure 6). This means that butyl or propyl borates may be useful additions for the selective separation of carbohydrates in aqueous buffer systems or hexamethyl sulfonates to prepare a strong cation exchanger for the aqueous-buffered separation of amines. Whole new avenues are open here.

RP-TLC TO HPLC CORRELATIONS

The selectivity of the KC_{18} plate correlates well to Whatman's Partisil 10 ODS-3, Waters's μ-Bondapak C_{18} HPLC columns, or any other ODS column that is fully capped with a 10 to 12% carbon

TABLE II. SURVEY OF REVERSE-PHASE HPTLC USING CHEMICALLY BONDED STATIONARY PHASES

Stationary Phase	Mobile Phase	Application
C_{-1} to C_{-18} (homemade)	Acetonitrile-0.01M KH_2PO_4 (40:60) Acetonitrile-0.01M $(NH_4)_2CO_3$ (30:70)	Hydroxybenzoates, substituted anilines and phenols
RP-8, -18(M)	Acetonitrile Methanol-water (97:3) Methanol-acetonitrile (90:10)	Headache pill, air-particulate extracts
RP-8 (homemade)	Acetonitrile, methanol	Alkylpyridines, polynuclear aromatic hydrocarbons
C_{18}(W)	Methanol-water (70:30)	Zeranol and its diastereoisomer
C_{18}(Q)	Methanol-acetone-water (20:4:3 and 20:20:10)	Chloroplast pigments
KC_{18}(W)	Methanol-0.5M NaCl(50:50)	Constituents of APC tablets
RP-8(M)	Methanol-water (var.) Acetonitrile-water (var.)	Phthalate esters, chloroanilines, polynuclear hydrocarbons

RP-18(M)	Methanol-water (85:15)	Polynuclear aromatic hydrocarbons
RP-2, -8, and -18(M), KC_{18}(W)	Methanol-water (var.)	Barbiturates, alkylbenzenes
RP-2, -8, and -18(M)	Ion-pair chromatography with tetraalkylammonium bromide	Food dyes
RP-2, -8, and -18(M)	Various mixtures of methanol, ethanol and other organic solvents with aqueous solutions	Alkylphenones, steroids, organic acids, fluoranthene
KC_{18}(W)	Methanol-water (80:20)	Aromatic hydrocarbons

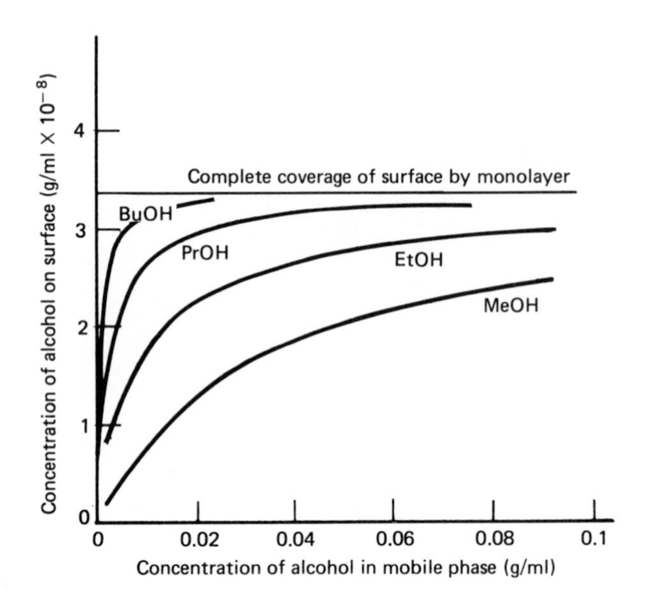

Figure 6. Adsorption isotherms for the C_1 to C_4 alcohols on ODS2 reverse phase.

load. The separation of some PAH on KC_{18} and the Whatman Partisil 5 ODS-3 column is identical in selectivity and very near in retention. This work was confirmed for seven aromatic hydrocarbons in an independent lab (2).

The operating range on RP-TLC for this correlation was very narrow. Using the formula $k' = 1 - R_f/R_f$, the useful R_f range is 0.1 to 0.5, which will give a column k' of 1 to 9.

DETECTION METHODS

Systems for detection create unique problems for reverse-phase TLC. The character of this media requires modification of some of our best reagents. It should be a persistent goal to work in the UV mode since lower detection levels require lower sample loads and consequently higher resolution possibilities. It is also much more stable for densitometric measurements and therefore leads to better quantitative procedures. The following is a list of reagents successfully used in our laboratories.

1. UV quench (254 nm excitation) sensitivity 500 ng.
2. Alcoholic sulfuric acid:
   a. 25% H_2SO_4 in methanol--after spraying plate, heat at 110°C for 1 minute. View under long-wave UV (366 nm). Detects sterols, sterol esters (negative when no ring unsaturation exists).
   b. 8 to 10% methanolic sulfuric acid--heat 2 minutes at 170°C after spraying plate. View under long-wave UV (366 nm). Detects bile acids, estrogens, sterols, esters. Sensitivity for bile acids: 10 to 20 ng.
3. Fluorescamine--0.01% in EDC--spray or dip plate, air dry using warm-air blower. View under long-wave UV (366 nm). Detects primary amines (amino acids, sulfonamides, amphetamines).
4. Marquis reagents--1% formaldehyde solution in concentrated H_2SO_4. Spray plate while hot. Induces fluorescence in numerous aromatic compounds. Plate does not wet uniformly, but procedure is usable for qualitative work. Sensitivity can be as low as 10 ng.
5. Iodoplatinate--regular aqueous reagent diluted with ethanol (2:1). Prepare fresh. Spray plate, allow spots to develop on standing. Sensitivity for morphine: 1 μg.
6. Diphenylcarbazone-mercuric sulfate (for barbiturates). (1) Spray heavily with diphenylcarbazone. (2) Spray with mercuric sulfate diluted 1:1 with methanol. Blue and purple bands are visible. Background not uniform, color of bands not stable, therefore not suitable for densitometry. Sensitivity: 1 to 3 μg.
7. Phosphomolybdic acid (5% in MeOH/IPOH, 1:2). Dip plate, heat to char at 110°C until spots develop, usually 10 minutes or longer. (Background discolors on prolonged heating.) Detects sterols, sterol esters. Sensitivity: 1 to 5 μg. Prewashing plate recommended when charring.
8. Dragendorff reagent (used as is)--spray plate. Detects alkaloids, nitrogen-containing compounds, polyglycols.
9. Bratton-Marshall reagent:
   a. 1% $NaNO_2$ in 4% aqueous HCl (v/v).
   b. 0.2% N-naphthylethylenediamine di HCl.

   Spray plate with a. till damp. Dry 1 minute in 110°C oven; cool. Spray lightly and uniformly with b. Detects compounds that can be diazotized--sulfonamides. Sensitivity: 25 ng (for sulfonamides).

10. Iodine vapors--place plate in chamber containing iodine crystals for 10 minutes. Brown bands on tan background. Detects lipids, unsaturates.
11. α-Cyclodextrin--1% in ethanol. Sprayed plate is air-dried, then placed in iodine chamber. Detects straight-chain lipids. Sensitivity: 5 to 10 μg.

SUMMARY

The reverse-phase TLC plate KC_{18} shows wide applicability to the separation of various classes of both nonpolar and polar compounds opening new opportunities for TLC. It is extremely simple to use and is very reproducible. Care must be exercised in spotting reverse-phase plates, taking into account the spotting volume and concentration. Up to 100% water can be used, providing 3% NaCl or NH_4A is added. KC_{18} plates provide the highest speed of analysis over the broadest range of mobile-phase compositions. Unique separations can be obtained employing selected solvent additions or adsorbed species. Detection systems must be largely lipophilic in nature.

REFERENCES

1. R. P. W. Scott and C. F. Simpson, unpublished.
2. U. A. Th. Brinkman and G. DeVries, J. of Chromatogr., 192, 331 (1980).

CHAPTER 2

Preparative TLC

Herman R. Felton

INTRODUCTION

Preparative thin layer chromatography (Prep-TLC) may be simply defined as the technique involved in separating and isolating relatively larger quantities of material than those normally used for analytical TLC. Usually the material to be isolated is for the purpose of (1) further analytical studies for identification, (2) use as purified material for biological activity studies or chemical purposes, or (3) obtaining pure standards for use as comparator materials for unknown mixtures.

Most Prep-TLC is performed on coatings thicker than the normal 100 to 250 μm used for analytical purposes. Layer thicknesses available are 500, 1000, 1500, and 2000 μm; the 1000 and 2000 representing some 90% of those most commonly employed. It should be observed that sometimes as much as 5 mg of sample may be separated on a 250-μm "analytical" plate. If that quantity is sufficient for the purpose at hand, an "analytical" plate can be used for Prep-TLC without suffering the slight degradation of performance ascribed to the thicker layer (but possibly suffering some "overload" effects). An analytical plate can be loaded to the point of "overload" (or even somewhat beyond) until spot distortion or tailing interferes with the desired separation. Analytical plates with the relatively inert "preadsorbent" layer can be loaded quite a bit higher than standard single-coated plates. In general, Prep-TLC can be used to obtain quantities of material from 10 to 500 mg. Of course, the maximum amount that can be handled satisfactorily depends on the specific separation. As a rule of thumb, the "capacity" of a Prep-TLC plate will increase as the square root of the thickness with little or no degradation of the separation, a 1000-μm layer having twice the load capacity of a 250-μm layer. If some deterioration of the separation can be tolerated, the load can be considerably increased beyond this point.

Column chromatography (CC), another technique, is often used for preparative-scale separations. With the newer small size and narrow cut adsorbent particles available for CC, excellent separations can be obtained. For a number of reasons, Prep-TLC is a preferred choice over CC. The conditions for Prep-TLC may be easily and quickly established on an analytical plate prior to running Prep-TLC. Separated zones, once located, may be easily removed from a Prep-TLC plate, whereas such removal is often difficult from a column. Reference compounds may be run simultaneously by TLC under identical conditions to assist in locating the desired material. Consistent quality Prep-TLC plates are commercially available in a wide variety of coating materials and are very moderate in cost.

Prep-TLC Plates can be obtained in the same 5 × 20, 10 × 20, 20 × 20, and 20 × 40 cm sizes as analytical plates. Since the load capacity is the single most important factor in Prep-TLC, it is not surprising that the 20 × 20 and 20 × 40 cm sizes are most commonly employed.

SAMPLE APPLICATION

Two factors related to sampling should be considered in achieving the high load desired for Prep-TLC. In analytical TLC normal sample sizes range from 2 to 10 µl. Larger volumes are applied to preparative plates. However, in normal analytical TLC the maximum concentration of components in the sample solvent should be of the order of 1 or 2%. A more concentrated sample solution may be used for Prep-TLC if sufficient information is at hand so that it is certain that the components are adsorbed on the coating and not deposited as crystals on the surface. This type of "crystal" overload can give badly distorted spots or bands since the component's kinetic rate of solution in the carrier liquid becomes limiting.

A simplified mechanistic picture of the TLC process may be in order here. The solvent used to dissolve the components of a sample for TLC should be fairly volatile and a "good" solvent for the materials. However, the solvent used as the carrier for a TLC separation should neither be a "good" solvent nor a "poor" solvent for the sample components. Too good a solvent will strip components easily from the adsorbent, and the components will move with the carrier front. Too poor a solvent will lose in the competition with the adsorbent's attraction for the sample, and nothing will move. Therefore, what is desired is a carrier of "mediocre" solvent strength. This mediocrity could imply that the carrier also is a "slow" dissolver of the sample. So as the carrier passes a compact deposited crystal, slow dissolving takes place. This causes a vertical broadening or smearing of the component distribution in the moving carrier. If the sample is deposited on the coating in such a way that it is either molecularly adsorbed or exists as agglomerations of only a few molecules, the solution kinetics is much faster. In addition, the ultimate saturation solubility of each sample component in this "mediocre" carrier also determines how much solvent must pass a given point in order to "pick up" the component.

The preferred method for putting a sample on a Prep-TLC plate is to apply it as a narrow streak across the plate. There are a number of devices, both manual and automatic, available commercially for applying such a streak. It is very desirable to have the streak as straight (horizontally) and as narrow (vertically) as is possible. Most available streakers can give a sample zone only 1 to 3 mm high. As generally operated, a streak is made across the plate starting at a point 2 to 3 cm from one edge and stopping about the same distance from the other edge. These clear areas are left partly because "edge effects" may make the solvent move either faster or slower at the edges than across the center 15 to 17 cm of the plate. After waiting for the sample solvent to completely evaporate from this applied streak, a second identical streak is laid down directly over the first deposition. This repeated streaking is done until the desired total volume of sample solution has been deposited. If the sample solvent is not removed between streaks, it will "wick" into the plate, resulting in a wide sample streak that is not uniform in concentration across its vertical dimension. For a similar reason, the streaking method used must not gouge or damage the adsorbent coating. Gouging interferes with the regular straight-line movement of carrier, making later component recovery difficult.

Some successful Prep-TLC work has been reported using the same manual techniques as in applying analytical spots. In these cases, a small (i. e., 2 to 4 μl) pipette is used to apply successive round spots immediately touching each other across a plate. Needless to say, this is time-consuming and tedious (100 μl takes 50 fillings of a 2-μl pipette) and requires great care. The irregularities that necessarily are a part of this method of sample application are magnified in the separation process and often can make isolation of developed material very difficult. Of course, if the separations are widely spaced, the hand "spot-streaking" method may be perfectly satisfactory.

With considerable practice, it is possible to streak a plate by hand satisfactorily using a syringe. The syringe is held nearly vertically and not quite touching the coating surface. Then, using the edge of a bench or some such guide to help steady the hands, the syringe is moved across the plate while the plunger is slowly pushed into the barrel.

Since "hand" methods require practice, skill, and care, it is obvious why a mechanical "streaking" device is the method of choice if many Prep-TLC plates are to be run.

Careful streaking need not be done when Prep-TLC plates equipped with the preadsorbent layer are used. In this case, the sample is applied to the relatively inert preadsorbent layer as a "crude" streak across the plate. Although this type of plate permits "sloppy" techniques, it still gives better results if the sample is applied carefully across the plate, since high local concentration can still give rise to some of the solution kinetic effects and overloading effects previously discussed.

DEVELOPMENT

Development of Prep-TLC plates is carried out in almost exactly the same manner as that used for analytical scale plates. Almost invariably the use of Prep-TLC is preceded by development on analytical plates that establishes solvent systems suitable for achieving the desired separation. With this knowledge of the system behavior it is reasonable to use the same solvent system. It is advisable to keep in mind the fact that Prep-TLC depletes the solvent in the developing chamber faster than analytical TLC. The carrier level should be periodically observed to be sure there is sufficient liquid present. If necessary, additional liquid may be added (carefully) to the reservoir during a Prep-TLC run. Some developing chamber systems are provided with the means (i. e., a hole in the cover) to accomplish this. Care should be taken not to create waves or severe disturbances during the solvent addition since the liquid front ascending the plate may be distorted.

It is also commonly known that differential adsorption occurs on TLC plates when mixed solvents are used. The solvent composition changes with respect to a given point on the plate as solvent moves past that point. Concomitantly, the solvent composition in the reservoir that supplies the plate is also changing. Therefore, it is good practice to start with a relatively large volume of solvent in the developing chamber when operating Prep-TLC with mixed solvents. As a guide at least 100 ml of solvent would be needed to run a 2000-μm, 20 × 20 cm plate the full length of solvent travel.

LOCATING THE DESIRED COMPONENT (OR COMPONENTS)

There are a number of methods available to locate components on Prep-TLC plates. Obviously, if the desired material or materials are colored, they can be located simply by "eye." If the compound desired fluoresces, its position on the plate can be determined by the use of an appropriate UV light source. Conversely, a Prep-TLC plate containing a fluorescent material will indicate the separated components as dark streaks or spots in the fluorescent background when examined under UV illumination. Most of the currently used fluorescent agents are nearly insoluble in the solvents later used for component elution and, therefore, will give little or no contamination of the isolated material.

The exposure of a developed TLC plate to iodine vapor will show the separated components from a large variety of chemical classes as either dark brownish zones or lighter zones on a tan background. The iodine vapor chamber technique, if applicable, is preferable to most of the available spray reagents since, in most cases, the iodine can be evaporated leaving the separated components unchanged. The zones containing the separated components must be marked before evaporating the iodine since the materials will again become invisible. This is generally done by outlining the zones with a scribed mark made with a sharp pointed object.

Of course, the most difficult instance is where none of the above techniques reveal the presence and/or location of the desired components. When performing "analytical TLC" the operator has a wide variety of "spray reagents" (approximately 100) available to "react" with the component in order to "make it visible". Generally, these reagents either produce a color with the compound sought or color the plate background leaving the material contrasting. However, many of these reagents react irreversibly with the material and, therefore, change its chemical nature. Obviously, using a chemical reaction is not desirable for Prep-TLC since the object of the technique is to obtain a "pure" fraction of the unchanged material. The following simple methods can be used to determine the location of the components desired. When the sample is "streaked" on the Prep-TLC plate, additional small samples are "spotted" in the unstreaked 2 to 3 cm margins on each side of the sample streak at the same vertical level. After the plate has been removed from the developing chamber and the carrier solvent evaporated, the streaked portion of the developed plate is masked (covered with an appropriately sized piece of glass or even cardboard) and the two outer edges carefully sprayed with an appropriate visualizing reagent. The desired component will then be visually located on each edge of the plate. Horizontal lines are scribed across the plate to outline the zone using the spray-developed spots as a guide. Prior to spraying the edges of the "masked" plate it is suggested that channels be scribed vertically on the plate between the streak and the spots to prevent the sprayed reagent from "wicking" into the masked area.

Another way of achieving the same end result would be to scribe two vertical channels after development through the original sample streak leaving the outer edge portions of the streaked sample outside the mask. This would use a small portion of the potentially recoverable material, but does ensure positive location of the separated component.

It is very important to remember that if heating is required for the specific visualization reagent, the chemical nature of the material to be collected may be changed. Preparative-scale TLC plates are available that achieve positive identification without the possibility of damaging the adsorbent surface when "masking" or spraying the plate. These plates are prescored 1 in. in from each side. As in the previously described technique, a sample is either "spotted" or, more generally, the "streak" extended to enter each prescored edge zone. After the plate is developed, the 1-in. edges are "snapped" free of the center portion. The separated narrow plates may then be sprayed, charred, reacted, and/or treated in any appropriate manner to provide visualization of the desired components. The visualized edge strips are placed next to the developed center section and the zones marked for collection. With these plates there is no possibility of sprayed reagent "wicking" into the desired component. Additionally, if heat is necessary for the visualization procedure, it is only the separated strip plates that are heated, and no chemical effect could be produced on the untreated central portion.

REMOVAL OF MATERIAL FROM THE PLATE

The desired substance, once located, may be removed from the plate in a number of ways. Generally, the simplest way involves physical removal of the adsorbent zone, eluting the substance out of the adsorbent with a solvent, separating from the residual adsorbent, and concentrating the solution. The most common procedure is first to outline the desired area by scribing a line around it through the adsorbent. Then the outlined layer of adsorbent is scraped off clean to the glass with a scraper or razor blade. The adsorbent so removed is put on a sheet of weighing paper or similar hard surface material and transferred to a small glass vial or tube having a solvent resistant cap. If a centrifuge is available, the container can be a centrifuge tube. One or two drops of water can now be added. The water displaces the compound of interest from the very active sites on the adsorbent.

The choice of solvent for eluting the compound off the adsorbent is quite important. One should use as polar a solvent as possible that is also a good solvent for the component. Several factors are worth mentioning at this point. Water is not really a desirable solvent, since it is most difficult to get rid of later when concentrating the eluent. Methanol is often recommended in the literature but suffers from two real drawbacks. Silica gel is noticeably soluble in methanol, a fact that is not generally appreciated (which incidentally, contributes to deterioration of HPLC columns), and, coincidentally, silica gel is the most commonly used adsorbent for Prep-TLC. This solubility is ascribed to the formation of methyl esters of silicic acid. Water also has this solubilizing effect, although to a lesser extent than methanol. The second objection to methanol (and water) is that the impurities normally present in silica gel such as Fe, Na, and SO_4 ions are quite soluble and will contaminate the isolated material after removal of the solvent. Depending on subsequent use and treatment (such as recrystallization), these impurities may or may not be bothersome. Acetone is quite good as a solvent for compound elution, and ethanol can also be recommended. If the solubility of the desired material allows the use of chloroform, it is strongly recommended that this solvent be used. There is virtually no solubility of the normal silica impurities in chloroform, and it is relatively easy to vaporize from the eluted solution obtained.

In the event that water is used to remove the desired compound from silica gel, a simple procedure may be tried. Shake the extracted solution (after extraction and filtering) with ethyl acetate. Both methyl silicate and silicic acid are virtually insoluble in ethyl acetate. Therefore, a considerable volume can be used to effect a two-phase separation of the desired material into the ethyl acetate layer (leaving the impurities in the water layer).

After the selected solvent is added to the tube containing the scraped adsorbent, the tube is shaken, either mechanically or by hand for a few minutes. Usually 5 minutes of agitation will suffice. The tube may then be centrifuged and the supernatent liquid carefully removed by pipet or decantation. If desired, another load of solvent can be added to the adsorbent and the two solutions combined. Often, unless dealing with a complete unknown, sufficient information is known about the desired compound to indicate the amount of solvent necessary to effect recovery. Alternatively, simple filtration can be used to separate the solution from the adsorbent particles. The filter medium should be capable of retaining 2-μm particles even though most silica gels for TLC are in the range of 5 to 20 μm in diameter.

There are alternatives to the hand-scraping method. Great care should be taken during hand scraping to avoid the problems of spillage and blowing. Obviously, the work should be done in a draft-free area to prevent blowing and loss of fine particles. The care needed can be tedious, as in the scraping of two adjacent zones, both of which are desired. There are collectors using vacuum suction that are designed for removing adsorbents. These collect the powder in filter thimbles that can be used once and discarded. These devices are fitted with a plastic tube suction end that can be guided over the adsorbent quite easily. With a little practice, the operator can become quite adept in the use of these collector systems. In some cases, a thin film of adsorbent coating may be left on the glass after passage of the suction collector. In this event, it is simple enough to loosen this small amount of material with a hand scraper and pick it up with another "pass" of the collector.

The thimbles may be directly extracted by using a Soxhlet device, or the powder collected can be carefully transferred to a vial for solvent extraction as previously described.

Once a solution of the material is obtained, the component is most commonly isolated by evaporating to dryness. Again care should be taken to ensure that heat does not decompose the collected sample or otherwise change its chemical nature. In the event that little is known about the substance, it may be good practice to evaporate off the solvent at low temperature in an inert gas system, such as nitrogen.

SUMMARY

Preparative thin layer chromatography is a simple and inexpensive technique for the isolation of desired compounds. This paper discusses some of the details involved in the manipulation of the method.

CHAPTER 3

Analysis of Polycyclic Aromatic Hydrocarbons Using Reverse-Phase TLC

Martin Gurkin

INTRODUCTION

Regulations in Europe require that drinking and process water for food-processing plants be analyzed for the following polycyclic aromatic hydrocarbons (PAHs) listed in Table I. Until recently, this was accomplished by a very time-consuming process utilizing two-dimensional thin layer chromatography (TLC) with aluminum oxide-cellulose acetate.

Recently, a change was proposed based on selected West German standards (DIN 38409--part 13). Water samples contaminated with less than 20% of the limit value stated in the regulation (total content 30 to 50 ng/l PAH) can now be designated as "not contaminated" by a preliminary test utilizing one-dimensional reverse-phase high-performance TLC (RP-HPTLC).

Water samples with higher concentrations (>20% of 30 to 50 ng/l) must be quantitatively assayed via the longer two-dimensional TLC procedure combined with subsequent photometric evaluation. The time and material required for the two-dimensional TLC method is 2 hr per sample/plate; the time required for RP-HPTLC is 25 minutes for 8 samples/plate.

Extraction

In a 250-ml volumetric flask add 1 ml of cyclohexane to the water sample, and shake intensively for about 5 minutes. Allow to stand for a short period of time (about 10 minutes). The ground glass stopper is exchanged for a microseparator (Figure 1). Distilled water is added through the left pipe junction of the microseparator. The cyclohexane phase rises in the other

TABLE I. POLYCYCLIC AROMATIC HYDROCARBONS WHICH MAY BE FOUND IN WATER

Chemical Designation	Chemical Formula	Molar Mass (g/mol)
Fluoranthene	$C_{16}H_{10}$	M = 202.16
Benzo[b]fluoranthene (3,4-benzofluoranthene)	$C_{20}H_{12}$	M = 252.32
Benzo[k]fluoranthene (11,12-benzofluoranthene)	$C_{20}H_{12}$	M = 252.32
Benzo[a]pyrene (3,4-benzopyrene)	$C_{20}H_{12}$	M = 252.32
Benzo[ghi]perylene (1,12-benzoperylene)	$C_{22}H_{12}$	M = 276.34
Indeno[1,2,3-cd]pyrene (2,3-o-phenylene pyrene)	$C_{22}H_{12}$	M = 276.34

ascending tube and can subsequently be removed simply with the use of a pipette.

Thin Layer Chromatography

Take 0.5 ml of the cyclohexane phase and pipette into a 10-ml test tube. The sample is concentrated to one small drop (about 10 to 15 μl) by evaporation with N_2, and the residue is applied (ϕ maximum: 5 mm) on the concentrating zone of the thin layer plate with the aid of a microliter syringe or capillary. Subsequently, the sample flask is rinsed once with about 20 μl of cyclohexane, and this is also applied to the plate. Samples, together with 2 standards (12 and 3 ng total of polycyclic aromatic hydrocarbons corresponds to 120 and 30 ng/l) can be analyzed on one thin layer plate.

The polycyclene standard solution is applied directly on the

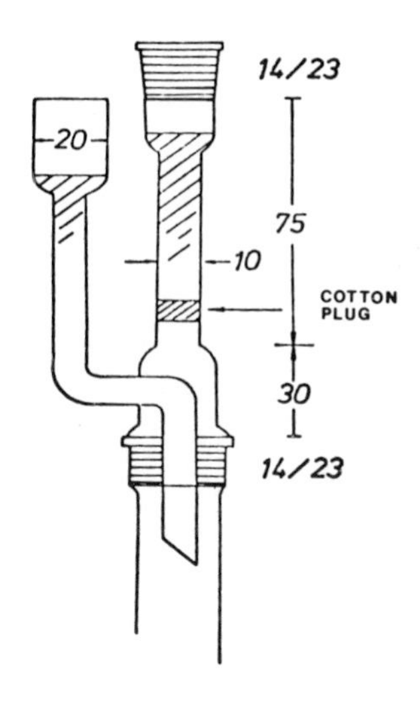

Figure 1. Microseparator.

concentrating zone of the thin layer plate. In order to obtain spot sizes that are useful for visual comparison standard, solutions must be diluted in a test tube with 0.5 ml of cyclohexane and concentrated by evaporation as described prior to application to a thin layer plate.

Development of the chromatogram is made in a chamber that contains acetonitrile, dichloromethane, and water (9:1:1) as the solvent. Migration time is 20 to 25 minutes.

Evaluation

After drying the plate under subdued light (5 to 10 minutes), evaluate the plate immediately under UV at 366 nm. The evaluation is made by visual comparison of the sample fluorescence with that of the standard solution.

Typically the materials separate as follows (Figure 2):

Origin.
Benzoperylene.
Indenopyrene.
Benzopyrene.
3,4- and 11,12-benzofluoranthene.
Fluoranthene.

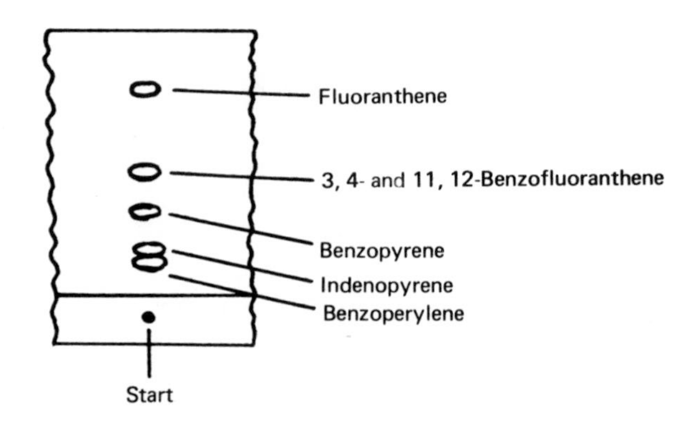

Figure 2. Separation under UV light.

If no fluorescent spot is visible or if it is not stronger than that of the standard solution (3 ng total of PAH), the total amount of polycyclic aromatic hydrocarbons is less than 50 ng/l.

If the intensity of the fluorescence of a sample reaches or surpasses that of the stock solution (3 ng = 30 ng/l), this sample must be regarded as positive and analyzed by other methods. This screening test can then be classified as follows:

Either the total amount of polycyclic aromatic hydrocarbons is less than 50 ng/l; or

The total amount of polycyclic aromatic hydrocarbons is more than 50 ng/l. Samples that indicate the presence of less than 50 ng/l PAH do not have to be tested further.

Goals

Distinguish between contaminated and noncontaminated drinking water/process water.

Detection limits 30 to 50 ng/1 total PAH.

EXPERIMENTAL

Apparatus

250 volumetric flask with ground glass stoppers.
1-ml volumetric pipette.
Microseparator.
Short test tube.
Nitrogen pressure bottle--with precision control.
RP-HPTLC--precoated RP-18 without fluorescent indicator.
10 × 20 cm--E. Merck Catalog No. 15037.
Development chamber 22 × 10 × 22 cm.
UV lamp, 366 nm.
Microliter syringe--10 µl, 50 µl.

Reagents

Cyclohexane (e.g., MCB OmniSolv).
Acetonitrile (e.g., MCB OmniSolv).
Dichloromethane (e.g., MCB OmniSolv).

Procedures

Extraction	Water 200 ml, cyclohexane 1 ml microseparator.
Concentrate	0.5 ml cyclohexane to one drop (10 to 15 µl) under N_2.
TLC	RP-18 HPTLC along with standards.
Develop	AcN/DCM/H_2O, 9:1:1.
Evaluate	

CONCLUSIONS

Reverse-phase high-performance thin layer chromatography (RP-HPTLC) with preconcentrating zone can be used to separate and semiquantitate lypophilic moities in drinking and process water containing 30 to 50 ng/1 polycyclic aromatic hydrocarbons.

CHAPTER 4

Automated HPTLC, A Cost Saver

Fred D. Sancillio

INTRODUCTION

In recent years a large percentage of scientific thought in pharmaceutical separations and nearly every industry involved in analytical chemistry has concentrated on high-performance liquid chromatography (HPLC). Although HPLC has been advantageous in many instances, other methods now available offer distinct advantages over HPLC in time, throughput, and overall cost.

One such technique, high performance thin layer chromatography (HPTLC), uses many of the same improvements that HPLC had made over conventional liquid-column chromatography; controlled particle size, flow rates, and spectroscopic measurement of the chromatogram. This paper describes the quantitative application of HPTLC to various analysis, reviews research in the automation of the method, and compares costs and benefits of this new technology with the older HPLC techniques.

ANALYSIS OF ERROR

The introduction of HPTLC to various analytical laboratories has only recently occurred. The renaissance of this thin layer technique has been brought about by the development of layers of superior resolution power and the use of sophisticated sample handling and analysis techniques. This new approach greatly reduces the risk of error, the cost, and the time usually associated with thin layer chromatography when large numbers of samples are to be analyzed, and is easily interfaced to modern laboratory automation systems for data handling.

Using computer technology one can produce accuracy and

reproducibility equal to any other chromatographic method. It is necessary, however, to realize the various components of error in HPTLC and introduce techniques that will minimize them. Error in all chromatographic methods can be attributed to several sources: sample size errors, detector instabilities, and chromatographic (separation) errors. Therefore, a simplistic equation may be written to describe a relationship between them. This equation assumes the form:

Total error = Volume error + Detection error + Chromatographic error

Several experiments were designed to determine the contribution of each to the total observed error in HPTLC analysis. To calculate "total error," a series of seven pharmaceutical drug standards was used. Each substance was chromatographed 10 times and a standard deviation calculated. Data obtained are listed in Table I. The average standard deviation for all seven substances was 1.6%; therefore, total error = 1.6.

Using a similar scheme and an internal standard it is found that the standard deviation is reduced by 0.81%; and so, volume error = 0.81.

Finally, to determine the error associated with the detector, several components were carefully scanned using a newly developed computer-controlled scanner. Each zone was centered and scanned 10 times and the standard deviation calculated. The results indicate a detection error of 0.17%, or detection error = 0.17.

One can now calculate the chromatographic error as follows:

Chromatographic error = Total error - Volume error - Detection error

Chromatographic error = 0.62%

Therefore, it can be reasonably assumed that by employing an internal standard and a carefully configured computer scanning method, routine analysis by HPTLC will yield standard deviations of around 0.62%.

INSTRUMENTATION

This simple analysis of error in HPTLC assumes the possibility of reproducing the analysis procedure. This is facilitated by using instrumental methods of sample application and detection (assuming that the plate quality is maintained).

To date, TLC analysis was viewed as a procedure comprising

TABLE I. CALCULATION OF TOTAL ERROR

Substance	Standard Deviation (%)
Pyrilameme maleate	1.5
Naphazoline hydrochloride	1.8
Phenylephrine hydrochloride	1.1
Acetominophen (APAP)	1.6
Caffeine	1.8
Aspirin	2.1
Salicylamide	1.6
Total error	1.6
Corrected error (internal standard)	0.79

several separate methods: sample application, development, and detection. Sample application and development were usually slow and highly inaccurate. Scanning of the developed chromatogram was manual and regarded as only marginally useful. In many instances the TLC plate would have to be scored with channels that helped the scanning device find its way along the plate surface.

In our laboratory, HPTLC uses automated sample application and scanning. The sample is applied with the CAMAG Linomat III sample application system.* The solution to be analyzed is accurately sprayed onto the plate surface in precise bands evenly spaced along the HPTLC plate. Forty samples and standards are easily placed on one 20 × 20 cm plate and automatically scanned. To facilitate the scanning of the chromatograms, a CAMAG-HPTLC scanner (1) is used. This instrument is a highly precise spectrophotometer especially designed for HPTLC applications.

AUTOMATION

In general automation in HPTLC has been lacking. Instruments

---

* CAMAG Model Number 27805 with 500-μl syringe.

were specifically designed for a function but never linked together into one system. At our laboratory HPTLC is beginning to become integrated into a system, controlled by the DS500C microcomputer.* The purpose of the microcomputer is not only to collect and process data but to control sample application, development timing, and appropriate centering and scanning of each band of the chromatogram. The DS500C system is depicted in Figure 1.

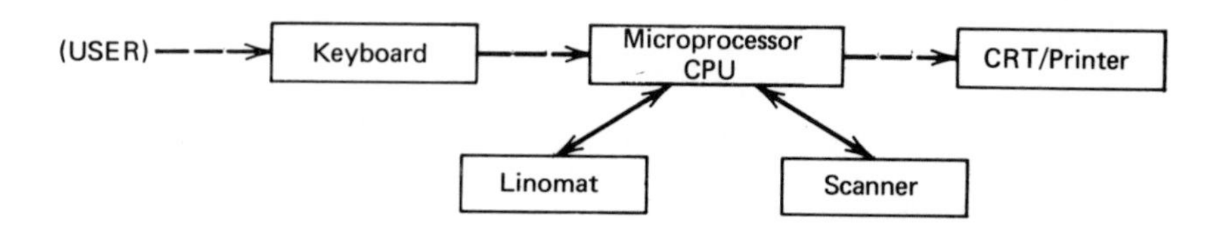

Figure 1. The DS500C system.

A user communicates to each instrument through the keyboard of the CPU, and the dialogues includes the plate number, sample identification, a sample order, and the number of points required for the standard curve. After the text has been entered, the DS500C searches for the sample identification in its library and will instruct the user of the solvent system to be used and prime the user to:

Place standard No. 1 in the Linomat.
Place standard No. 2 in the Linomat.
Place sample No. 1 in the Linomat. . . .

As this proceeds, the CPU will automatically advance the Linomat stage to the appropriate position, spray a precise alliquot of sample onto the plate surface, and await the introduction of the next sample into the Linomat syringe.

After the samples are applied, the user removes the plate and begins development of the chromatogram in the CAMAG Linear

* Available through Applied Analytical Industries, Inc.

Developing Chamber. While developing one plate, the user is free to begin scanning or preparing a second plate. Many plates may be prepared in this fashion and scanned together later.

After development the user returns to the keyboard and instructs the system to scan the plate (now identified by the plate number assigned during preparation). He need not enter any other parameter. The CPU searches its files for specific parameters for the experiment, sets the wavelength to be scanned, and proceeds to instruct the scanner to move to the first sample band, center and scan it, and then move to the second band and again center and scan. The data are transmitted from the scanner's analog output to an A/D converter and collected for analysis by the DS500C. At the conclusion the system will analyze the data, calculate the chromatographic error and standard deviation, and display them on the CRT, where the user may either direct the system to recalculate or transmit the report of the analysis in a user definable format, to the printer/plotter.*

In this fashion hundreds of different methods may be stored on file in the computer and accessed when needed by the user. Data obtained from the experiments may also be filed onto cassette tape files for future reference.

TIME AND COST SAVINGS

As compared to conventional TLC techniques, HPTLC offers considerable time savings as well as increased reliability and precision (2). Published data indicate that Linear HPTLC can reduce analysis time by at least 54%. In a recent publication, Janchen reports that samples analyzed by conventional TLC methods require 4.9 minutes per sample as compared to only 2.4 minutes per sample in linear HPTLC. Both of these time requirements are significantly less than that required by other chromatographic techniques. In fact, the average time required to perform an HPLC analysis in our laboratory is 13 minutes. A savings of 10.6 minutes per sample is realized. When one compares methods, however, the error encountered in conventional TLC reduces its utility. But using HPTLC similiar magnitudes of error are encountered, and the time-saving may contribute substantial cost reductions in routine applications.

---

* The system described in this section is under development by Applied Analytical Industries, Inc. Write for additional information.

Finally, there are many hidden costs in HPLC that are not a factor in HPTLC. These are listed in Table II. It is shown that each analysis performed by HPLC will cost approximately $2.65; this is compared to 30 cents per analysis performed using HPTLC (solvent and plate cost only). This amount translates to more than $10,000 per year in a moderately busy laboratory. This figure alone justifies the purchase of the most sophisticated systems, with a payback of less than three years.

TABLE II. HIDDEN COSTS IN HPLC ANALYSIS

Item	Cost Per Analysis
Column cost (HPLC) (300 analyses/column)	$1.16
Solvent cost (HPLC grade)	$1.20
Solvent storage and handling	$0.09
Fittings and hardware	$0.20
Total	$2.65

CONCLUSION

The future of HPTLC is clear. It is an analytical method that should be considered an alternative to HPLC in routine applications. It is rapid and inexpensive and easily automated. It is clearly the method of choice for laboratories performing a high volume of analysis routinely and where cost, speed, and ease of use are all critical criteria.

REFERENCES

1. D. Janchen and H. R. Schmutz, High Performance Radial Chromatogr. and Chromatogr. Commun., 2, 133 (1979).
2. D. Janchen, Kontakte, 3, 9 (1979).

CHAPTER 5

Automated TLC Detection, Analysis, and Characterization of Steroids

Thomas B. Roos

INTRODUCTION

During the nearly 30 years since the development of paper chromatography of steriods (1), numerous improvements in technique have greatly increased the speed and precision of chromatographic analysis. Thin layer, vapor-phase and high-performance liquid chromatography have all improved both the convenience and precision of analyses. Most recently these procedures that utilize heated gases, vapors, or liquids have become increasingly popular, using computer-controlled elution schedules and electronic data collection. Planar techniques, whether paper or thin layer, remain in wide use for sample preparation and verification (2, 3), but have not lent themselves to automation of data collection. In the course of developing procedures for gathering data, we have found that these planar techniques yield information in a form suitable for rapid, direct, and automated analysis, saving both the expense and effort of VPC and HPLC techniques and avoiding the problems of isotope disposal or immunological interpretation.

This paper describes the development of an interface between a plate scanning reflectance spectrophotometer and a high-speed minicomputer that permits rapid determination and recording of voltages generated by photons reflected from a moving thin layer chromatogram. It works equally well with photographs (e. g., autoradiograms), paper chromatograms, or electropherograms (paper or gel). Data may be collected after either a one- or two-dimensional separation; it is stored initially in a data matrix that can be readily transformed into an R_f matrix. These data can be analyzed for qualitative identification of solutes, by comparison of migratory properties against a library of standard values, or for quantitative presence, by integration and comparison against an internal standard and a standard curve. Moreover, they may be used to reconstruct either linear arrays, representing a single scan, or rectangular fingerprint matrices, suitable to mathematical comparison of complex fingerprints. The speed of data collection permits repetitive analyses, using different frequencies of incident light, that allow easy determination of spectrometric ratios for solutes on the plate. Finally, the procedure leaves the solutes unaltered on the plate, available for subsequent chemical or physiological characterization.

EQUIPMENT

Detection

A Zeiss TLC scanning spectrophotometer serves as a source of monochromatic light in the range of 180 through 1000 nm. Light from the monochromator (Figure 1f), focused on a slit of variable width onto the moving chromatogram, reflects from the plate toward a photocell (Figure 1h) oriented normally to the reflected light. In common operation, the current in the photocell shows as either absorbance or percent reflectance from a meter (Figure 1i) operating over a 10-mV range. A parallel linear recorder (Figure 1j can be used to record these voltage changes and the integrated area swept out by the moving recording pen. Mechanical recorders impose a serious time delay, however. Since using the computer system we have dispensed with the strip recorder as imprecise and slow.

In the present system, a preamplifier (Figure 1e) driven by the computer's power supply amplifies the spectrophotometer output to ±5 V, using two variable potentiometers. Signals from the preamplifier enter the memory of the computer (Figure 1a) as elements of a vector and show as points on a Tektronix oscilloscope (model 5110). Output and gain can be separately adjusted to yield a desirable image, but since subsequent analyses utilize ratios, the absolute values have no influence on ultimate comparisons. The characteristics of the oscilloscope screen (Figure 1d) impose a practical limit of 1024 on the number of data points collected from the chromatogram and stored in computer memory.

The electronic system requires about 40 sec to scan 10 cm, rather than the 5 minutes needed for a mechanical scan. Slowing the scan rate yields no improvement in reliability or precision. More rapid scanning forces both the drive motor and the experimenter beyond their limits of reliable operation.

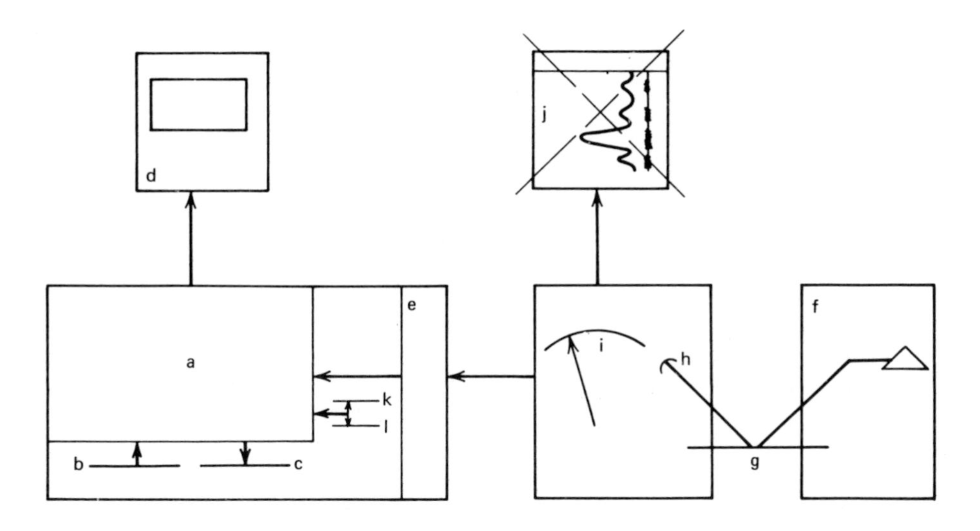

Figure 1. Schematic diagram of data collection apparatus. (a) Computer (with random access memory), (b) disk with system and operating programs, (c) disk with data, (d) oscilloscope, (e) preamplifier, (f) monochromatic light sources (adjustable), (g) object plate (motor driven, computer controlled), (h) photosensor, (i) voltage meter (scaled for absorbance and transmittance), (j) linear recorder and integrator, (k) disk with information on analyzed standards, (l) disk with information on analyzed samples.

Object

The scanning table (Figure 1g) of the Zeiss TLC spectrophotometer will accept any planar object. Its movable head permits objects of different thicknesses to be scanned. We have used thin layer gel plates (either prepared or made from slurries), photographs, and dried acrylamide or starch gels.

The monochromator allows specification of chromophores to be detected in the object plane. Since steroids with ultraviolet chromophores absorb energy at maxima of between 250 and 290 nm (4), we have located and specified molecules absorbing energy at 254 and 280 nm, characteristic of androgens or corticoids and estrogens, respectively.

The object X-Y moving table operates through a distance limited by mechanical pressure switches at a location of X specified mechanically by a rack and pinion gear. In the standard TLC spectrophotometer the X location must be manually specified before each scan. For one-dimensional thin layer work, lines inscribed 10 mm apart delimit vertical channels from the lower edge of the plate to the terminating band, 10 cm above the origin. Leaving the side channels empty to avoid edge effects, there remain 18 channels for reference standards, blanks, and samples. An 8-mm wide aperture in the light path restricts illumination to 1 channel at a scan.

For two-dimensional work the pinion gear is attached by a friction belt to a stepping motor, activated on command by a signal from the processor. An _on_ signal from the computer moves the reading strip a distance equal to the width of the source aperture. It can be modified to change precision in spot location. The _off_ signal comes from contact made between the object mover and one of the mechanical switches on the spectrophotometer frame. Apertures and increments can be adjusted from 4 to 16 mm in 4-mm steps. Four millimeters makes a good reproduction of any reflecting surface, including a photograph. Alternate sweeps occur in bustrophedon, changing direction with each advance along the abscissa.

In order to provide an internal marker for the origin and terminus, the object must have a frame. Since the photocell responds quickly to changes in incident light, a light-absorbing rim around the scanning area works well. From the beginning of the traverse to the object area, the photocell receives minimal stimulation. Current in the cell rises quickly as the object comes into the photocell field. Both uni- and bidimensional scans use this masking signal, imposed by attaching matte black paper or, with glass chromatoplates, scraping the gel away to expose the underlying black metal platform.

Processor

A New England Digital Company computer, model Able 60B with 24K, 16-bit word random access memory, serves as the central processor. It has 16 ports, software addressable for either input or output. Four disk drives permit extensive subsidiary storage for data and processed information, as well as for programs and subsidiary routines. The computer has an easily accessible board rack, which contains slots that allow adding up to 64K of memory and additional function boards. For spectrophotometric automation we use an analog/digital converter, an X/Y interface board, and a scientific timer.

Four disk drives interface with the processor under control from the monitor. The compiler reads automatically from a disk in one drive unit when the instrument is turned on. Additional programs can be stored on the same disk for data collection, processing, and analysis. One 8-in. floppy disk can hold all of the programs (Figure 2) needed to analyze a planar object as a matrix of 1000 × 50 eight-bit words (Figure 1b). Of the other disk drives, one (Figure 1c) permits keeping a permanent, magnetic record of an object that can later be reconstituted on a printer as a picture, density stippled. Four 256-word sectors suffice to contain the data and their appropriate labels for visualization. The other two disks store 16 words of processed information about samples and standards, respectively.

Data Collection

Data collection depends on the timer that controls the interval between acceptances of data from the photomultiplier. Amplified analog voltages from the phototube are converted to digital values, stored in RAM, and simultaneously displayed on the oscilloscope. Since the timer and memory operate in parallel, data accumulate in a buffer file during the specified collection period. Voltages collected at intervals of about 6 μsec are summed for t sec, where t is an experimenter specified time: the window during which the scan is run. Every 5 sec an average is computed and stored in RAM (Figure 1a) as an eight-bit integer.

Our solvent systems yield good separation of steroids in a space of 10 cm. We collect data over a length of 11 cm, however, in order to allow room for the margin described above. At a scan rate of 27.5 mm/sec we gather voltages for 40 sec. Approximately 1000 eight-bit data points are gathered per traverse in a one-dimensional scan, that is, 20 points per second. The reading frequency is thus about eight points per millimeter. Each point, however, represents the average of the nearly 100 readings made at 5-μsec intervals during the traverse. In consequence, even at high sensitivity, background noise remains small: far less than found in strip recordings. For two-dimensional objects, disk storage capacity precludes using a 1000 × 1000 matrix, instead we use approximately 1000 points in Y and 12 to 50 in X. A single disk can thus contain the results from three objects scanned at one frequency or one scanned at three frequencies.

COMPUTATION

Data Storage

The same program (Figure 2, SPECCOPY) that collects data from the plates processes it into a form suitable for future use. Before starting to read the object, the experimenter stipulates whether the collection will be one or two dimensional, automatically or manually driven. Values entered from the monitor keyboard label the mode of collection, experiment, object code, chromatographic system, sample source and number, frequency of incident light, and scan duration. These values are prefixed to the collected data along with the actual numbers of points collected.

As an option, the collection program permits noise suppression and image enhancement. Values for tolerable noise and signal amplification are prefixed to the data vector, but not used until later when the data are analyzed. During analysis, points that lie within the limits of acceptable noise are used to compute a 10-point running average. Points that differ from their neighbors by more than the noise threshold are read individually.

If the initial scan on the oscilloscope satisfies the experimenter the program can be directed to continue, deleting those values of the vector that correspond to the leading and trailing margin. These cropped vectors are stored in consecutive 4-sector bands on a data disk (Figure 1c).

Should the data fail to satisfy the experimenter, they may be recollected from the object using the same or different noise-suppression or image-enhancement values. Alternatively, they may be retained in raw form but subjected to new suppression or enhancement values. A repetition mode allows sequential scanning with the result summed to reinforce signals and destructively interfere with noise. Such recompilation and repetitive scanning can only be accomplished in a system, like the one described, in which the separated solutes remain unchanged by their detection and description.

Background Computation

All background computations (Figure 2, SPECBACK) occur in RAM, averaging the strips on a disk identified as blanks by the appropriate label. The program for computing backgrounds excludes data from channels containing samples or standards and those with inappropriate background measurements.

The background vector for each set of experimental parameters contains 101 data points, rather than the approximately 1000 in

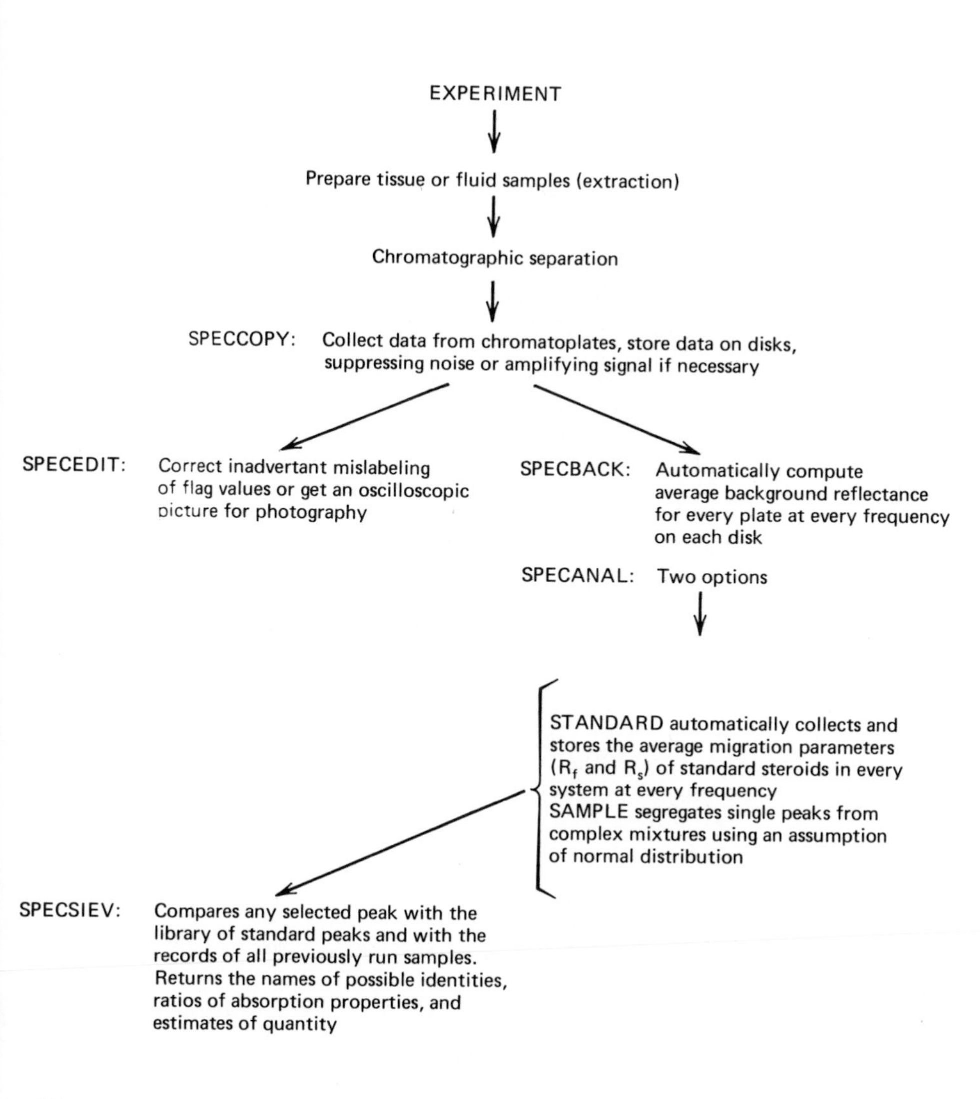

Figure 2. Flow diagram for automated steroid hormone analysis, showing steps in data preparation, collection, and analysis.

data vectors. This reduction permits more efficient storage without significant loss of accuracy. Background points are stored on the data disk (Figure 1c) and then automatically subtracted in RAM from data in all subsequent analyses: the stored background and data values themselves remain unchanged.

Analysis

Standard solutes and samples are analyzed similarly (Figure 2, SPECANAL), by the same program, but differ in the converted data that they store. Both locate and characterize the solute peaks present in a chromatogram or gel by a common statistical procedure. Pertinent labels and parameters for each peak are stored on command as sixteen 16-bit words on one of two disks: Standard (Figure 1k) or Output (Figure 1l). Since each disk sector contains eight records, a 5-in. disk can carry information on 1400 peaks.

a. Scan Reduction

A data vector or matrix is called from the data disk (Figure 1c) and transferred to RAM. Raw data are then converted into hR_f values and reduced by the amount of background appropriate to that location. Finally, the minimum value for the modified curve is subtracted from each hR_f point to produce a zero baseline. In the course of these calculations, the raw data curve and then the hR_f curve (1001 points, 0 to 100 by tenths) appear in sequence on the oscilloscope screen (Figures 3a, 3b, 3c, 3d).

b. Transformation

A new threshold for peak smoothing and image amplification can be introduced to treat the raw data. We have not found this option necessary in any work yet done.

Outpeak from this part of the analysis reports the total area under the (zero minimum) curve, the locations (hR_f) of the maximum and minimum and an arbitrary numerical value of the maximum (Figure 4).

c. Peak Identification and Isolation

Values for three parameters define the conditions for holding portions of the chromatogram for storage: a minimal variance, a minimum value, and a maximal deflection. Peak isolation depends on estimating variances and then subtracting the normalized curve from the data curve (see below). Spurious curves with low variance can be generated. Setting minimal values for variance and detection reduces artifacts. The maximal value can be used for quantitation when the qualitative identification of a peak is known and can be compared against the reflectance properties of an authentic, measured standard run in another strip on the plate.

Beginning with the maximum point (X_m), a normal distribution around that point is estimated from up to 14 adjacent points (X_i). The variance of each point is estimated from the relation:

$$\hat{\sigma}_i^2 = X_i \cdot \exp(t^2/2) = \frac{d_1^2 - d_2^2}{\ln[X(d_1)/X(d_2)]}$$

$$= \text{abs}\left\{\frac{(X_m - X_i)^2 - (X_m - X_i \pm 1)^2}{2 \cdot \ln[(X_i \pm 1 \cdot Y_i)/(X_i \cdot Y_{i\pm 1})]}\right\} \tag{1}$$

Where X_m: hR_f of maximum point.
X_i: hR_f of test point.
$X_{i\pm 1}$: hR_f of next point above (if $X_m - X_i < 0$) or below (if $X_m - X_i > 0$) the point X_i.
Y_i: value of Y at X_i.
$Y_{i\pm 1}$: value of Y at X_{i+1} or X_{i-1}.

Estimates alternate above and below the maximum until reaching either 14 values or the baseline. A geometric mean of the variance is then calculated for use in computing a theoretical curve:

$$\hat{Y}_i = Y_m \cdot \exp -\left(\frac{(x_i - x_m)^2}{2 \cdot \sigma_m^2}\right) \tag{2}$$

Where Y_m: deflection of Y from 0 at X_m.
σ_m^2: geometric mean variance for the peak at X_m.

A new RAM vector is produced by subtracting the peak computed by Equation (2) from the zero minimum peak. The new vector then serves as a base for locating and characterizing another

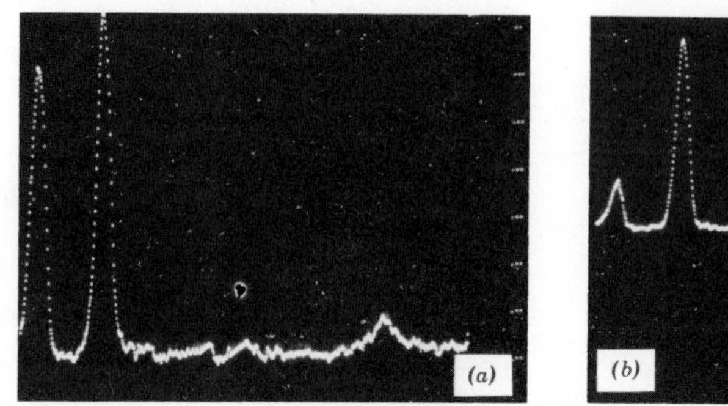

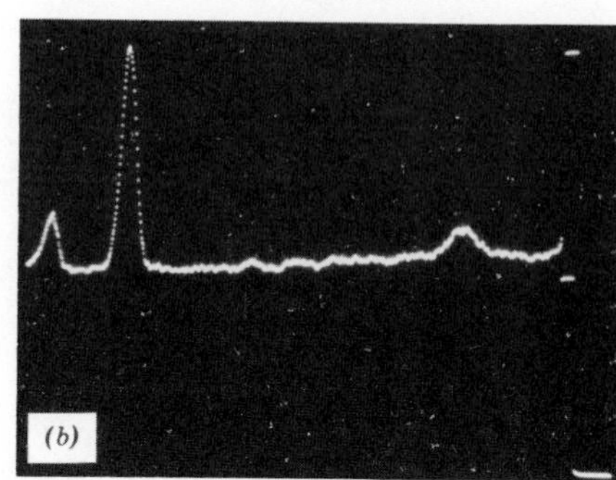

(a) Trace of original data at 254 nm, (b) 280 nm;

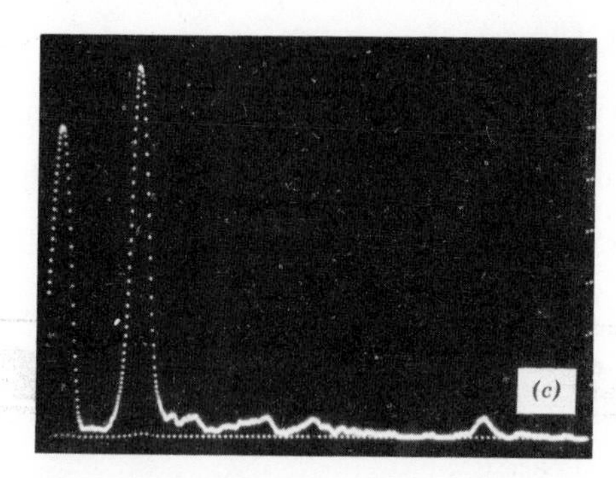

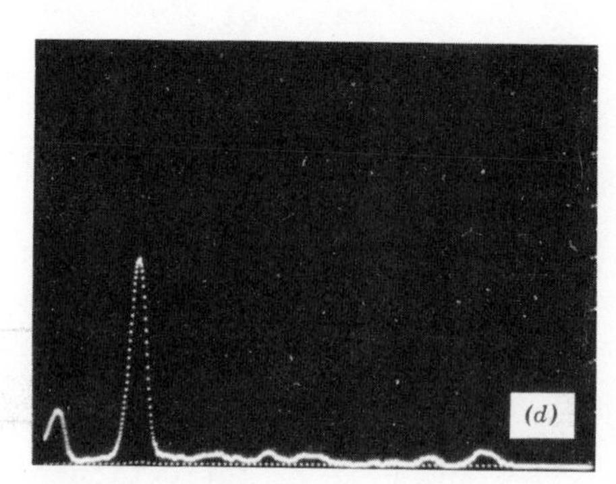

(c) trace of data after subtraction of background and conversion to hR_f at 254 nm, (d) 280 nm;

Figure 3. Photographs of the oscilloscope screen during the use of SPECANAL on steroids derived from a 13-day-old rat embryonic adrenal tissue (equivalent of 10 glands), 254 and 280 nm.

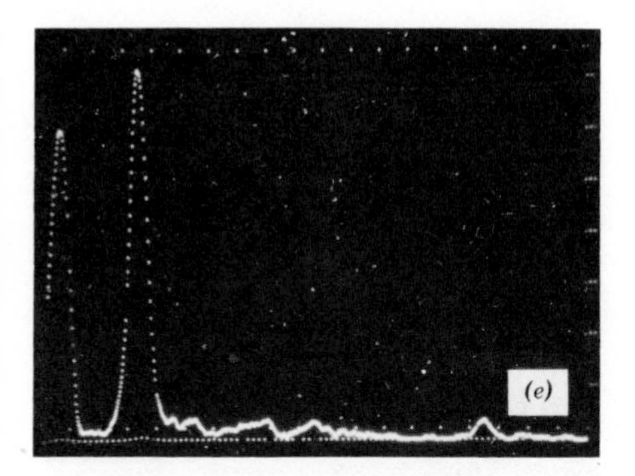

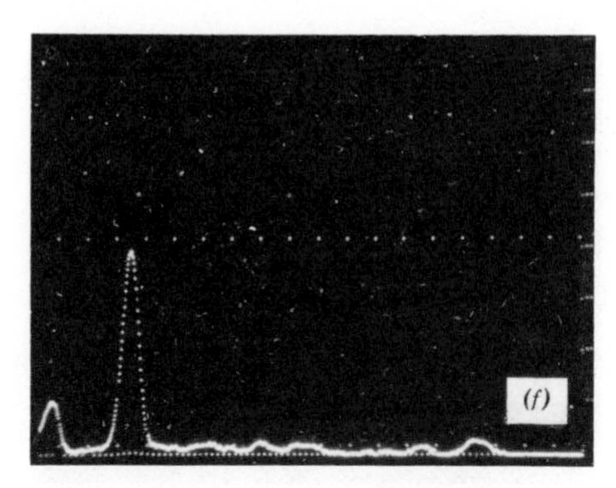

(e) hR_f curve showing the maximum deflection (upper dotted line) and minimum value for detection (lower dotted line) at 254 nm, (f) 280 nm;

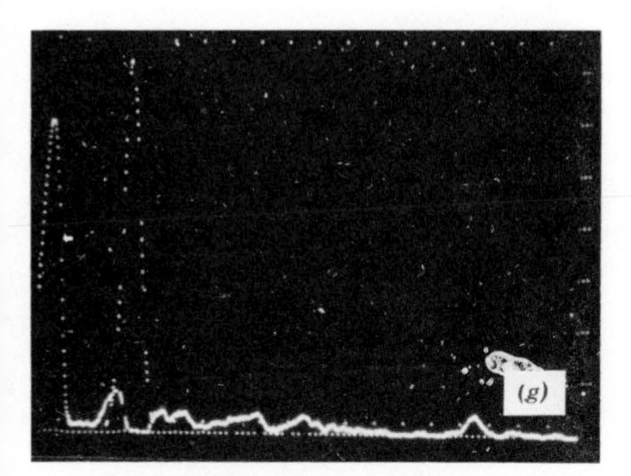

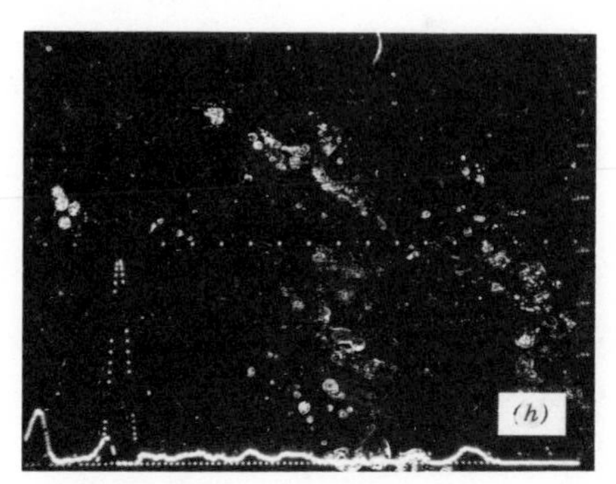

(g) hR_f curve after isolation of the first peak (internal standard) at 254 nm, (h) 280 nm;

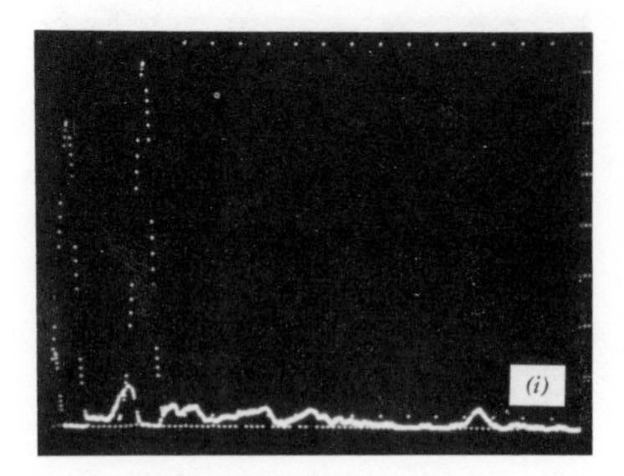

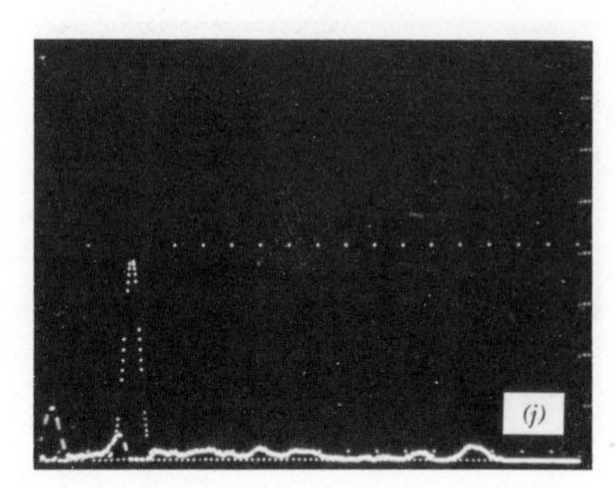

(i) hR_f curve after isolation of the first embryonic steroid peak at 254 nm, (j) 280 nm;

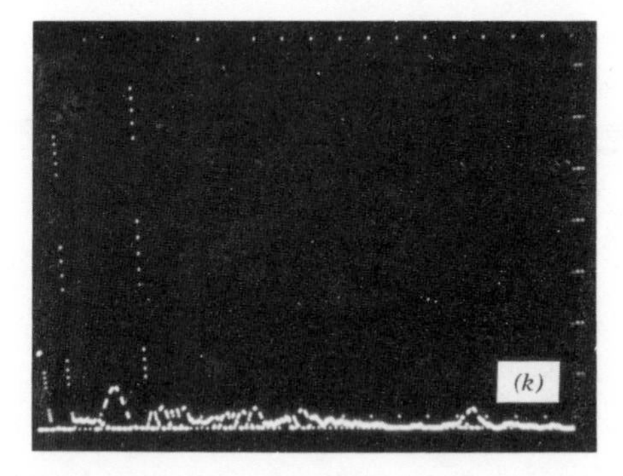

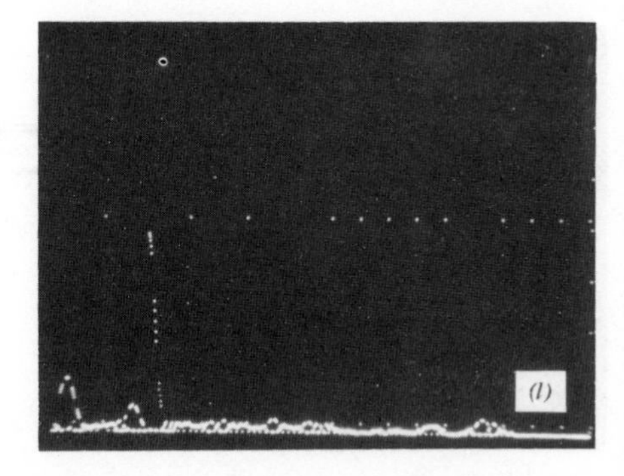

(k) hR_f curve after isolation of all peaks to the level of the minimal value for detection at 254 nm, (l) 280 nm.

peak, and then continues until the remaining maximum falls below the previously defined minimal value for detection. Minima can be set at any level: in practice, values of 40 to 100 (about 2% of the maximum) yield good results. As peaks disappear, artifacts are introduced by the subtraction. These have low variance estimates and can be excluded automatically by choosing a minimal acceptable variance value, usually near 0.65 (Figures 3e through 3l, 4).

Each accepted computed peak, its hR_f, area, and variance, is saved and reviewed. If an internal reference is present in a scan, a value of R_s is also computed. The area under each peak is computed relative to the aggregate curve excluding the referent, if present. Table I shows the values stored for each peak, both standard and sample. Standard and sample properties are stored on separate disks (Figures 1k and 1l, respectively).

TABLE I. VALUES STORED FOR EACH CHROMATOGRAPHIC PEAK

	Sample	Standard
1.	Sector occupation signal	
2.	Experiment number	0
3.	Sample or standard identification	
4.	System and illumination code	
5.	0	Number of replicates in mean
6.	0	Standard deviation for hR_f
7.	Code for internal referent	
8.	Peak number	
9.	hR_f	Mean hR_f
10.	Maximum deflection for sample	Standard deviation for R_s
11.	Relative maximum for curve	Mean maximum value at mean hR_f
12.	Standard deviation of variant estimates	Mean variance
13.	Total area under curve	
14.	Area of peak relative to total	
15.	Area of computed peak relative to local area	
16.	R_s	Mean R_s

```
In which sector are the data ?10037
Constants are: Strip: 04010 Sample: 01323 System: 07254; OK? YES

Identify the lowest frequency used: ?254

Will you smooth the curve? NO

The total area under the curve is +139137.0.                              A
The maximum point is 01879 at 00202. The minimum is at 00420
Enter a value for the minimal accepable variance ?.65                    B
Enter values for maximum deflection and minimum detection ?1900.65
Are these satisfactory? YES
                                                                          C
Peak # 00001:  hR(f) = +22.49442 (+1794.181)   var. = +2.032982 D
   Will you stop with  00001 peaks? NO
Peak # 00002:  hR(f) = +8.797325 (+1506.817)   var. = +2.225443 E
Peak # 00003:  hR(f) = +6.458795 (+253.4412)   var. = +.9848793
Peak # 00004:  hR(f) = +20.15589 (+221.5438)   var. = +1.861814
Peak # 00005:  hR(f) = +4.342983 (+121.0000)   var. = +.3458656
Peak # 00005:  hR(f) = +28.17371 (+105.1999)   var. = +.6831616
Peak # 00006:  hR(f) = +26.61469 (+98.29998)   var. = +.1906851
Peak # 00006:  hR(f) = +31.51446 (+92.39998)   var. = +4.171926
Peak # 00007:  hR(f) = +44.76614 (+91.09089)   var. = +1.156250
Peak # 00008:  hR(f) = +82.51668 (+90.00000)   var. = +1.507868
Peak # 00009:  hR(f) = +52.89531 (+84.90907)   var. = +1.336080 F


The total area under the curve is +54936.00.                              A
The maximum point is 01041 at 00203. The minimum is at 00113
Enter a value for the minimal accepable variance ?.65                    B
Enter values for maximum deflection and minimum detection ?1050.45
Are these satisfactory? yes
                                                                          C
Peak # 00001:  hR(f) = +22.53051 (+985.9089)   var. = +1.576843 D
   Will you stop with  00001 peaks? no
Peak # 00002:  hR(f) = +8.768032 (+254.0908)   var. = +1.809542
Peak # 00003:  hR(f) = +20.42175 (+148.3125)   var. = +1.077297 E
Peak # 00004:  hR(f) = +5.549388 (+74.00000)   var. = +1.233478
Peak # 00005:  hR(f) = +81.35403 (+70.45453)   var. = +1.176573
Peak # 00006:  hR(f) = +44.67258 (+56.79998)   var. = +1.867161
Peak # 00007:  hR(f) = +83.35181 (+55.09090)   var. = +.4248460
Peak # 00007:  hR(f) = +36.84794 (+49.44443)   var. = +.4590633
Peak # 00007:  hR(f) = +35.84904 (+48.62500)   var. = +.1770210
Peak # 00007:  hR(f) = +50.94338 (+47.00000)   var. = +1.465304
Peak # 00008:  hR(f) = +26.19311 (+46.50000)   var. = +.3077578
This peak (+42.90908) below minimum: +45.00000                           F
   Will you change the minimum? no
```

Figure 4. Computer listing of peaks isolated at 254 and 280 nm from a sample of 13-day-old rat embryonic glands (same sample as shown in Figure 3). Letters show the point at which each of the photographs of the oscilloscope screen were taken.

Comparisons

(Figure 2, SPECSIEV).

a. Sieving

A sample, designated by its experiment code and strip number, is retrieved to RAM from the sample disk (Figure 11) and compared against R_s, variance, and spectrometric properties of standards and, on demand, of other samples. Comparison depends on a multiple sieve that excludes matches between any selected sample peak and a standard peak or other sample peak when:

1. R_s or hR_f of the selected sample deviates from a standard peak by more than three standard deviations of the standard estimate or from another sample by more than 0.1 units.
2. Variance estimate for the sample differs from that of the comparison peak by less than three standard deviations.

The sieve program lists each peak of the selected sample separately, with its migration values (hR_f and R_s), its computed variance, its height, its relative area, and its spectrometric ratio at any other illumination frequencies used. It appends a list of all standards (and samples) with compatible values (Figure 5).

b. Interference

Two samples or a sample and standard, either one or two dimensional, are reconstructed from the values of their individual peaks. Matrix subtraction of the hR_f vector or matrix permits direct, two by two comparison in the form of a new, result matrix. Deviations from zero, or a selected detection threshold, identify regions of dissonance between the comparisons.

Sequential matching of samples against a common standard or test sample produces a table of dissonances that can be compared further by successive matrix subtractions. The process allows precise quantitative characterization of complex two-dimensional chromatographic or electrophoretic objects and then both direct and indirect comparisons.

254 NM

```
Peak # 00001, hR(f) = +22.49442
     ( 00092)  R(s) =  +1.000000
     Maximum = +1794.000, Variance = +2.032982
     Variance of variances is +.0153429
     The ratio of computed area (+57292.71) for peak # 00001
        to the total curve is +41.17718 %
        to segment +18.48551 - +26.39197   is +93.64766 %
     Will you store these data? NO
   No data stored for peak # 00001

Peak # 00002, hR(f) = +8.797325
     ( 00092)  R(s) =  +.3910890
     Maximum = +1507.000, Variance = +2.225443
     Variance of variances is +2.451031
     The ratio of computed area (+50312.36) for peak # 00002
        to the total curve less standard is +36.16056 %
        to segment +4.677058 - +12.80623   is +96.15359 %
     Will you store these data? NO
   No data stored for peak # 00002

Peak # 00003, hR(f) = +6.458795
     ( 00092)  R(s) =  +.2871286
     Maximum = +253.0000, Variance = +.9848793
     Variance of variances is +.0153631
     The ratio of computed area (+5405.123) for peak # 00003
        to the total curve less standard is +3.884776 %
        to segment +4.454341 - +8.351891   is +21.08082 %
     Will you store these data? NO
   No data stored for peak # 00003
```

280 NM

```
Peak # 00001, hR(f) = +22.53051
     ( 00092)  R(s) =  +1.000000
     Maximum = +986.0000, Variance = +1.576843
     Variance of variances is +.0091457
     The ratio of computed area (+27831.50) for peak # 00001
        to the total curve is +50.66168 %
        to segment +18.97890 - +25.97113   is +91.59618 %
     Will you store these data? no
   No data stored for peak # 00001

Peak # 00002, hR(f) = +8.768032
     ( 00092)  R(s) =  +.3891625
     Maximum = +254.0000, Variance = +1.809542
     Variance of variances is +.0218402
     The ratio of computed area (+7555.047) for peak # 00002
        to the total curve less standard is +13.75270 %
        to segment +5.660375 - +11.76470   is +91.38800 %
     Will you store these data? no
   No data stored for peak # 00002

Peak # 00003, hR(f) = +20.42175
     ( 00092)  R(s) =  +.9064037
     Maximum = +148.0000, Variance = +1.077297
     Variance of variances is +.0013374
     The ratio of computed area (+3322.287) for peak # 00003
        to the total curve less standard is +6.047669 %
        to segment +18.31298 - +22.41953   is +19.72737 %
     Will you store these data? no
   No data stored for peak # 00003
```

Figure 5. Full detail of data printed on the computer terminal and stored for retrieval on the first three peaks (including the internal standard, Prednisolone) at 254 and 280 nm.

RESULTS OF THE SYSTEM IN USE

The automated TLC analytical system has been used to separate steroids from embryonic and fetal rat tissues, using standard techniques of tissue incubation and solute extraction (5). Only minute tissue quantities are available, but previous work has shown that many, different steroids appear in incubation media (6). Moreover, the constellation of steroids changes during development and differs among steroidogenic tissues.

Standard Library

A library of steroid hormones and metabolic intermediates with ultraviolet chromophores was assembled using purified recrystallized solutes. Table II shows the average hR_f and R_s properties for a representative set of standards, along with a statistical measure of their errors of estimate. There is no correlation between distance migrated and standard deviation of either hR_f or R_s. Thus, although the standard deviation for hR_f measurements (1.359 ± 0.554) greatly exceeds that for R_s measurements (0.062 ± 0.046), both approximate 1.5% of the usable abscissa. We use R_s whenever possible, however, in order to guard against systematic error.

Compilation of the standard library is based on a table of biochemical conversion (Figure 6) showing the oxidative transformations between cholesterol and the steroid hormones. Table III lists the steroids included in Figure 6.

Embryonic Rat Steroids

Work now in progress shows that the rat embryo, age 13 days, secretes a number of steroidal solutes, three common to all embryos and produced during incubation of both gonadal and adrenal rudiments. Preliminary results show that the most abundant has properties different from any of the standards yet included in the library. It thus resembles a hydrophilic, progestane derivative (R_s 0.226 ± 0.015 compared with prednisone, 0.352 ± 0.018 compared with prednisolone, ratio of reflectance 280:254 about 0.2). The next most abundant, present in much lower quantity, migrates similarly to both cortexone and estrone in two solvent systems, but its reflectance properties resemble estrone, with relatively high absorption at 280 nm (about 130%). The last presumptive steroid migrates very rapidly, faster than any of the tested standards, and has the optical properties of

TABLE II. COMPARISON OF STANDARD PEAK PARAMETERS, SYSTEM 7

Standard	hR_f	R_s	$\hat{\sigma}^2$	280/254
Progesterone	70.78 ± 0.95	3.140 ± 0.113	5.42	9.99%
Cortexone	63.36 ± 0.97	2.793 ± 0.103	3.54	14.65%
Aldosterone	32.02 ± 0.85	1.430 ± 0.024	2.46	9.98%
Corticosterone	44.15 ± 0.96	1.946 ± 0.041	2.90	15.85%
17-α-OH progesterone	61.32 ± 1.93	2.786 ± 0.084	3.73	11.56%
Cortisol	27.25 ± 0.77	1.203 ± 0.013	2.67	19.79%
Cortisone	41.24 ± 1.28	1.838 ± 0.029	2.62	19.68%
Prednisolone	22.27 ± 0.74	1.000 ± 0.000	2.62	57.02%
Testosterone	54.01 ± 0.88	2.417 ± 0.044	3.15	11.32%
Androstendione	69.21 ± 2.62	3.152 ± 0.082	4.81	10.72%
Estrone	63.65 ± 2.34	2.886 ± 0.124	2.79	155.30%
Estradiol	50.90 ± 1.92	2.291 ± 0.059	2.89	142.59%
Estriol	18.56 ± 1.11	0.848 ± 0.024	2.13	159.44%

an estrane derivative. We shall report on this work in greater detail in a subsequent publication.

Figure 3 shows a scan of 13-day old rat embryonic adrenal extract. The first scan (Figures 3a and 3b) show the raw data, after trimming the vector, as displayed on the oscilloscope at the beginning of SPECANAL. It appears at the position marked A in Figure 4. After background subtraction and conversion to hR_f, the curve reappears as shown (Figures 3c and 3d) and the terminal requests values as shown at position B in Figure 4.

The program disassembles the curve, peak by peak, as shown in the list, Figure 4. Each isolated peak disappears from the data vector and the oscilloscope, leaving only the trace of its recomputed curve behind (Figures 3e through 3l). When the variance fails to reach the prescribed minimum (0.77 in the example shown), the curve, although subtracted, is not retained as a record.

Detailed information on individual peaks is given seriatim after the initial listing (Figure 5). This provides a literal

```
Load a STANDARD disk in R0 and an OUTPUT disk in R1.  Press RETURN when ready to proceed.
 61 standards entered
What Sample will you test?  1323
Will you compare #1323 with other Samples as well as with Standards?  NO
Records read:  1153, saved:  230
Sieve records:  648  Each record sounds a bell as read.
 00000 records checked.
 00100 records checked.
 00200 records checked.

For sample 1323 on strip 4010 in system 7 at 254 nm:
   Peak #3:  R(s) = +.2970296 relative to Prednisolone     hR(f) = +6.681512     Variance = +1.003284
      Its height = +368.9999,  Its area = +5.822031% of the total
      Its parameters are compatible with no samples yet run

For sample 1323 on strip 4010 in system 7 at 254nm:
   Peak #4: R(s) = +.8861384 relative to Prednisolone      hR(f) = +19.93318     Variance = +2.550388
      Its height = +195.9999, Its area = +4.709694% of the total
      The ratios for maxima and variance are:
         280:254 nm +1.63265     +.4640292
      Its parameters are compatible with sample(s):

      Estriol,65, Standard, Peak #1:-
         hR(f) = +18.55656
          R(s) = +.8476864 relative to Prednisolone
      variance = +2.133937
```

```
For sample 1323 on strip 4010 in system 7 at 254nm:
   Peak #5:  R(s) = +.1930692 relative to Prednisolone     hR(f) = +4.342983     Variance = +.8460110
      Its height = +123.9999, Its area = +1.631214% of the total
      Its parameters are compatible with no samples yet run

For sample 1323 on strip 4010 in system 7 at 280nm:
   Peak #2:  R(s) = +.3891625 relative to Prednisolone     hR(f) = +8.768032     Variance = +1.882958
      Its height = +253.9999, Its area = +13.99084% of the total
      The ratios for maxima and variances are:
         280:254nm +.3062085     +.9517584
      Its parameters are compatible with no samples yet run


For sample 1323 on strip 4010 in system 7 at 280nm:
   Peak #3:  R(s) = +.9014776 relative to Prednisolone     hR(f) = +20.31076     Variance = +1.183454
      Its height = +125.9999, Its area = +5.337832% of the total
      Its parameters are compatible with sample(s):

      Estriol, 65, Standard, Peak #1:-
            hR(f) = +18.68992
             R(s) = +.8644978 relative to Prednisolone
         variance = +2.939606

For sample 1323 on strip 4010 in system 7 at 280nm:
   Peak #4:  R(s) = +.2463054 relative to Prednisolone     hR(f) = +5.549388     Variance = +.9431056
      Its height = +74.99998, Its area = +2.665048% of the total
      Its parameters are compatible with no samples yet run
```

Figure 6. Detail of comparison generated in the program SPECSIEV for 13-day-old rat embryonic adrenal glands. The portion shown gives information on the sample illustrated in Figures 3 through 5.

TABLE III. KEY TO THE NUMERICAL CODE FOR STEROIDS

1*	Cholesterol
2*	20-α hydroxy cholesterol
3*	20-α, 22-χ dihydroxy cholesterol
4*	Pregnenolone
5*	17-α, 20-α dihydroxy cholesterol
6*	17-α, 20-α, 22-χ trihydroxy cholesterol
7*	22-β hydroxy cholesterol
8*	20-α, 22-β dihydroxy cholesterol
9*	Pregnenolone sulfate
10	Pregnendione
11	Progesterone
12	Cortexone (Desoxycorticosterone, DOC)
13	18 hydroxy cortexone
14	18 hydroxy corticosterone
15	Aldosterone
20*	21-hydroxy pregnenolone
21*	Pregnentriolone
22	11-β hydroxy progesterone
23	16-α hydroxy progesterone
24	Corticosterone
25	17-α, 20-α dihydroxy progesterone
30*	17-α hydroxy pregnenolone
31	17-α hydroxy pregnendione
32*	3-β, 17-α, 21-pregnentriol
33	17-α hydroxy progesterone
34*	3-β, 11-β, 17-α, 21-pregnentetraol
35	21 desoxy cortisol
36	11 desoxy cortisol
37	Cortisol (hydrocortisone)
38	Cortisone
39	Dihydrotestosterone
40*	Androstendiol-17-acetate
41	Testosterone acetate
42	Testosterone
43	Androstendione
45	11-β hydroxy testosterone
46	11-β hydroxy androstendione
47	11-oxo androstendione (adrenosterone)
48	19-hydroxy androstendione

Table III Continued

49*	Estrone
50	Androstendiol
51	17-α androstenolone
52	δ-1 testosterone
53	19 hydroxy δ-1 testosterone
54	19 hydroxy testosterone
55	19 oxo testosterone
56	19 carboxy testosterone
57	19 nor-testosterone
58	17-β estradiol
60	Dehydroepiandrosterone
61	Dehydroepiandrosterone sulfate
62	16-α hydroxy dehydroepiandrosterone
63	16-α, 19 dihydroxy dehydroepiandrosterone
64	16-α hydroxy estrone
65	Estriol
66	16-α hydroxy dehydroepiandrosterone sulfate
67	16-α, 19 dihydroxy dehydroepiandrosterone sulfate
68	16-α hydroxy estrone sulfate

record of the information stored in the sample disk.

Comparison of sample data requires a serial examination of the standard (and sample) disks. Figure 5 shows the use of the program SPECSIEV on the data shown in Figure 3.

An entire run, from removal of tissue, through extraction, chromatography, and preliminary identification, requires no more than 6 hr. The sample, still present and intact on the chromatoplate, remains available for further study.

ACKNOWLEDGMENTS

I thank Marie T. Girard for her invaluable help in chromatography and David vonLehn and Michael Shapiro for their contributions to computer programming and interface design.

REFERENCES

1. I. E. Bush, J. Biochem., 50, 370 (1952).
2. G. Cavina, G. Moretti, R. Alimenti, and B. Gallinella, J. Chromatogr., 175, 125 (1980).
3. B. F. Maume, C. Millot, D. Mesner, D. Patourauz, J. Doumas, and E. Tomori, J. Chromatogr., 186, 581 (1979).
4. G. D. Fassman (Ed.), CRC Handbook of Biochemistry and Molecular Biology, 3rd ed., Lipids, Carbohydrates, Steroids, 1975, p. 531.
5. G. F. Cortland and M. H. Kuizenga, J. Biol. Chem., 116, 57 (1936).
6. T. B. Roos, Endocrinology, 81, 716 (1967).

CHAPTER 6

The Role of Solvent Classes in Chromatographic Selectivity

Dexter Rogers

INTRODUCTION

The solvent strength concept was developed as a means for understanding and predicting solute mobility in adsorption chromatography (1). Experimentally, solvent strength is the free energy of adsorption of a pure solvent onto a sorbent of standard surface activity, which is normalized to a unit of surface area. Originally, solvent strength constants were obtained for alumina, but corresponding values for other sorbents could be recalculated from the alumina data by using appropriate conversion factors. Solvent and solutes were assumed to be competing for common sites on the sorbent surface by some common mechanism of sorption. The relative strength of solvent binding could then account for the relative displacement or mobility of the solute. The practicality of the solvent strength concept was easily perceived from its explanatory and predictive capability for both development (thin layer) and elution (liquid-solid) chromatography.

In a previous article (2), the solvent strength concept was applied to derive a formula to equate solute mobility relative to the solvent front (R_f) with solvent strength (E^0). The derivation $R_f = -a + bE^0$ was tested over a solvent strength range of 0.00 to 0.50 for a number of solutes using several general-purpose sorbents, including alumina and several grades of silica gel. It was also observed that at least two classes of solvents existed. Within each solvent class the derived formula held. The existence of more than one solvent class challenged the premise upon which the solvent strength concept was based and suggested that the solvents and solutes might be interacting at different binding sites by different mechanisms.

This article is concerned with the theory and application of the solvent strength concept and with the method of classifying the solvents according to their solute partitioning characteristics, also developed by Snyder (3). An understanding of solvent classes should lead to a better theory for adsorption chromatography and for more successful applications.

For thin layer chromatography, solute mobility depends on the competition between solute and solvent for common binding sites on the sorbent surface. It is the following equilibrium, which is being considered:

$$S_s + nM_l = S_l + nM_s$$

where S is the solute, M is solvent (or mobile phase), s is the sorbent (or solid) phase, l is the liquid phase, and n is the average number of solvent molecules per displaced solute molecule. It is assumed that no alteration occurs to the surface activity of the sorbent during the chromatographic process.

R_f = solute displacement relative to the solvent front.
ΔE = net energy change for solute displacement by solvent.
$R_f = \Delta E = -E_{S,l} + nE_{M,l} \quad -E_{S,s} + nE_{M,s}$

Liquid-Phase Energy	Sorbent-Phase Energy
Can be neglected for nonsolvating solvent because these energy terms should be negligible or self-canceling.	$E_{S,s}$ = a constant (a) $E_{M,s}$ = the independent variable, which is normalized to a unit of surface area (= E^0, solvent strength).

Therefore, between 0 and 1, $R_f = -a + bE^0$, where b is the slope.

A typical result is shown in Figure 1. Clearly, solvent strength determines solute mobility. However, two solvent classes were indicated by the two curves shown in this figure. These solvent classes are called "inert" and "solvating" solvents. The solvating solvents provided for relatively greater solute mobility than the inert solvents. For each solvent class, this formulation held true. This formulation should be able to furnish a direct and quantitative means for relating appropriate physical measures of solvent character of chromatographic behavior,

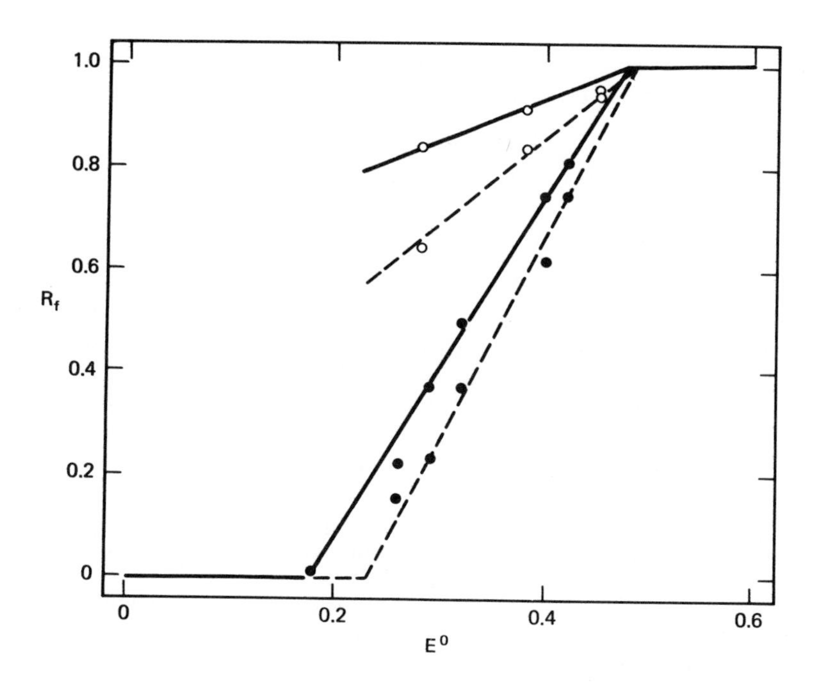

Figure 1. Correlation of solute mobility (R_f) and solvent strength (E^0) by solvent class. Solute: tetraethylthiuram disulfide. (Solid circles, inert solvents; open circles, solvating solvents. Solid lines, Whatman LK5D silica gel; dashed lines, Whatman LK6D silica gel. Redrawn from reference 2.)

provided the reality of solvent classes can be established. Solvent class is the subject of this paper.

SOLVENT POLARITY AND SELECTIVITY: THE SNYDER SUBGROUPS

In order to investigate the validity of the concept of solvent class, various solvent characteristics were compared with solvent strength values. These included solvent polarity and selectivity. Polarity (P') is defined as relative chromatographic strength, which should make it analogous to solvent strength. Polarity is used in the sense of organic chemistry of functional groups rather than physical chemistry of charge dissymmetry. Polarity is better expressed in terms of the Hildebrand solubility parameter instead of dipole moment. Polarity is calculated in the following way:

K_g experimental distribution coefficients determined for ethanol (e), dioxane (d), nitromethane (n), toluene (t), and n-octane (o).

K_g' adjusted distribution coefficients, which are calculated as follows:

$K_g' = K_g \cdot V_s$, where V_s is solvent molar volume.

K_g'' polar distribution coefficients, for which the effects due to solute molar size, solute/solvent dispersion interaction, and solute/solvent induction resulting from solvent polarizability have been eliminated as follows:
$\log K_g'' = \log K_g' - \log K_\nu$, where K_ν is the adjusted distribution coefficient for a hypothetical n-alkane with the same molar volume as solute x. $\log K_\nu = (\bar{V}_x/163) \log K_0$, where K_0 is the adjusted distribution coefficient for n-octane.

P' $= \Sigma \log K_g''$, for ethanol, dioxane, and nitromethane.

The correlation of polarity and solvent strength is shown in Figure 2. Generally, the individual points correlated reasonably well for both inert and solvating solvents. Alcohols and amines, however, formed a third solvent class called the "hydroxylic" solvents. A listing of solvents by solvent classes and Snyder subgroups (3) is given in Table I.

A relationship has been claimed for solvent strength and dielectric constant or dipole moment (5). Although this relationship has since been denied, the belief persists (3). An attempt was made to rationalize this discrepancy by correlating solvent strength and dielectric constant (ε) by solvent class (Figure 3). The results were encouraging, although exceptions were still noted. The line with the least slope accommodated the inert solvents, except for inert hydrocarbons with strongly nucleophilic groups, such as carbonyls, sulfones, nitrile, and nitro and dichloro derivatives. Monohalogenated hydrocarbons lay above this line. Ethers and amines functioned as inert solvents in terms of dielectric constant. Solvating solvents, except for ethers and esters, and the inert hydrocarbons with strongly nucleophilic groups formed a second grouping. Hydroxylic solvents, except for amines, and pyridine formed a third grouping. In a similar fashion, dipole moment (μ) correlated with solvent strength by solvent class (Figure 4). For the present, a direct role of dielectric constant or dipole moment in the determination of solvent strength remains to be elucidated.

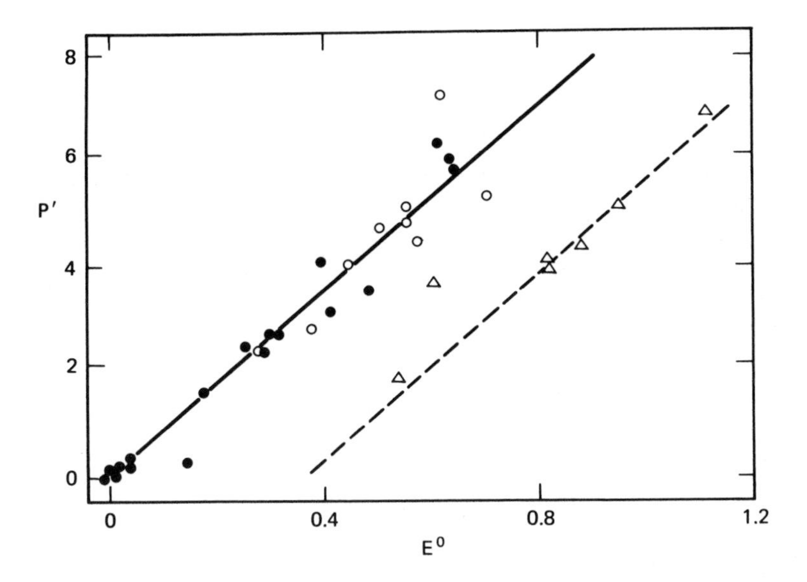

Figure 2. Correlation of solvent polarity (P′) and solvent strength by solvent class. (Solid circles, inert solvents; open circles, solvating solvents; triangles, hydroxylic solvents. Data from reference 3.)

By definition, solvent strength and molecular area (A_s) are related inversely: $E^0 = S^0/A_s$, where S^0 is sample adsorption energy. However, the plot of $1/A_s$ versus E^0 did not suggest any relationship other than a horizontal line with zero slope and much scatter. If anything, this plot suggested that the molecular contact areas between solvents and sorbent might be characteristic of solvent class. Their molecular areas, in 8.5-Å^2 units, are inert solvents, 6.0 ± 1.9 (N = 25); solvating solvents, 5.0 ± 0.6 (N = 12); and hydroxylic solvents, 7.9 ± 0.2 (N = 9).

In addition to polarity, Snyder (3) classified solvents according to their solvent selectivity (X_i). Selectivity is defined as the ability of the solvent to engage in hydrogen bonding [as proton donors (X_d) or proton acceptors (X_e)], dipole interactions (X_n), or dispersion interactions (X_t). Selectivity constants are calculated as follows:

$$X_e\text{, proton acceptor activity} = \frac{\log K''_{\text{ethanol}}}{P'}$$

TABLE I. COMPILATION OF SOLVENT PROPERTIES

Solvent Class and Group, and Member Solvents[a]	$E^{0\,b}$	$P'^{\,c}$	ε^{d}	μ^{e}	$A_s^{\,f}$	$X_e^{\,g}$	$X_d^{\,h}$	$X_n^{\,i}$
Inert								
VII Pentane	0.00		1.84	0.00	5.9			
Hexane	0.00	0.1	1.88	0.09	6.8			
Isooctane	0.01	0.1	1.94	0	7.6			
Decane	0.04	0.4	1.99	0	10.3			
Cyclohexane	0.04	0.2	2.02		6.0			
Cyclopentane	0.05		1.97	0.00	5.2			
Diisobutylene	0.06				7.6			
1-Pentene	0.08		2.02	0.34	5.8			
1-Hexene			2.05	0.34				
Carbon disulfide	0.15	0.3	2.64	0.06	3.7			
Carbon tetrachloride	0.18	1.6	2.24	0	5.0			
Amyl chloride	0.26		6.6	1.94	4.2			
Butyl chloride	0.26		7.39	1.90				
Xylene	0.26	2.5	2.27	0.02	7.6	0.27	0.28	0.45
Isopropyl chloride	0.29		9.82	2.02	3.5			
Toluene	0.29	2.4	2.38	0.31	6.8	0.25	0.28	0.47
Propyl chloride	0.30		7.7	1.97	3.5			
Fluorobenzene		3.2				0.24	0.32	0.45
Chlorobenzene	0.30	2.7	5.62	1.54	6.8	0.23	0.33	0.44
Bromobenzene		2.7	5.40	1.55		0.24	0.33	0.43
Iodobenzene		2.8				0.24	0.35	0.41
Benzene	0.32	2.7	2.28	0	6.0	0.23	0.32	0.45
Ethyl bromide	0.37		9.39	1.90	3.4			
Chloroform	0.40	4.1	4.81	1.15	5.0	0.25	0.41	0.33
Nitrobenzene		4.4	34.82	4.03		0.26	0.30	0.44
Nitropropane	0.53		23.24	3.54	4.5			
Nitroethane		5.2	28.06	3.60		0.28	0.29	0.43
Aniline	0.62	6.3	6.89	1.51	6.7	0.32	0.32	0.36
Nitromethane	0.64	6.0	35.87	3.56	3.8	0.28	0.31	0.40
Ethoxybenzene (phenetole)		3.3	4.22	1.36		0.28	0.28	0.44

X_t[j]	E_T[k]	(^7Li)[l]	(^7Li)[m]	A_N[n]	A_N[o]	(^{19}F)[p]	(^{19}F)[q]	AN[r]	DN[s]
	51.96			15.134	15.219	-46.29	14.32	0.0	0.0
	50.87			15.289	15.374	-45.73	13.96		
	50.61			15.331	15.404	-46.39	13.96	8.6	0.0
0.13									
--				15.347	15.461	-46.06	13.85		
0.18									
0.17	48.62			15.472	15.563	-45.80	13.53		
0.18									
0.19									
0.16	49.73			15.404	15.532	-46.12	13.71	8.2	0.1
0.15	48.13			15.863	15.775	-45.90	12.81	23.1	
0.17								14.8	4.4
0.16									
0.13									
0.14	48.39			15.759	15.858	-45.10	12.34	20.5	2.7
0.16									

Table I (continued)

Solvent Class and Group, and Member Solvents[a]	E^{0}[b]	P'[c]	ε[d]	μ[e]	A_s[f]	X_e[g]	X_d[h]	X_n[i]
Methoxybenzene (anisole)		3.8	4.33	1.25		0.27	0.29	0.43
Phenyl ether		3.4	3.69	1.16		0.27	0.32	0.41
V Methylene chloride	0.42	3.1	8.93	1.14	4.1	0.29	0.18	0.53
Ethylene chloride	0.49	3.5	10.36	1.86	4.8	0.30	0.21	0.48
VIb Benzonitrile		4.8	25.20	4.05		0.31	0.27	0.42
Acetonitrile	0.65	5.8	37.5	3.44	10.0	0.31	0.27	0.42
Tris-Cyanoethoxy-propane		6.6				0.32	0.27	0.41
Oxydipropionitrile		6.3				0.31	0.29	0.40
Tetrahydro-thiophene-1, 1-dioxide		6.9				0.33	0.28	0.39
Propylene carbonate		6.1				0.31	0.27	0.42
Benzyl ether		4.1				0.30	0.28	0.42
Solvating								
I Isopropyl ether	0.28	2.4	3.88	1.22	5.1	0.48	0.14	0.38
Ethyl ether	0.38	2.8	4.34	1.15	4.5	0.58	0.13	0.34
Ethyl sulfide	0.38		5.72	1.61	5.0			
Tetramethyl-guanidine		6.1				0.47	0.17	0.35
Hexamethylphos-photriamide		7.4	30.	4.31		0.47	0.17	0.37
III 2,6-Lutidine		4.5				0.45	0.20	0.36
2-Picoline		4.9	9.8			0.44	0.21	0.36
Pyridine	0.71	5.3	12.4	2.37	5.8	0.41	0.22	0.36
Quinoline		5.0	9.00	2.18		0.41	0.23	0.36
Tetrahydrofuran	0.45	4.0	7.58	1.75		0.38	0.20	0.42

X_t^j	E_T^k	$(^7Li)^l$	$(^7Li)^m$	A_N^n	A_N^o	$(^{19}F)^p$	$(^{19}F)^q$	AN^r	DN^s
0.14									
0.19									
0.20				15.752	15.775	-45.56	13.28	20.4	
0.20	48.44			15.655	15.709			16.7	0.0
0.15								15.5	11.9
0.13	48.97	2.90	2.15	15.666	15.761			18.9	14.1
0.16									
0.17									
0.16									
0.15								18.7	15.1
0.17									
0.07									
0.09	51.58			15.334	15.421	-46.09	14.09	3.9	19.2
0.11									
0.12								10.6	38.8
0.11									
0.12									
0.13	47.98	-2.26	-3.02	15.608	15.663	-45.45	13.40	14.2	33.1
0.13									
0.14	48.72	1.06	0.00	15.373	15.474	-45.80	13.93	8.0	20.0

Table I (continued)

Solvent Class and Group, and Member Solvents[a]	E^0[b]	P'[c]	ε[d]	μ[e]	A_s[f]	χ_e[g]	χ_d[h]	χ_n[i]
Tetramethylurea		6.0	23.06	3.47		0.42	0.19	0.39
Dimethylacetamide		6.5	37.78	3.72		0.41	0.20	0.39
Dimethylformamide		6.4	36.71	3.86		0.39	0.21	0.41
_N_-Methyl-2-pyrrolidone		6.7	32.0	4.09		0.40	0.21	0.39
Methylformamide		6.0	182.4	3.86		0.41	0.23	0.36
Methoxyethanol		5.5	29.6	2.08		0.38	0.24	0.38
Diethylene glycol		5.2	31.69	2.31		0.44	0.23	0.33
Triethylene glycol		5.6	23.69	5.58		0.42	0.24	0.34
Dimethylsulfoxide	0.62	7.2	46.68	3.90	4.3	0.39	0.23	0.39
Butyl ether		2.1	3.08	1.18		0.44	0.18	0.38
Nonylphenolethylate						0.38	0.25	0.40
VIa Tricresyl phosphate		4.6	6.9			0.36	0.23	0.41
Methylisobutylketone	0.43		13.11		5.3			
Butanone	0.51	4.7	18.51	2.76	4.6	0.35	0.22	0.43
Acetophenone		4.8	17.39	2.96	4.2	0.33	0.26	0.41
Acetone	0.56	5.1	20.70	2.60		0.35	0.23	0.42
Cyclohexanone		4.7	18.3	3.01		0.36	0.22	0.42
Ethyl acetate	0.58	4.4	6.02	1.88	5.7	0.34	0.28	0.43
Methyl acetate	0.60		6.68	1.61	4.8			
Butyrolactone		6.5	39.	4.12		0.34	0.26	0.40
Bis-(2-ethoxyethyl) ether		4.6				0.37	0.21	0.43
Dioxane	0.56	4.8	2.21	0.45	6.0	0.36	0.24	0.40
Formylmorpholine		6.4				0.36	0.24	0.39
Cyanomorpholine		5.5				0.35	0.25	0.40
Hydroxylic								
II Octyl alcohol		3.4	10.34	1.76		0.56	0.18	0.25
Triethylamine	0.54	1.9	2.42	0.87	7.5	0.56	0.12	0.32
Isoamyl alcohol	0.61	3.7	14.7	1.82	8.0	0.56	0.19	0.26

X_t[j]	E_T[k]	(^7Li)[l]	(^7Li)[m]	A_N[n]	A_N[o]	(^{19}F)[p]	(^{19}F)[q]	AN[r]	DN[s]
0.14									
0.13								13.6	27.8
0.14		-0.32	-0.17					16.0	26.6
0.15								13.3	27.3
0.11								32.1	
0.11									
0.09									
0.12									
0.14	47.26	1.29	1.29	15.692	15.771	-44.81	13.40	19.3	29.8
0.11									
0.13									
0.15									
0.13									
0.15									
0.12	49.14	-1.23	-1.07	15.527	15.621	-45.43	13.65	12.5	17.0
0.14									
0.13		0.43	-0.56					9.3	17.1
								10.7	16.5
0.16									
0.13									
0.14	50.07			15.452	15.539	-45.63	13.69	10.8	
0.15									
0.16									
0.07									
0.05								1.4	61.0
0.04				16.004	15.961				

Table I (continued)

Solvent Class and Group, and Member Solvents[a]		E^0[b]	P'[c]	ε[d]	μ[e]	A_s[f]	X_e[g]	X_d[h]	X_n[i]
	Diethylamine	0.63		3.58	1.11	7.5			
	Isobutyl alcohol		4.1	17.93	1.79		0.56	0.20	0.24
	Butyl alcohol		3.9	17.51	1.75		0.59	0.19	0.25
	Isopropyl alcohol	0.82	3.9	19.92	1.66	8.0	0.55	0.19	0.27
	Propyl alcohol	0.82	4.0	20.33	3.09	8.0	0.54	0.19	0.27
	Ethyl alcohol	0.88	4.3	24.55	1.66	8.0	0.52	0.19	0.29
	Methyl alcohol	0.95	5.1	32.70	2.87	8.0	0.48	0.22	0.31
IV	Ethylene glycol	1.11	6.9	37.7	2.28	8.0	0.43	0.29	0.28
	Benzyl alcohol		5.7	13.1	1.66		0.40	0.30	0.30
	Acetic acid		6.0	6.15	1.66		0.39	0.31	0.30
	Formamide		9.6	111.	3.37		0.36	0.33	0.30
VIII	Dodecafluoroheptanol		8.8				0.33	0.40	0.27
	Tetrafluorapropanol		8.6				0.34	0.33	0.30
	m-Cresol		7.4	11.8	1.54		0.38	0.37	0.25
	Water		10.2	80.37			0.37	0.37	0.25

[a]Symbols used in Figures 1 through 14: solid circles, inert solvent open circles, solvating solvents; open triangles, hydroxylic solvents. Numbers in parentheses are literature citations.

[b]E^0, solvent strength (1).

[c]P', polarity (8).

[d]ε, dielectric constant (4).

[e]μ, dipole moment (4).

[f]A_s, molecular area, in 8.5-Å^2 units (3).

[g]X_e, proton acceptor activity (3).

[h]X_d, proton donor activity (3).

[i]X_n, strong dipole character (3).

X_t[j]	E_T[k]	(^7Li)[l]	(^7Li)[m]	A_N[n]	A_N[o]	(^{19}F)[p]	(^{19}F)[q]	AN[r]	DN[s]
0.02									
0.07									
0.03				15.973	16.044	-45.17	12.61	33.5	
0.06									
0.08	47.20			16.030	16.075	-45.02	12.40	37.1	
0.06	47.03	0.85	0.86	16.210	16.199	-44.79	12.03	41.3	19.0
0.13				16.364	16.298	-44.34	11.94		
0.11									
0.08								52.9	
0.20								39.8	
0.07									
0.08									
0.08									
0.07	0.00	0.00						54.8	18.0

[j] X_t, dispersion interaction (3).

[k] E_T, electronic transition energy, kilocalories per mole (6).

[l] δ, ^{7}Li-NMR chemical shift, $LiClO_4$ (7).

[m] δ, ^{7}Li-NMR chemical shift, LiBr (7).

[n] A_N, ^{14}N-ESR hyperfine splitting constant, di-*tert*-butyl nitroxide (8).

[o] A_N, ^{14}N-ESR hyperfine splitting constant, 4-amino-2,2,6,6-tetramethyl-piperid-1-yloxy (8).

[p] δ, ^{19}F-NMR chemical shift, 2-fluoropyridine (9).

[q] δ, ^{19}F-NMR chemical shift, 3-fluoropyridine (9).

[r] AN, acceptor number (10).

[s] DN, donor number (10).

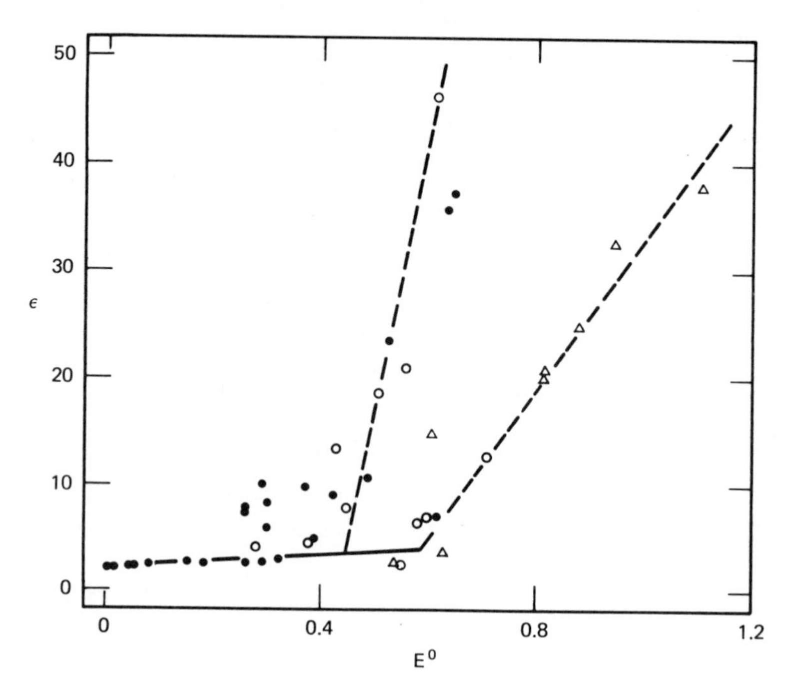

Figure 3. Correlation of dielectric constant (ε) and solvent strength by solvent class. (Same symbols as Figure 2. Data from reference 4.)

$$X_d\text{, proton donor activity} = \frac{\log K''_{dioxane}}{P'}$$

$$X_n\text{, strong dipole character} = \frac{\log K''_{nitromethane}}{P'}$$

$$X_t\text{, dispersion interaction} = \frac{\log K''_{toluene}}{P'}$$

For solutions of polar solutes, dispersion interactions (X_t) are considered to be relatively unimportant. This is seen in Figure 5. Although adsorption energy related to solvent strength, the proportion of dispersion interaction decreased with increasing solvent strength, while the proportion of specific interactions increased. For these reasons, solvent selectivity is thought to

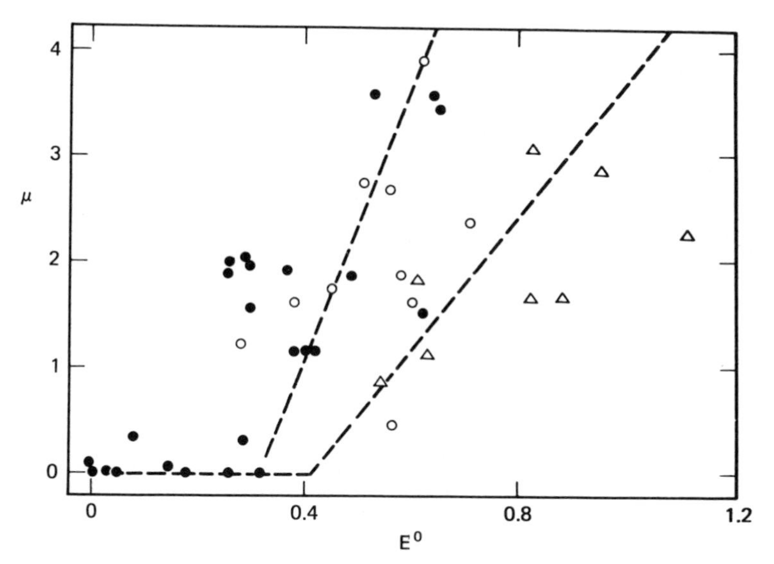

Figure 4. Correlation of dipole moment (μ) and solvent strength by solvent class. (Same symbols as Figure 2. Data from reference 4.)

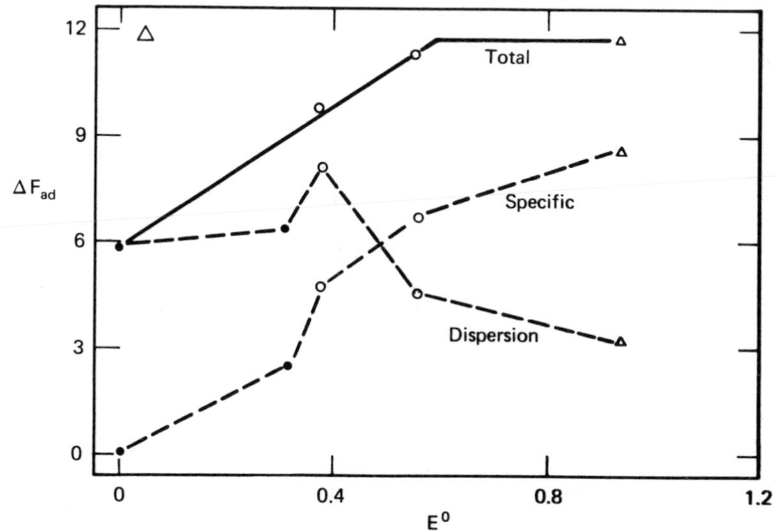

Figure 5. Partition of free energy of adsorption (ΔF_{ad}) between dispersion and specific energies as a function of solvent strength. (Same symbols as Figure 2. Data from reference 1.)

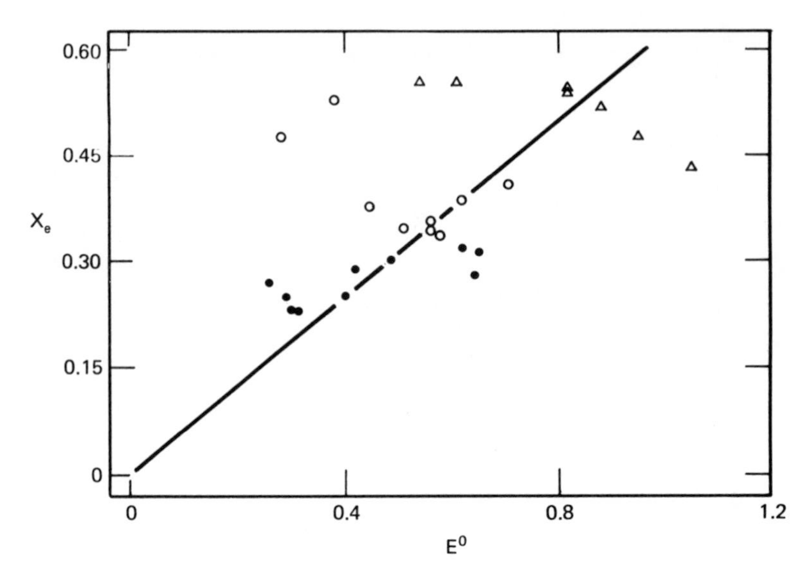

Figure 6. Correlation of proton acceptor activity (X_e) and solvent strength by solvent class. (Same symbols as Figure 2. Data from reference 3.)

be restricted to hydrogen bonding and dipole interaction. Although the data are limited, solvating solvents appeared to exhibit relatively greater specific interaction than inert solvent, and hydroxylic solvents relatively lesser amounts of specific interaction.

Proton acceptor activity (X_e) correlated reasonably well with solvent strength (Figure 6). Exceptions were noted for Snyder groups I (ethers), II (alcohols and amines), and IV (ethylene glycol). These groups were distributed across the correlation curve roughly according to their relative polarity. The ethers were displaced somewhat further than indicated by their polarity.

Proton donor activity (X_d) grouped the solvents into three solvent classes when correlated with solvent strength (Figure 7). With certain exceptions, each of the solvent classes was grouped into characteristic curves. The exceptions were the more polar of the inert solvents. Methylene chloride and ethylene chloride (group V), aniline (group VIb), and acetonitrile and nitromethane (group VII) were grouped with the solvating solvents.

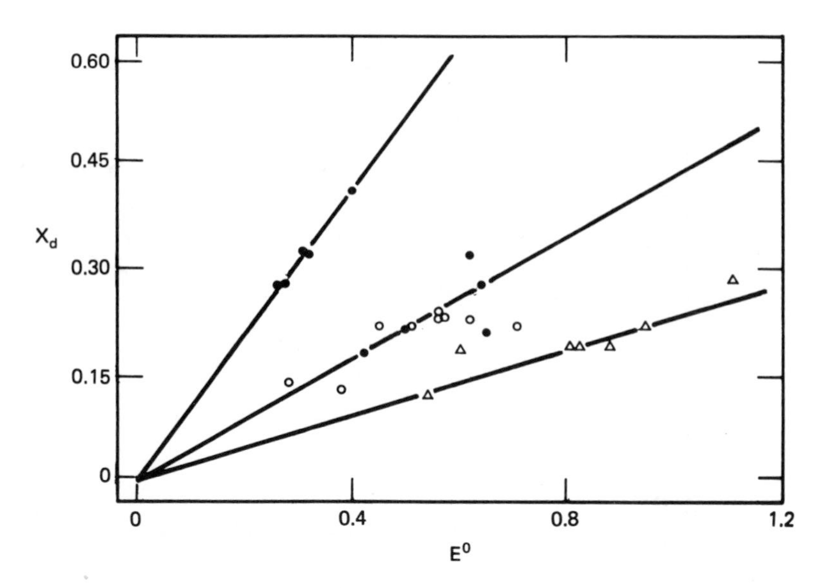

Figure 7. Correlation of proton donor activity (X_d) and solvent strength by solvent class. (Same symbols as Figure 2. Data from reference 3.)

Strong dipole character (X_n), with some exceptions, grouped the solvents into three solvent classes when correlated with solvent strength (Figure 8). The exceptions were the inert solvents, ethylene chloride (group V), acetonitrile (group VIb), and nitromethane, chloroform, and aniline (group VII), and the hydroxylic solvent, triethylamine (group II). These exceptions were found among the solvating solvents. The solvating solvent, isopropyl ether (group I), was grouped with the inert solvents.

Dispersion activity (X_t), with a few exceptions, grouped the solvents into three solvent classes when correlated with solvent strength (Figure 9). The exceptions were the inert solvents, acetonitrile (group VIb), and aniline and nitromethane (group VII). These exceptions were found among the solvating solvents.

In summary, proton acceptor activity (X_e) correlated with solvent strength, except for ethers and hydroxylic solvents. The latter solvents deviated from the correlation curve according to their relative polarity. Each of the selectivity constants--strong dipole character (X_n), proton donor activity (X_d), and dispersion interaction (X_t)--grouped the solvents into three solvent classes. Their relative contributions were approximately 45, 30, and 15,

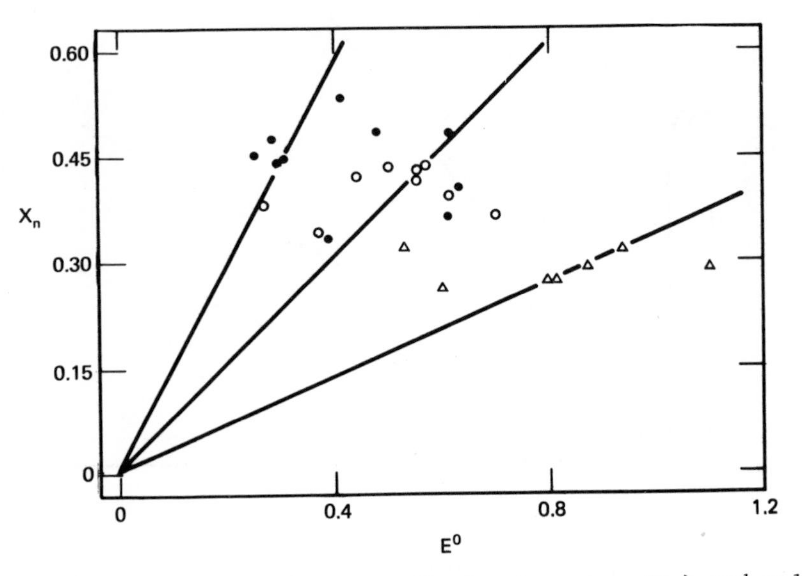

Figure 8. Correlation of strong dipole character (X_n) and solvent strength by solvent class. (Same symbols as Figure 2. Data from reference 3.)

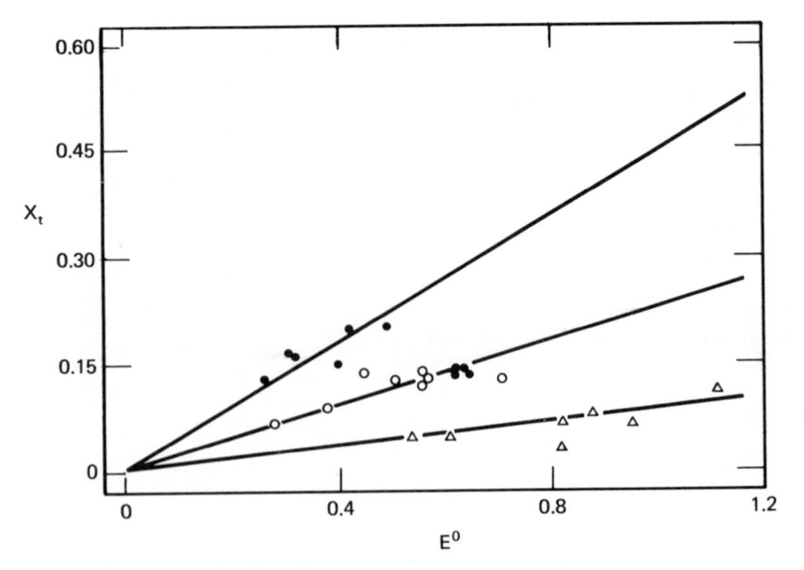

Figure 9. Correlation of dispersion forces (X_t) and solvent stren by solvent class. (Same symbols as Figure 2. Data from reference 3.)

respectively. The usual practice of overlooking the contribution of dispersion activity (X_t) may be justified because X_t seems to be analogous to both X_n and X_d, and because it provides only a minor contribution to the total selectivity. In effect, X_t seems to reinforce the effects of X_n and X_d without altering the qualitative picture or distorting their quantitative contribution to selectivity. Plots of X_e and any one of X_n, X_d, or X_t group the solvents into their proper classes and subgroups, although there is some overlapping observed for the X_e-X_t plot. Plots of any two combinations of X_n, X_d, and X_t can also group the solvents into their proper solvent classes and subgroups, although there is appreciable overlapping within some groups.

The triangular plot of X_e, X_d, and X_n values is shown in Figure 10. This plot showed the concordance of solvent classes and Snyder subgroups. The Snyder subgroups formed subdivisions of solvent classes. All data points fell within their expected class and group, except for tetrahydrothiophene-1,1-dioxide and acetophenone (at the boundary between groups VIa and VIb), and dodecafluoroheptanol (at the boundary between groups VII and VIII). The following generalizations have emerged:

1. The transition from inert to solvating and hydroxylic classes occurs at 0.33 proton acceptor activity (X_e) units.
2. The transition from solvating to hydroxylic classes occurs at 0.325 strong dipole character (X_d) units.
3. Within the inert class of solvents, the transitions from Snyder groups VII to VIb to V occur with increasing strong dipole character (X_n) or decreasing proton donor activity (X_d) or both. The boundaries are above 0.285 proton acceptor activity (X_e) units at 0.30 proton donor activity (X_d) units (group VIb) and at 0.23 proton donor activity (X_d) units (group V).
4. Within the solvating class of solvents, the transitions from Snyder groups I to III to VIa occur with decreasing proton acceptor activity (X_e) or increasing proton donor activity (X_d) or both, while maintaining a constant strong dipole character (X_n). The boundaries are at 0.40 proton acceptor activity (X_e) units (group I), 0.365 proton acceptor activity (X_e) units (group III), and 0.33 proton acceptor activity (X_e) units/0.37 strong dipole character (X_d) units (group VIa).
5. Within the hydroxylic class of solvents, the transition from Snyder groups II to IV to VIII occur with decreasing proton acceptor activity (X_e) or increasing proton donor activity (X_d) or both, while maintaining a constant

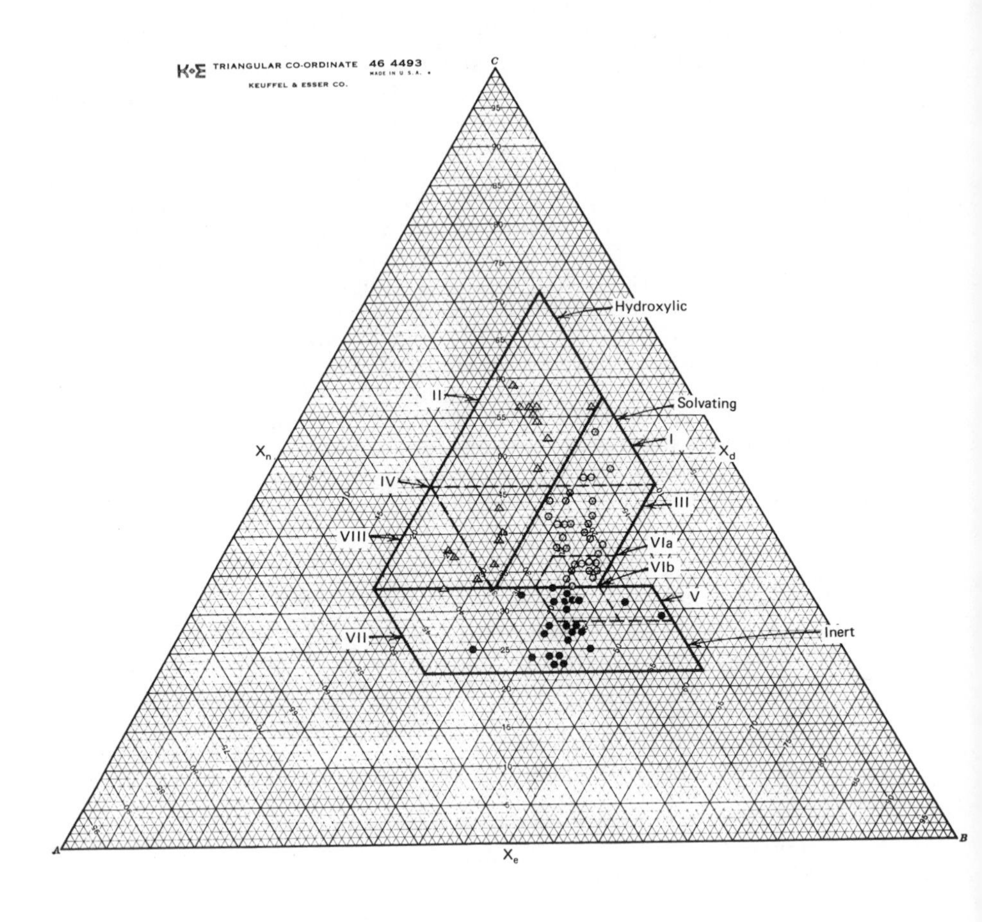

Figure 10. Triangular plot of X_e, X_d, and X_n constants by solvent class and Snyder group. (Same symbols as Figure 2. Data from reference 3.)

strong dipole character (X_n). The boundaries are at 0.46 proton acceptor activity (X_e) units (group II), below 0.345 proton donor activity (X_d) units (group IV), and above 0.345 proton donor activity (X_d) units (group VIII).

Sorption can be considered to involve ionic bonding and hydrogen bonding. Ionic bonds are characterized by sharp charge separation, and chromatography involving ionic bonding should be very sensitive to solvents. The ionizing process would be dependent on both the dielectric constant of the medium, which indicates the ease of forming charged products, and solvation, which stabilizes the charged products. Dipole moment may not be as appropriate as dielectric constant for characterizing solvents, because dipole moment assesses the net molecular polarity when polar bonds are opposing and mutually canceling.

Cations are electrophiles, which are deficient in electrons, and they are solvated by compounds containing unshared electron pairs. These solvents, such as water, ammonia, alcohols, ethers, carbonyls, organic acids and esters, and sulfur dioxide and dimethylsulfoxide, function as Lewis bases. Their basicity can be assessed in terms of proton acceptor activity (X_e). Anions are nucleophiles, which contain a surplus of electrons, and they are solvated by solvents containing hydrogen attached to a strongly electronegative element by means of a highly polar bond. These solvents, such as water, alcohols, and amines, function as Lewis acids. Their acidity can be assessed in terms of proton donor activity (X_d) and strong dipole character (X_n). The latter parameters are related inversely. Dispersion forces (X_t) would not be especially significant for polar compounds.

Hydrogen bonds are characterized by diffuse charge separation, and chromatography involving hydrogen bonds should be sensitive to solvents, but less dependent than chromatography involving ionic bonds. In addition to solvents suitable for ionic bonds, polar aprotic solvents, such as dimethylsulfoxide, tetramethylenesulfone (sulfolane), dimethylformamide, and hexamethylphosphotriamide, can serve for chromatography involving hydrogen bonds. Polar aprotic solvents have high dielectric constants, but lack the capability to form hydrogen bonds with solutes. Consequently, these solvents would be effective for reactions involving diffuse charge separations, and they might provide solute selectivity different from hydrogen bonding solvents.

SPECTROSCOPIC EVIDENCE FOR SOLVENT CLASSES

Empirical measures of solvent polarity have been attributed to medium effects on the electronic transition energies of model chromophores, and on NMR chemical shifts and ESR hyperfine splitting constants for various isotopic species. For adsorption chromatography, these measures could prove useful for classifying solvent properties to confirm the concept of solvent classes.

The correlation of electronic transition energy (E_T) and solvent strength is shown in Figure 11a. This correlation grouped the solvents into inert and solvating classes with some exceptions. The inert solvents, acetonitrile and nitromethane, were grouped among the solvating solvents, and the solvating solvents, dimethylsulfoxide and tetrahydrofuran, were found among the inert solvents. The third class of solvents, the hydroxylic class, could not be established with the limited number of data points available. Similar results were obtained with a Lewis acid probe. The correlation of ^{7}Li-NMR chemical shift and solvent strength is shown in Figure 11b. Although the number of data points was limited, the solvents could be grouped into inert, solvating, and hydroxylic classes. In this comparison and in others to be discussed, dimethylsulfoxide behaved as an inert solvent, although it was classified as a solvating solvent. Lewis acid probes would seem to be appropriate probes to simulate the surface of the silica gel with which solvent classes are especially evident.

In contrast, Lewis base probes provide a pattern of response different from those given by the Lewis acid probes. The correlation of ^{14}N-ESR hyperfine splitting constants (A_N) and solvent strength is shown in Figure 12. These probes arranged the solvents into a solvating class and a combined inert-hydroxylic class. Of possible interest is the fact that the curves for these two classes of solvents intersected at -0.25 solvent strength units, the low end of the solvent strength scale provided by the fluoroalkanes.

The correlation of ^{19}F-NMR chemical shift and solvent strength also grouped the solvents into a solvating class and a combined inert-hydroxylic class of solvents (Figure 13). Because 3-^{19}F-pyridine involves a less complicated mechanism of interaction with the medium than 2-^{19}F-pyridine, it should provide more reliable data. In these comparisons made with Lewis base probes, dimethylsulfoxide exhibited a considerable degree of inert solvent character.

Finally, the correlation of acceptor number (AN) and solvent strength is shown in Figure 14a. Acceptor number is a measure of

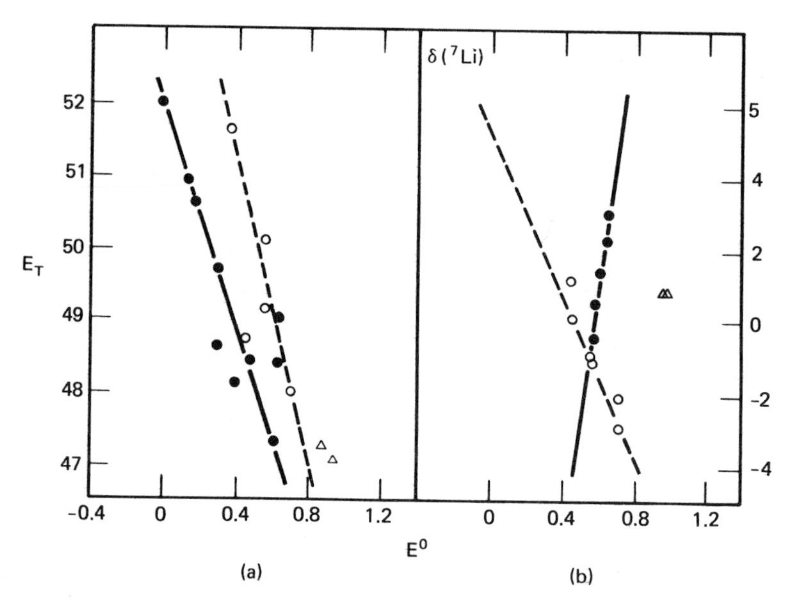

Figure 11. Correlation of (a) electronic transition energy for phenol blue (E_T) or (b) ^{7}Li-NMR chemical shift (δ) with solvent strength by solvent class. (Same symbols as Figure 2. Data from references 6 and 7.)

the electrophilic character, or localized electron deficiency in the solvent molecule, its Lewis acid function. However, AN grouped the solvents into an inert class and a combined solvating-hydroxylic class, a different arrangement from those discussed previously. Exceptions were the inert solvents, acetonitrile and nitromethane, which were found in the solvating-hydroxylic class. The correlation of donor number (DN) and solvent strength is shown in Figure 14b. Donor number is a measure of nucleophilic character, or localized electron surplus in the solvent molecule, its Lewis base function. However, DN grouped the solvents into inert and solvating classes of solvents, like a Lewis acid probe. Exceptions were the inert solvents, acetonitrile and nitromethane, which possessed some solvating solvent character. Not enough data points were available to decide the arrangement of the hydroxylic solvents.

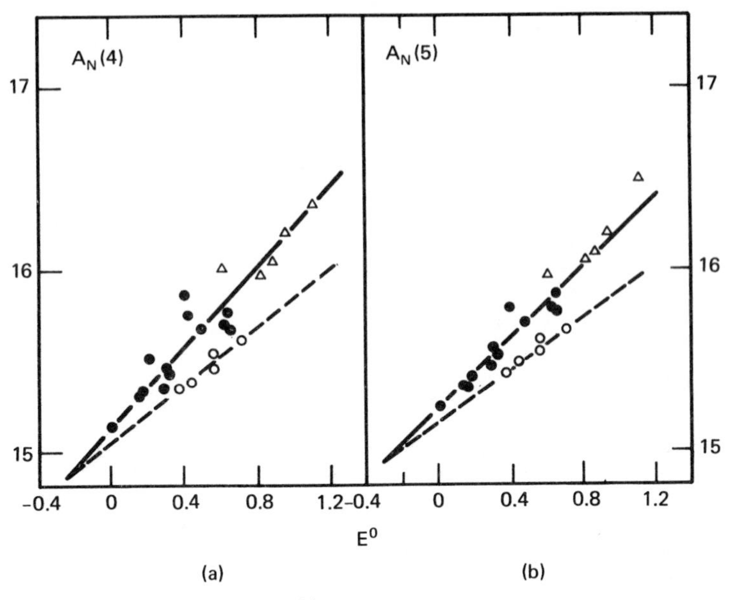

Figure 12. Correlation of ^{14}N-ESR hyperfine splitting constants (A_N) for (a) di-tert-butyl nitroxide or (b) 4-amino-2,2,6,6-tetramethylpiperid-1-yloxy and solvent strength by solvent class. (Same symbols as Figure 2. Data from references 6 and 8.)

A recapitulation of the spectroscopic properties of inert solvents, acetonitrile and nitromethane, and solvating solvents, dimethylsulfoxide and tetrahydrofuran, revealed an interesting ambivalence among these solvents. This comparison is shown in Table II. Acetonitrile exhibited significant degrees of solvating solvent character (40 to 100%) by all the solvent parameters tested, and nitromethane exhibited solvating solvent character by some, but not all, of the solvent parameters tested. On the other hand, dimethylsulfoxide and tetrahydrofuran exhibited inert solvent character to varying degrees depending on the solvent parameter used. Assuming the validity of the data, solvent character must be a complex relationship of several physical measures of solvent character. This complexity was also evident in Snyder's three selectivity criteria.

HETEROGENEOUS SURFACE STRUCTURE OF AMORPHOUS SILICA GEL

For polar sorbents, especially silica gel, the principal mode of sorption is hydrogen bonding between the sorbent, which functions

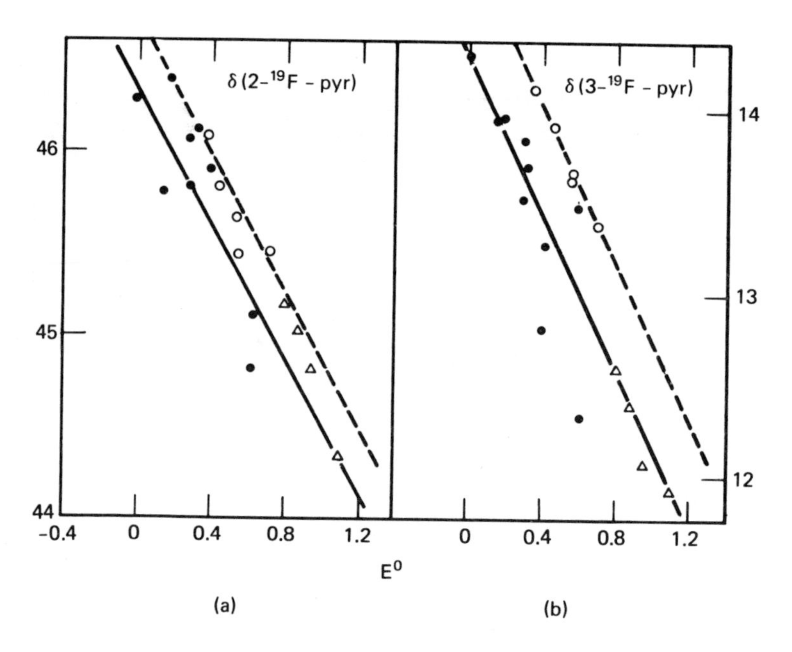

Figure 13. Correlation of ^{19}F-NMR chemical shift (δ) for (a) 2-fluoropyridine or (b) 3-fluoropyridine and solvent strength by solvent class. (Same symbols as Figure 2. Data from reference 9.)

as a Lewis acid (electron acceptor), and the solutes and solvents, which function as Lewis bases (electron donors). Hydrogen bonding is defined as the interaction between weak acids and bases without proton transfer. For a given sorbent, the stronger the basicity of the solute or solvent, the stronger will be its binding. Additional electron donor groups may enhance binding by rendering the solute or solvent more basic. For polar sorbents, all other mechanisms of sorption seem to be relatively insignificant.

The surface structure of polar sorbents and the preparative treatments that influence the surface structures are crucial to an understanding of sorption. In the case of silica gel, the constellation of hydroxyl (silanol) groups determines both strength and selectivity of sorption. Surface area can be altered by controlling the acidity of precipitation and by heat treatment of the moist gel. The activity of the silica gel surface can be enhanced by dehydration and moderated by subsequent controlled rehydration. Variation in the activity of silica gel was apparent in Figure 1. Whatman LK5D was appreciably (27%) more

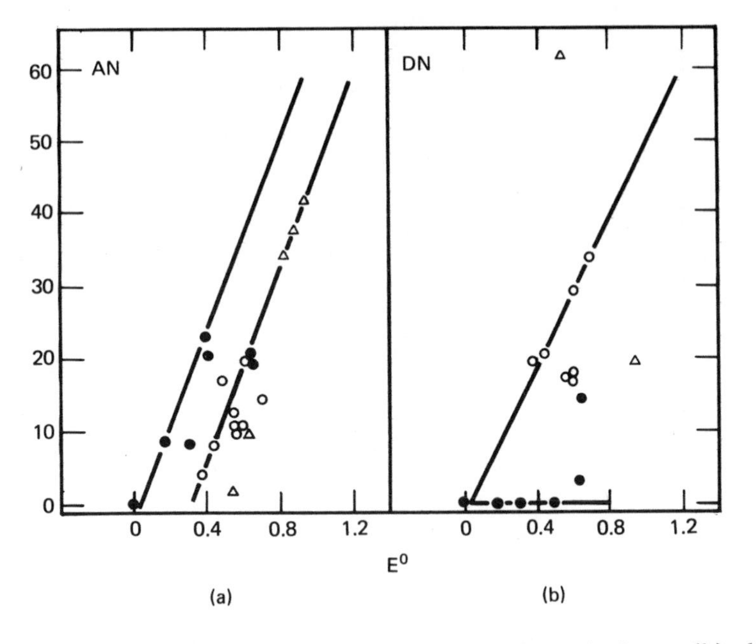

Figure 14. Correlation of (a) acceptor number (AN) or (b) donor number (DN) and solvent strength by solvent class. (Same symbols as Figure 2. Data from reference 10.)

sensitive to the solvating solvents than was Whatman LK6D.

For crystalline silica, all surface hydroxyl groups are equivalent because they are well-separated from each other and unable to interact. These hydroxyl groups are called "free" hydroxyls. For amorphous silica gel, the surface hydroxyl groups are predominately "reactive" and "bound" hydroxyl groups because they are close enough to interact with other hydroxyl groups. For wide-pore silica gel, the predominant hydroxyl group is considered to be the free hydroxyl group. For narrow-pore silica gel, the predominant hydroxyl groups are apparently reactive and bound hydroxyl groups. The binding strengths of these hydroxyl groups vary accordingly: bound < free < reactive.

TABLE II. THE SOLVENT CHARACTERISTICS OF SOME NONCONFORMING SOLVENTS[a]

	Inert Solvents		Solvating Solvents	
Solvent Parameters	Acetonitrile	Nitromethane	Dimethylsulfoxide	Tetrahydrofuran
Snyder subgroups (3)	VIb	VII	III	III
E^0, solvent strength	0.65	0.64	0.62	0.45
E_T, Electronic transition energy	Solvating	Solvating	Inert	Inert
δ, ^{7}Li-NMR chemical shift	(No data)	(No data)	Inert	Solvating
A_N, ^{14}N-ESR hyperfine splitting	0.4 Solvating	Inert	0.8 Inert	Solvating
δ, ^{19}F-NMR chemical shift	(No data)	Inert	0.3 Inert	Solvating
AN, Acceptor number	Solvating-Hydroxylic	Solvating-Hydroxylic	Solvating-Hydroxylic	Solvating-Hydroxylic
DN, Donor number	0.4 Solvating	0.1 Solvating	Solvating	Solvating

[a]Data from Figures 10 to 14.

The hydroxylated surface of crystalline silica has the following structure:

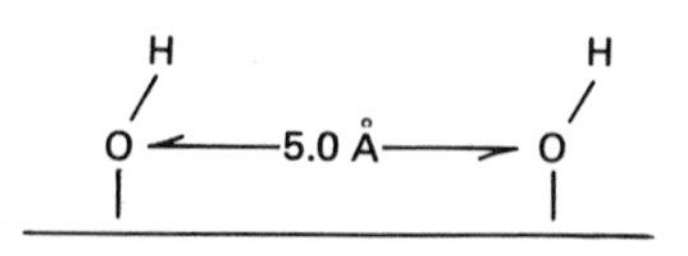

In contrast, the hydroxylated surface of amorphous silica gel has the following structures:

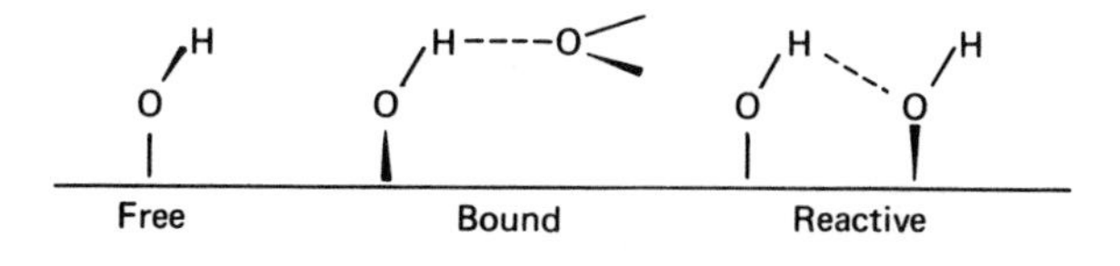

The selective sorption of unsaturated and polar solutes or solvents occurs at the reactive hydroxyl groups. These groups are also selective binding sites for polyfunctional solutes or solvents, whereas monofunctional solutes and solvents are bound equally strongly to free and reactive hydroxyl groups. Water is selectively bound to the reactive hydroxyl groups. The characteristic water content of activated silica gels is a gauge of their surface activity as well as the remaining reactive hydroxyl groups. Because of its selectivity, amorphous silica gel serves as a practical means for resolving mixtures of polar solutes.

SOME SECONDARY SOLVENT EFFECTS

Secondary solvent effects can occur in the solvent phase and influence solute sorption by controlling the availability of the sorbing material. Hydrogen bonding is one mechanism that may be involved. The strength of the hydrogen bond will depend on the basicity of the proton acceptor (amines) or the acidity of the proton donor (alcohols, phenols) or both, and the geometry (linearity) of the hydrogen bond thus formed. Other types of chemical interactions may also be involved. Examples of these reactions include (1) Schiff bases formed from aldoses (but not ketoses) and primary (but not secondary or tertiary) amines--the cyclic Schiff bases thus formed do not bind as strongly as the free aldoses and the ketoses; (2) acidic adducts formed from *cis*-diols and boric acid, which bind strongly to anion exchange sorbents and less strongly to silica gel; and (3) the polar adduct formed from decaborane and ether, which binds more strongly to silica gel than the nonpolar decaborane.

Secondary solvent effects can also involve the sorbent phase. These effect include the phenomena called (1) the basic eluent anomaly, and (2) the ether anomaly. The basic eluent anomaly refers to the enhanced binding of both a secondary amine and an aromatic amine or phenol to alumina from basic solvents. The sorbent concentrates the solutes in an array favorable for hydrogen bonding. The hydrogen-bonded pair is held more strongly to the sorbent than either of the solutes separately. The ether anomaly describes reduced sorption of compounds with strongly bonding functional groups from ether solutions. Ethers behave as stronger solvents because they have easily accessible functional groups, which bind to the strong binding sites on silica gel and exclude the polar solutes from their usual binding sites.

THE ETHER ANOMALY

For mixed function molecules, Snyder (3) has predicted that the more polar functional group will predominate selectivity. Alcoholic groups would be expected to dominate ether groups, for instance. This prediction does not prove to be true because methyoxyethanol and di- and triethylene glycols behave more like ethers than alcohols. In general, solvating solvents (ethers included) are chromatographically more efficient than indicated by their solvent strength values (Figures 1 and 2). The concept of solvent classes was introduced when deviations were noted from the expectations of the solvent strength theory. These deviations suggested that the solvating solvents were interacting more strongly or more specifically with the sorbent surface. Possibly, ethers interact preferentially with reactive hydroxyl groups of silica gel.

A physical characteristic of solvating solvents, including ethers, is their relatively small molecular contact areas, which could fit the reactive hydroxyl sites. In this case, structural effects ought to be noted. A comparison of various ethers is shown in Table III. Replacing the hydrogen atoms of ethyl ether with bulkier substituents reduced both their basicity (proton acceptor activity, X_e) and relative position in the solvent class/ group diagram (Figure 10). Likewise, the introduction of hydroxyl

TABLE III. INFLUENCE OF SUBSTITUENTS ON THE SOLVENT PROPERTIES OF SOME ETHERS[a]

Solvent Class	Group	Basicity (X_e)	Substituted Ethers
Solvating	I	0.53	Ethyl ether
		0.48	Isopropyl ether
		(0.46)	
	III	0.44	Butyl ether, diethylene glycol
		0.42	Triethylene glycol
		0.38	Tetrahydrofuran, methoxyethanol
		(0.375)	
	VIa	0.37	bis-(2-Ethoxyethyl) ether
		0.36	Dioxane, cyanomorpholine
		0.35	Formylmorpholine
		(0.33)	
Inert	VIb	0.32	tris-(Cyanoethoxy) propane
		0.31	Oxydipropionitrile
		0.30	Benzyl ether
		(0.285)	
	VII	0.28	Ethoxybenzene (phenetole)
		0.27	Methoxybenzene (anisole), phenyl ether

[a]Data from Table 1.

groups, structural rigidity (by cyclization), or aromatic rings reduced both basicity and relative position in the solvent class/group diagram. Ultimately, ethers become inert solvents through substitution. It remains to be seen whether there is interference with polar solutes by ethers at common binding sites on the sorbent surface.

DISCUSSION

This paper and others cited in references 1 and 3 have shown that the solvent exhibits a primary effect, which is determined by solvent strength and modified by solvent class, and a secondary effect, which is dependent on various solvent/solute/sorbent interactions. The primary effect determines solute mobility, while the secondary effect contributes to solute selectivity. The secondary effect is associated with solvent class in an as yet incompletely understood manner. Solvent class is determined primarily by proton acceptor activity (X_e) and secondarily by both proton donor activity (X_d) and strong dipole character (X_n). Adsorption chromatography is formulated as:

$$\log K^0 = \log V_a + \alpha(S^0 - A_s \cdot E^0)$$

where K^0 is the adsorption isotherm, V_a the adsorbent surface volume, α the adsorption surface area, S^0 the sample adsorption energy, A_s the sample molecular size, and E^0 the solvent strength. This formulation assumes that solvent/solute interactions are not significant, and solute (or solvent)/sorbent interactions are at least functionally equivalent, if not identical. This formulation is satisfactory for weak to moderately strong solvents. As solvent strength increases, the contribution of both liquid-phase and sorbent-phase interactions becomes increasingly more significant. These secondary effects can be evaluated by adding the term Δ_{eas}, secondary activity effects, to the formulation above. A chemical evaluation of Δ_{eas} will be the target of future work.

Solvent/sorbent interactions seem to be especially significant for solute selectivity. These interactions may involve localized sorption at stronger binding sites. Secondary effects imply heterogeneity of binding sites or binding mechanisms. A general-purpose sorbent appears to function as if its surface were heterogeneous owing to the presence of various types of binding groups. In particular, silica gel appears to provide the structural heterogeneity needed to account for the functional heterogeneity that is implicit in the solvent class concept and in solute selectivity. It is here that the inherent character of the solvents comes into play. Solvent classes represent one aspect of chromatographic selectivity, which can serve as a basis for a more comprehensive understanding of selectivity in general. The elucidation of solvent character by physical methods offers the promise for understanding solvent interactions more explicitly.

REFERENCES

1. L.R. Snyder, Principles of Adsorption Chromatography, Marcel Dekker, New York, 1968.

2. D. Rogers, Amer. Lab., 12, 49 (1980).
3. L. R. Snyder, J. Chromatogr. Sci., 16, 223 (1978).
4. J. A. Ruddick and W. A. Bunger (Eds.), Solvents: Physical Properties and Methods of Preparation, 3rd. ed., Wiley-Interscience, New York, 1970.
5. J. Jacques and J. P. Mathieu, Bull. soc. chim.,(1946), 94.
6. O. Kolling, Anal. Chem., 49, 591 (1977).
7. G. E. Maciel, J. K. Hancock, L. F. Lafferty, P. A. Mueller, and W. K. Muskie, Inorg. Chem., 5, 554 (1966).
8. B. Knauer and J. Napier, J. Am. Chem. Soc., 98, 4395 (1976).
9. C. Giam and J. Lyle, J. Am. Chem. Soc., 95, 3235 (1973).
10. U. Mayer, Pure Applied Chem., 51, 1697 (1979).

CHAPTER 7

Radioscanning of TLC

Heinz Filthuth

INTRODUCTION

In this article we describe some new developments in the detection technique of radiochromatography and electrophoresis.

The classical method is autoradiography, the exposure of thin layer plates or electrophoresis gels to X-ray film. After relatively long exposure times, days or weeks, the radioactive regions (spots) can be identified on the film. Their relative intensities can be evaluated by photometric methods from the X-ray film, which is rather limited because of the small dynamic range of the film. A far more precise method is the technique of liquid scintillation counting. The identified radioactive regions have to be removed from the plates, gels, or paper, and the substances have to be separated from the carrier material and introduced into the scintillation liquid.

This method is time-consuming, demands extensive human labor, and requires the operation of the liquid scintillation counter, which is costly and has the problem of producing radioactive and poisonous waste.

About 10 years ago the thin layer chromatography (TLC) scanner was introduced. It replaced autoradiography, when a spatial resolution of only a few millimeters was required. With a methane flowing through counter and a small entrance window of about 2 × 20 mm the surface of the TLC plate is scanned in several hours or a day, depending on the radioactivity in the plate.

In particle physics new experimental techniques and methods have been developed over the past 10 years to detect particle trajectories and to identify these particles.

With multiwire spark chambers and multiwire proportional chambers one detects particle trajectories in space to a very high precision and can determine the origin of their emission with a precision of about 100 μm. These experimental setups at accelerator laboratories are very large, complicated, and expensive, measured on the scale of "normal laboratories." In the last years several of the experimental techniques of high-energy physics have been applied to the fields of medicine and biology, like the cut scanner, the positron camera, and multiwire proportional chambers to study p.ex. molecular structure by X-ray scattering.

The aim of this work was to design a detector, relatively simple and not too expensive, to measure the distribution of radiochromatograms and electropherograms, replacing the old TLC scanner. It is described in the following article on the (TLC) linear analyzer and the beta camera.

THE (TLC) LINEAR ANALYZER

For the detection of "one-dimensional" radiochromatograms it is natural to use the technique of position-sensitive counters, which are widely used in particle physics.

The linear analyzer (Figure 1) introduced here is basically designed for the same applications as the TLC scanner, but there is a fundamental difference: The TLC scanner only counts that part of the chromatogram that is directly below the slit diaphragm with its typical width of 2 mm, and the distribution curve is produced by counting the individual slit areas sequentially.

The linear analyzer, in contrast, counts the whole trace simultaneously, using a position-sensitive detector. The main advantage of the linear analyzer, therefore, is that its sensitivity or speed is greater than that of the TLC scanner by a factor of approximately 100.

Basic Principle

The heart of the linear analyzer is a position-sensitive proportional counter. β-particles emitted from the chromatogram plate enter the counter from the bottom through a 250-mm-long and 15-mm-wide <u>open</u> entrance window. Therefore, the counter can

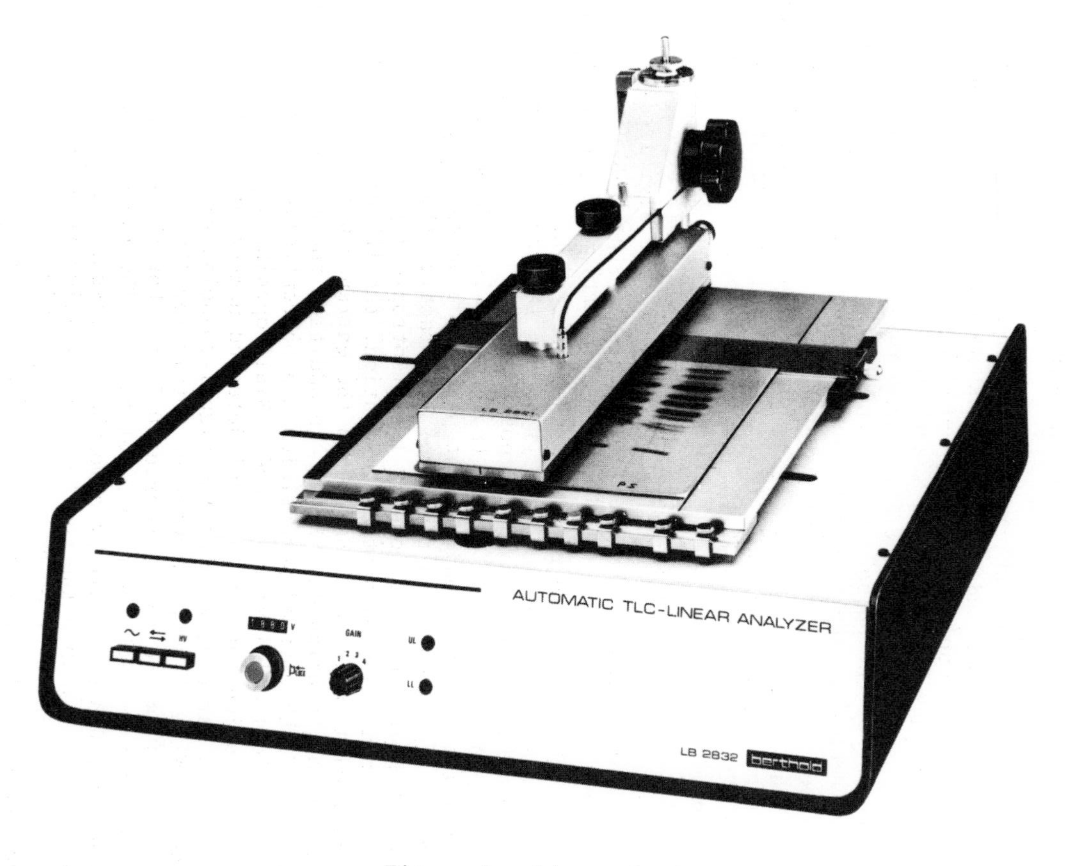

Figure 1. Linear Analyzer LB 283.

detect very low energy β-particles, as emitted from ^{3}H and ^{125}J. The counting wire is stretched along the center line of the counter. As the detector operates as a proportional counter, the electron avalanche produced by gas amplification of the primary ionization is limited to locations above those points where particles are emitted from the TLC plate.

The count pulses, which are proportional in height to the electron avalanches and in time to the position of the incident particles, are coupled electromagnetically into a delay line, which also runs over the full length of the counting chamber. These pulses pass along the delay line in both directions, and the arrival of the pulses at each end of the delay line is electronically recorded. The time difference between the arrival of two associated pulses is a direct measure of the location of an incident particle (see Figure 2). After preamplification the pulses are shaped in the time analyzer and converted into the fast output pulses, "start" and "stop," the stop pulse being delayed by an additional fixed delay time T_D equal to the delay time T_D of the delay line of the counter. The stop-start time difference is digitized in the time digital converter, and the digitized signal is received by the data acquisition system 3500 (Figure 3).

The time difference is T(stop) - T(start) = $2T_1$, where T_1 is the time the stop pulse runs from the position X_1 to 0. Time differences from 1 to 1000 nsec (1 nsec = 10^{-9} sec) are measured, corresponding to 0.25 to 250 mm positions.

The inherent resolution of the counter is 0.25 mm; this is the resolution for a nondivergent particle beam entering the counter perpendicular to the wire. In practice the β-sources from TLC plates radiate more or less isotropically.

To obtain a good spatial resolution, the TLC plate is positioned as close as possible to the entrance window, the counter height being only 5 mm. In addition the accepted opening angle θ of the emitted β-particles can be controlled electronically. No mechanical collimator is used. Small opening angle θ means high resolution but some loss of detected number of particles per minute via large θ, whereas the signal-to-noise ratio does not change. With this arrangement we obtain for ^{14}C sources a resolution of 1 mm and for ^{3}H sources 0.5 mm. For ^{3}H the problem of a divergent particle beam does not exist. The range of ^{3}H β's being only 0.5 mm in air at atmospheric pressure, they do not enter directly into the counter. β-particles from ^{3}H and ^{125}I produce a thin layer of negative (and positive) charges by primary ionization at the surface of the TLC plate. These negative charges, electrons, are drifted along the electric field lines of the

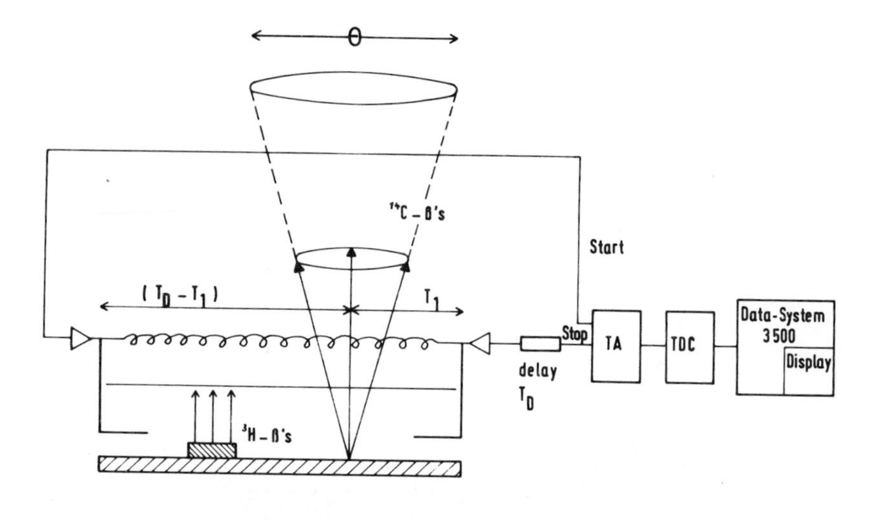

L .... X_1 .... 0 : X

$2 \cdot T_D$ .... $2 \cdot T_1$ .... 0 : ΔT

$$T_{start} = T_D - T_1$$

$$T_{stop} = T_1 + T_D$$

$$\Delta T = T_{stop} - T_{start} = 2 \cdot T_1$$

$$T_D = 500 \text{ nsec}$$

$$L = 250 \text{ mm}$$

$$0{,}25\text{mm} = 1 \text{ nsec}$$

Figure 2. Principle of position-sensitive proportional counter with delay line read out.

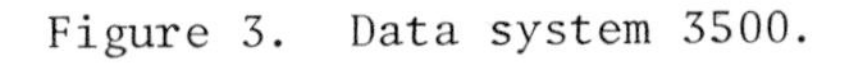
Figure 3. Data system 3500.

counting wire into the counter volume and are detected. Therefore, one obtains very sharp contours from ^3H sources, the resolution being 0.5 mm.

DATA SYSTEM 3500

The digitized start-stop signals are introduced into the data system 3500. A TLC plate with up to 10 chromatogram tracks can be measured fully automatically. The position of each track is manually preselected on the plate carriage. For each track the title, preselected measuring time or counts, memory group, are typed into the data memory by means of the keyboard of model 3500. After the measurement of a track is completed, the data are stored in the memory, the carriage receives a signal from the system 3500 and moves to the next preselected position, and the new measurement begins.

After data acquisition of a track (or several tracks), the user can call the data from the memory via the keyboard or light pen, being displayed at the television screen of the system. He can demand a fully automatic analysis of the data. The program finds all the peaks, subtracts background, and gives a full report of the data analysis on a hard copy, as peak position, width, counts, percentage areas (see Figure 4), and a graphic plot of the measured chromatogram, the ROI's* being labeled. The position scale can be automatically calibrated in any units, as millimeters or R_f values.

Of course, the data can be analyzed manually as well by interaction of the user with the data system 3500 via its keyboard and light pen.

The following analysis functions are available:

1. Calibration.
2. Background subtraction.
3. Smoothing (3,5,7,9 point).
4. Integration.
5. Addition of any 2 graphic displays.
6. Subtraction of 2 graphic displays.
7. Normalization.

---

*ROI = region of interest = region of integration.

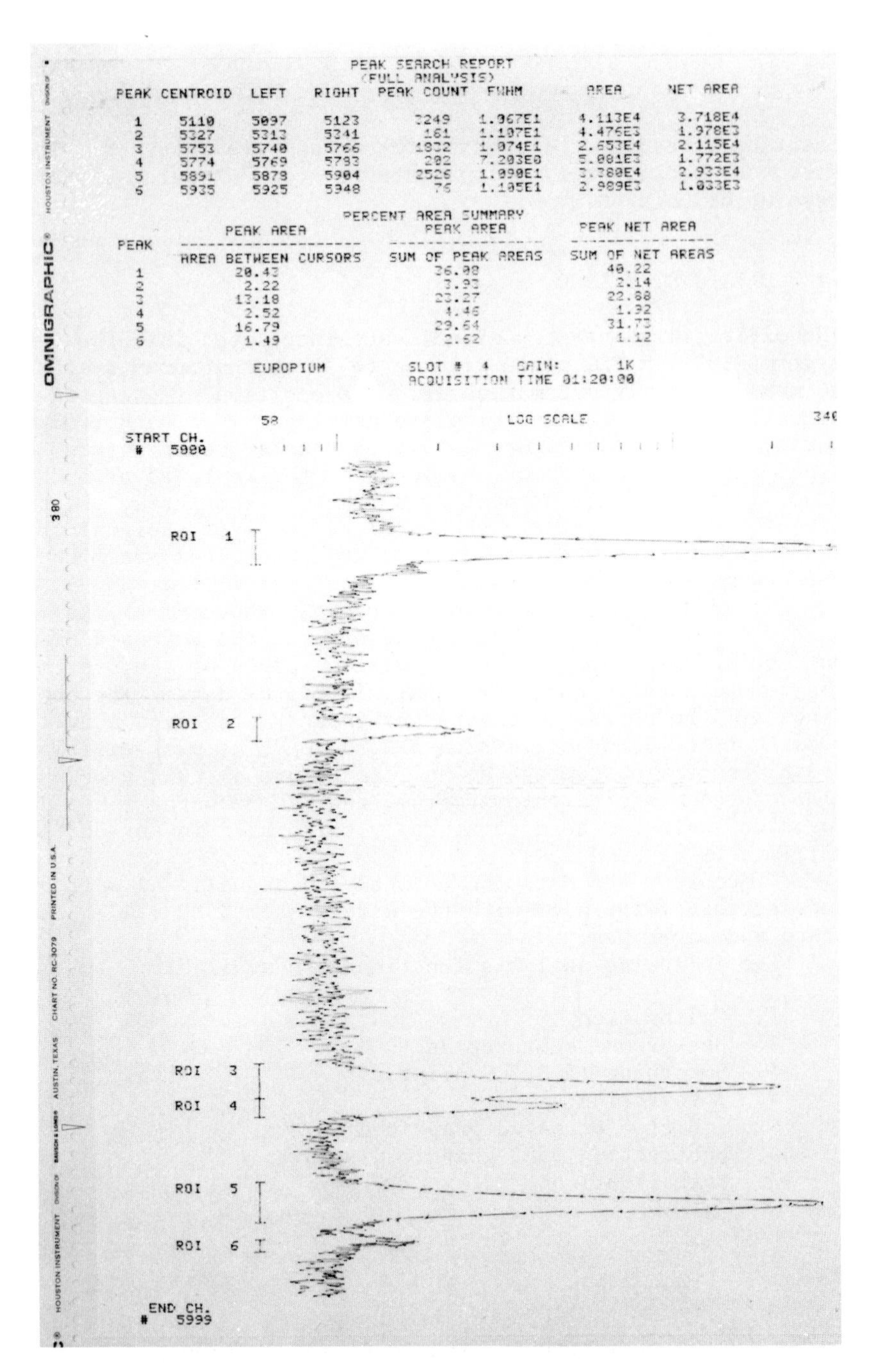

PEAK SEARCH REPORT
(FULL ANALYSIS)

PEAK	CENTROID	LEFT	RIGHT	PEAK COUNT	FWHM	AREA	NET AREA
1	5110	5097	5123	3249	1.967E1	4.113E4	3.718E4
2	5327	5313	5341	161	1.107E1	4.476E3	1.978E3
3	5753	5740	5766	1832	1.074E1	2.653E4	2.115E4
4	5774	5769	5793	202	7.203E0	5.081E3	1.772E3
5	5891	5879	5904	2526	1.990E1	3.380E4	2.933E4
6	5935	5925	5948	76	1.105E1	2.989E3	1.033E3

PERCENT AREA SUMMARY

PEAK	PEAK AREA / AREA BETWEEN CURSORS	PEAK AREA / SUM OF PEAK AREAS	PEAK NET AREA / SUM OF NET AREAS
1	20.43	36.08	40.22
2	2.22	3.93	2.14
3	13.18	23.27	22.88
4	2.52	4.46	1.92
5	16.79	29.64	31.73
6	1.49	2.62	1.12

EUROPIUM SLOT # 4 GAIN: 1K
ACQUISITION TIME 01:20:00

Figure 4. Hard copy of analysis report with system 3500.

PERFORMANCE OF THE LINEAR ANALYZER LB 283

The linear analyzer has the following properties:

1. Open Entrance Window

250 × 15 mm. For measurements of radioisotopes other than ^{3}H and ^{125}I, the window can be closed by a thin foil of 0.15 mg/cm^2. (In this case the resolution due to multiple scattering of the β-particles is only slightly worse.) Figure 5.

2. Background Radiation

60 cpm (counts per minute)/250 mm = 0.25 cpm (counts per minute)/mm or 1 cpm (counts per minute)/4 mm.

3. Sensitivity

The detection efficiency for low-energy β-particles is practically 100%. But due to their adsorption in the thin layer chromatogram surface only a small fraction enters the counter. For a thin layer of 20 mg/cm^2 the following is true:

^{3}H 0.4% cpm/dpm (disintegration per minute). 250 dpm detected in 20 minutes.

^{14}C 3% to 15% depending on "electronic collimation," 50 dpm detected in 20 minutes (see Figures 6 and 7).

4. Resolution

^{3}H 0.5 mm (for a source covered with 1/2-mm plastic strip).

^{14}C 1.0 mm (for a source covered with 1-mm plastic strip) (see Figures 8, 9, and 10).

5. Quantitative Measurements

With the present system one can perform quantitative measurements and therefore overcome the measurement with the liquid scintillation counter. This is demonstated in Figures 11, 12, and 13 for ^{3}H and ^{14}C. The results show that within about 10% of the measured counts per minute are proportional to the disintegrations per minute put onto the thin layer gel. Of course the chromatogram should have a uniform distribution in depth of the silica gel or polyacryl gel. Very thin layers have all the advantages to result in a precise quantitative measurement.

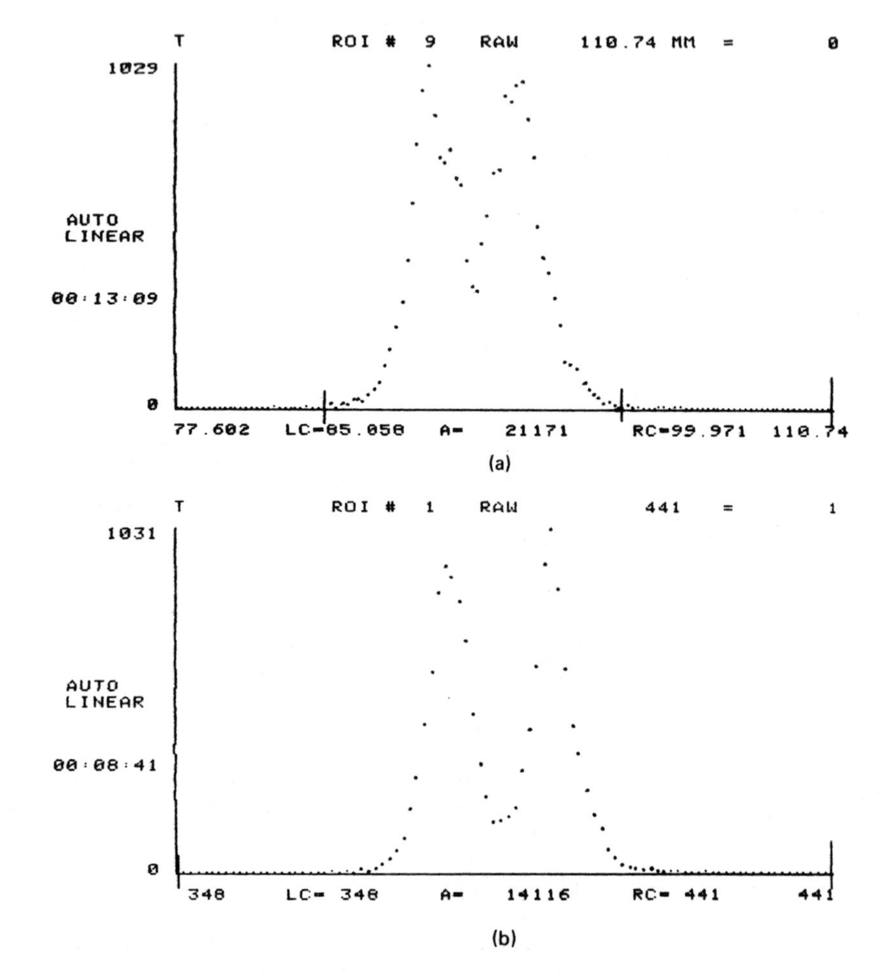

Figure 5. Effect of 0.15 mg/cm^2 foil on resolution.
(a) two ^{14}C sources, 2 mm separated, without foil,
(b) with foil.

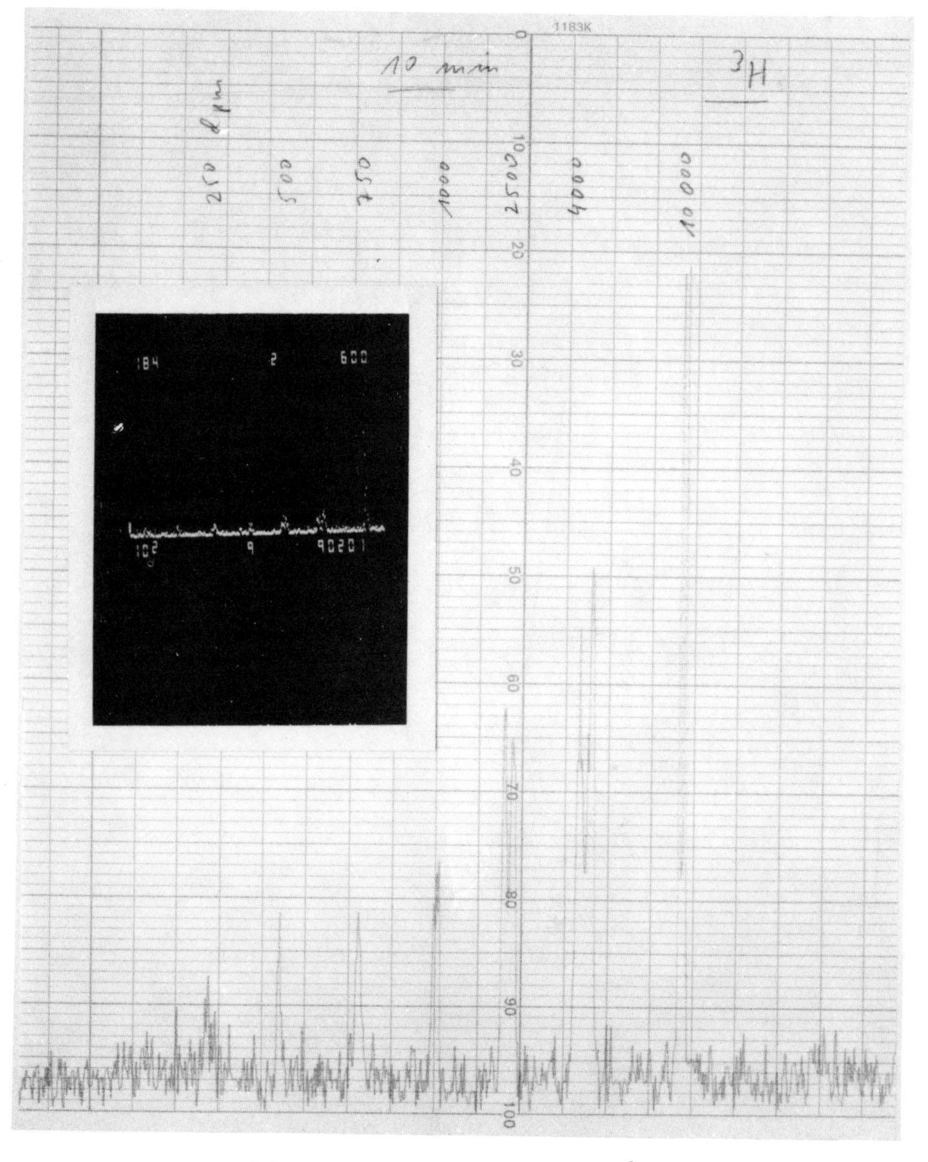

Figure 6. Sensitivity for ^{3}H.

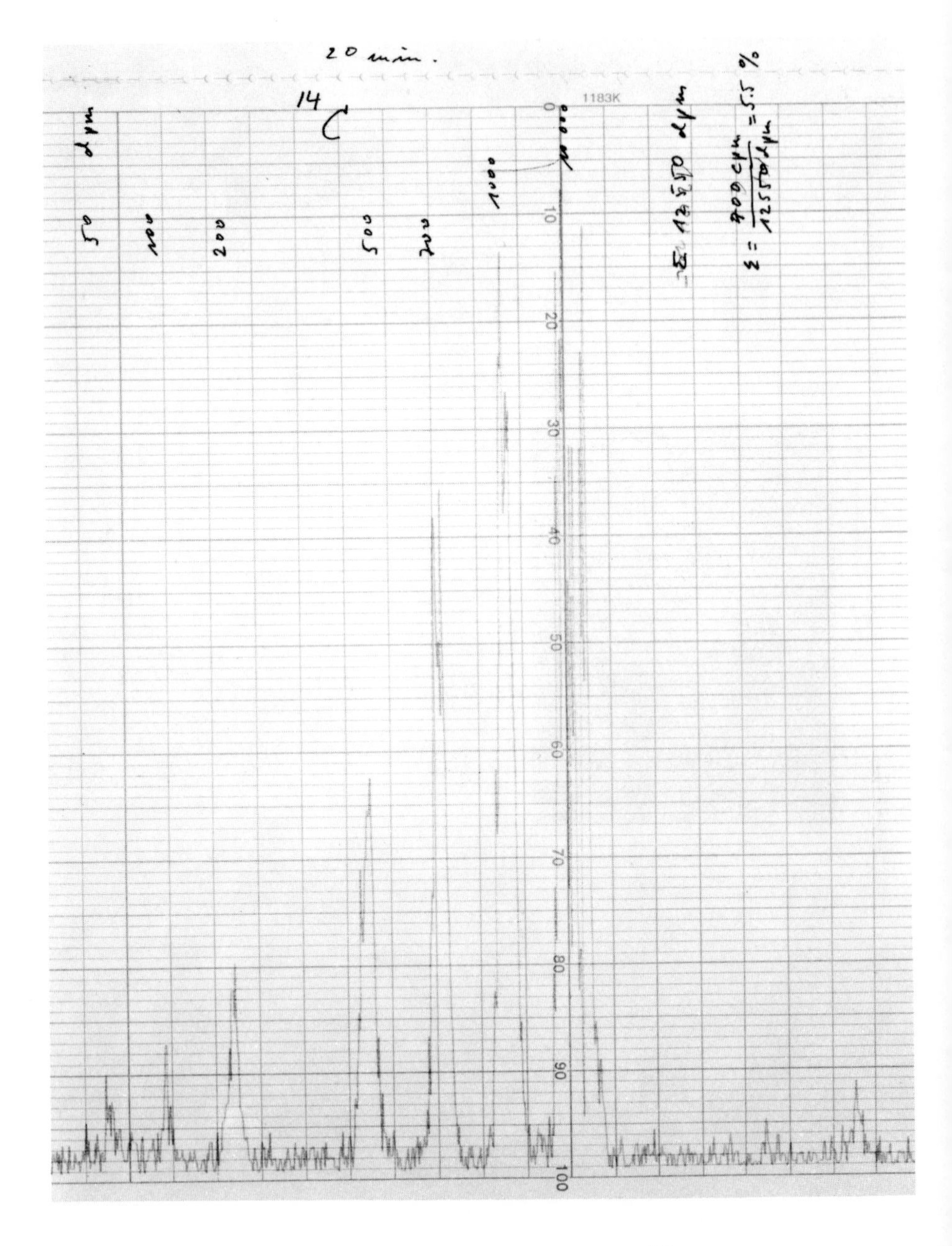

Figure 7. Sensitivity for ^{14}C.

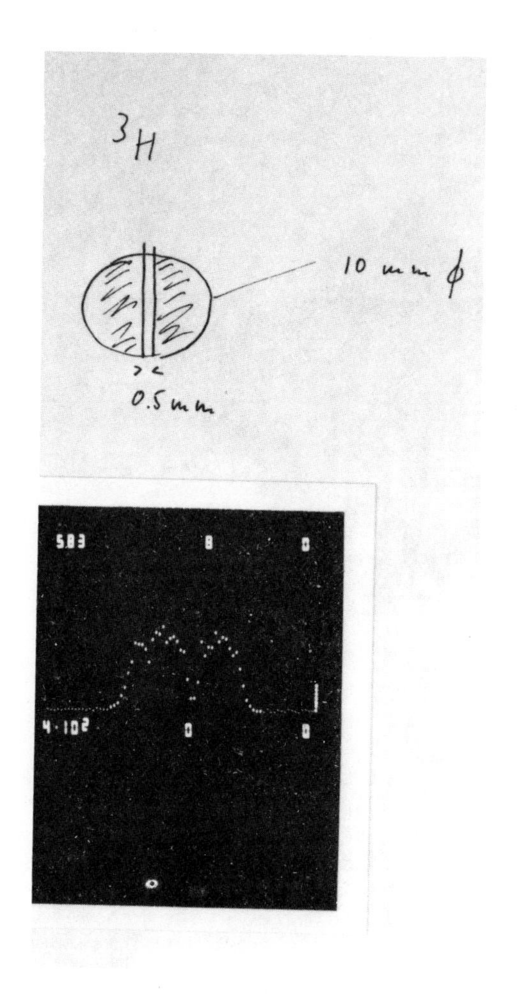

Figure 8. Resolution for ^{3}H. The ^{3}H source was covered with 1/2-mm plastic strip.

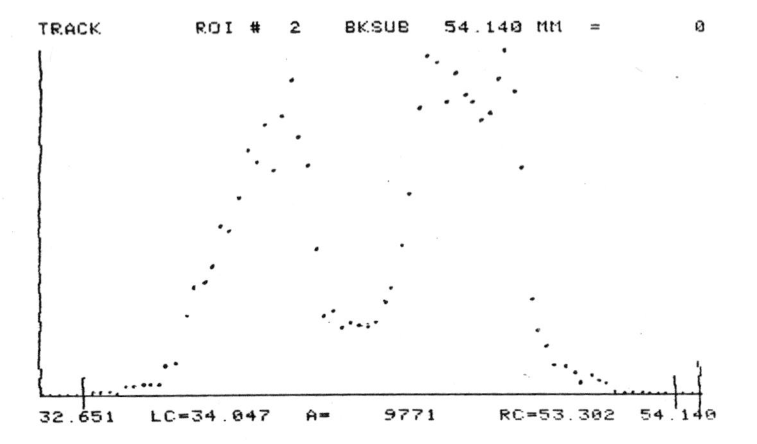

Figure 9. Resolution for ^{14}C. The ^{14}C source was covered with 1-mm plastic strip.

Figure 14 and Table I show the results of a quantitative measurement of a double-labeled ^{3}H and ^{14}C chromatogram. Each substance was labeled with ^{3}H and ^{14}C in the ratio 4.3 in total 13,000 dpm ^{3}H and 3000 dpm ^{14}C. In measuring (1) ^{3}H + ^{14}C with the open counter and (2) to measure only ^{14}C by covering the plate with a foil of 0.4 mg/cm^2, one can deduce the ^{3}H to ^{14}C ratio for each peak. For the sum over all peaks a ratio was measured (after all corrections): $^{3}H/^{14}C = 4.5$ to be compared with 4.3, which was put onto the plate (see Table I).

Figures 15 to 18 demonstrate the performance of the linear analyzer LB 283.

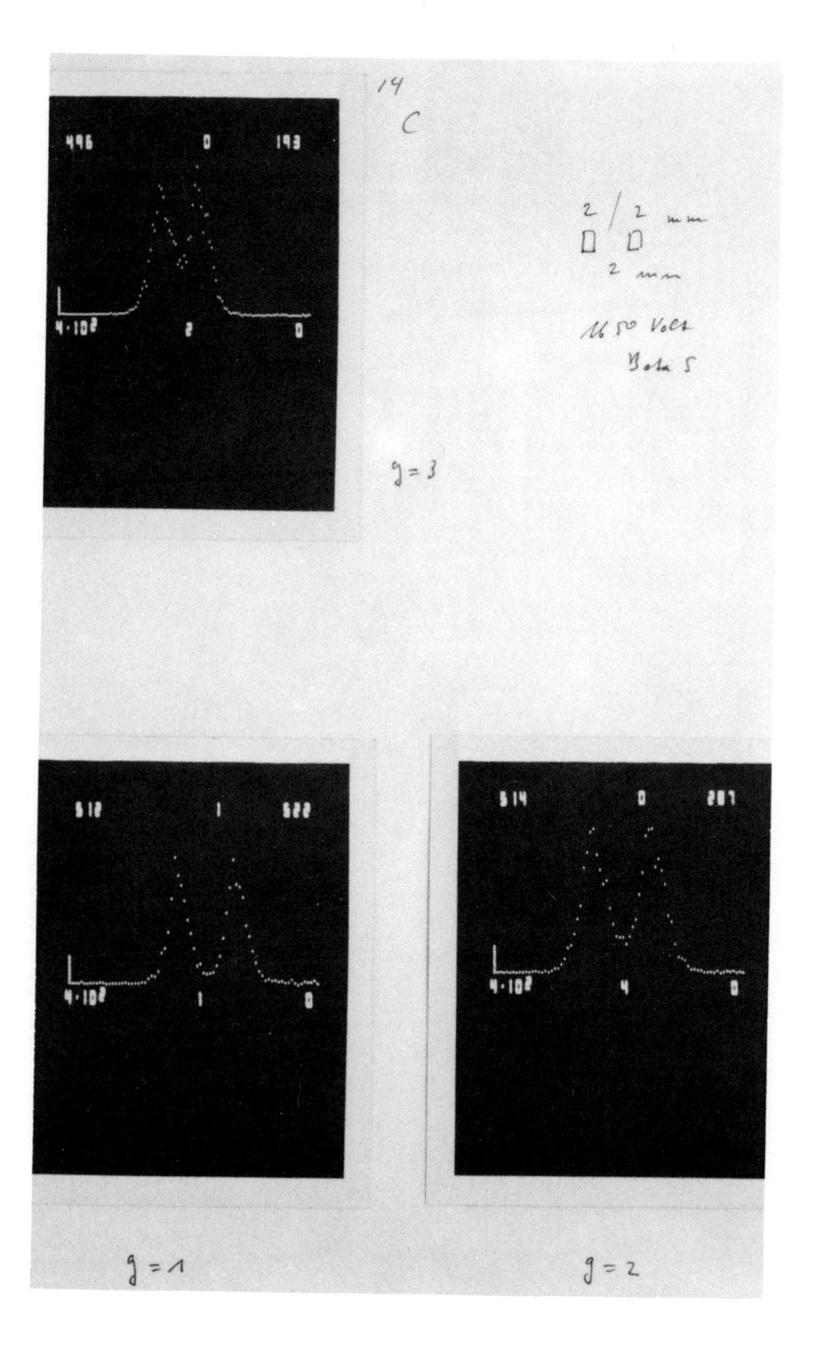

Figure 10. "Electronic Collimation." Two ^{14}C sources, separated by 2 mm, are measured with different settings of g. Setting g = 1 gives the best resolution, but the smallest accepted opening angle.

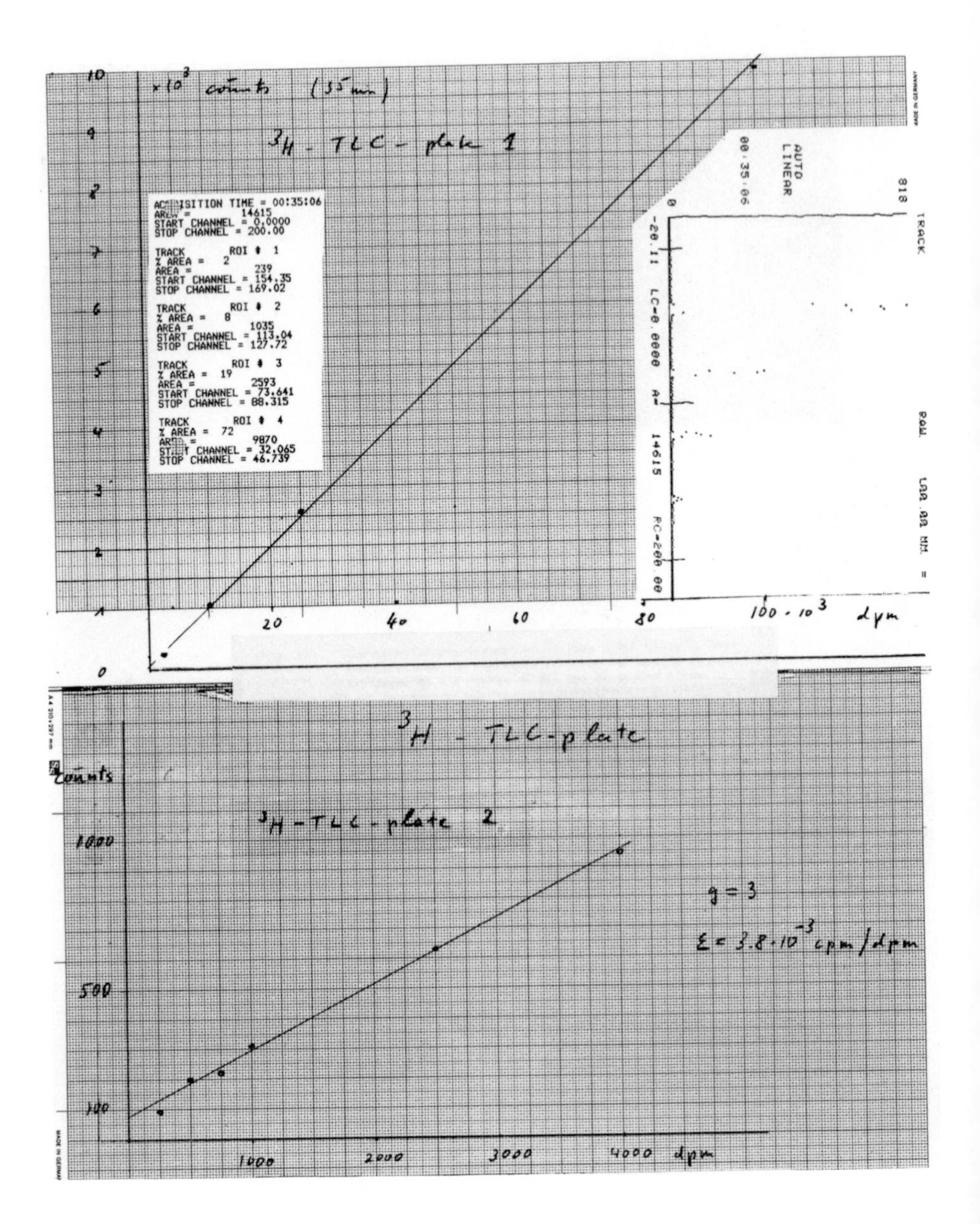

Figure 11. Top: Quantitative measurement of ^{3}H from a TLC plate 1. Certainly within 10% the measured number of counts is proportional to the number of disintegrations per minute put onto the plate. Bottom: From TLC plate 2.

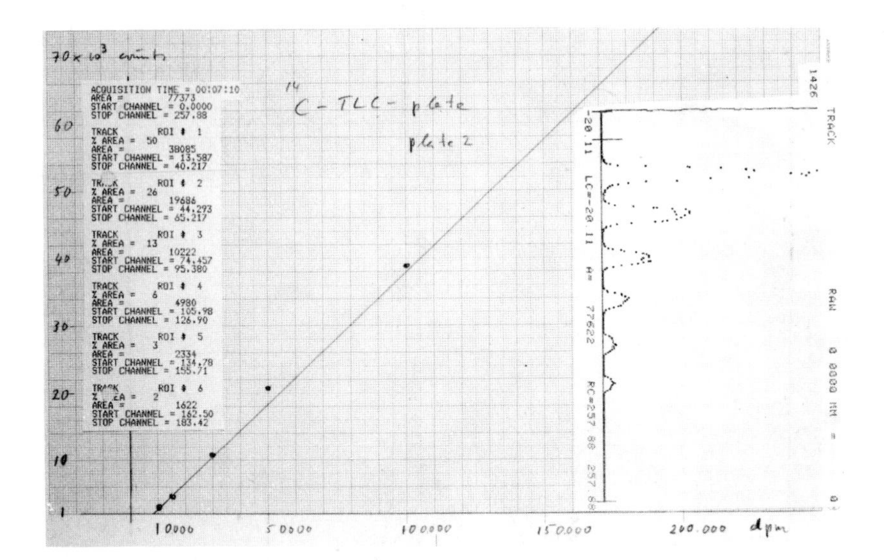

Figure 12. Quantitative measurement of ^{14}C from TLC plate 3.

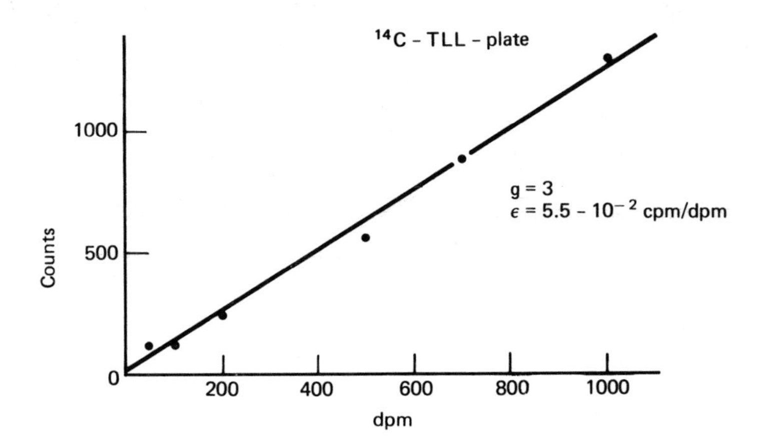

Figure 13. Quantitative measurement of ^{14}C from TLC plate 4.

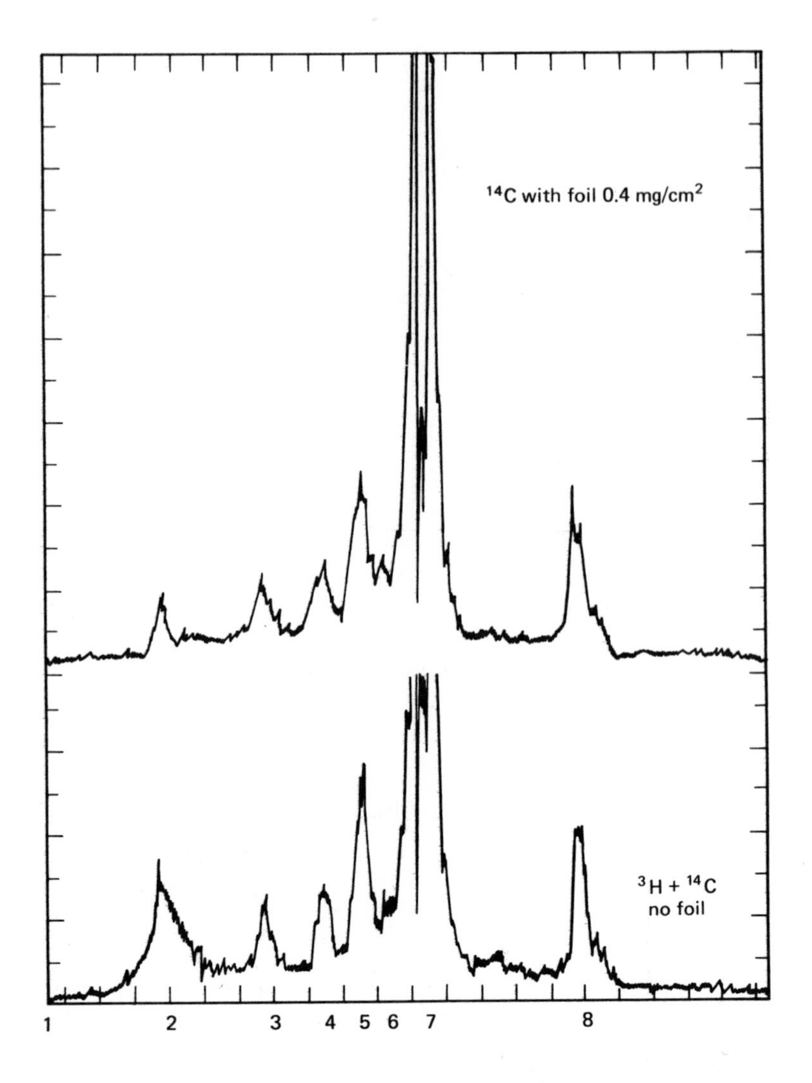

Figure 14. Quantitative measurement of a double-labeled ^3H and ^{14}C chromatogram.

TABLE I. RESULTS OF THE MEASUREMENTS FROM FIGURE 14

ROI	$^3H + ^{14}C$ No Foil	^{14}C 0.4-mg Foil	R $^3H + ^{14}C/^{14}C$	I_H/I_c
2	1110/min	322/min	3.45	16.9
3	560/min	338/min	1.55	3.4
4	595/min	334/min	1.78	5.0
5	942/min	573/min	1.64	4.0
6	263/min	188/min	1.44	2.6
7	5685/min	3513/min	1.62	3.9
8	783/min	522/min	1.50	3.0
Sum	9938/min	5790/min	1.72	4.5

$I_H/I_c = (\Sigma R - 1) \cdot \frac{(\eta_c)}{(\eta_H)}$, where $\eta_c = 2.9\%$ (detection efficiency ^{14}C), $\eta_H = 0.38\%$ (detection efficiency ^3H), and $\Sigma = 0.93\%$ (absorption ^{14}C in 0.4 mg/cm^2).

$I_H/^{14}C = \frac{(13,000 \text{ dpm})}{(3,000 \text{ dpm})} = 4.3$ put on plate.

BETA CAMERA, LB 292

A multiwire spark chamber of high sensitivity and resolution has been developed to measure two-dimensional chromatograms and two-dimensional radioactive distributions, p. ex. tissue sections (Figure 19). The chamber can detect 250 dpm ^3H in about 20 minutes (i.e., 0.1 nCi), and the resolution is better than 1 mm for ^3H and ^{14}C. The spark chamber is located directly above the chromatogram plate. Sparks are produced between the anode and the cathode of the chamber by the emitted β-particles from the chromatogram plate. These sparks are generated exactly above the source of β radiation, that is, above the spot of the chromatogram containing the radioactive-labeled substances. With aid of a Polaroid camera the spatial distribution of the sparks is recorded on Polaroid film. Projecting the image of the distribution from the Polaroid film onto the chromatogram plate with the correct magnification, the interesting spots, that is, the separated

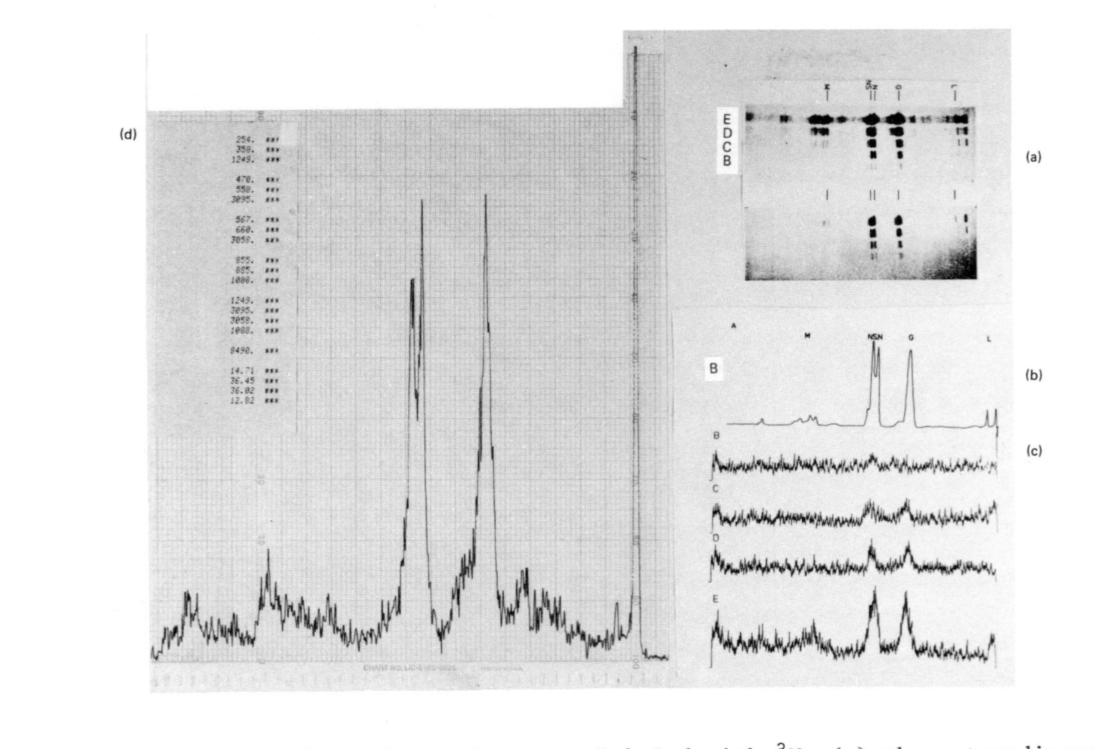

Figure 15. Measurements of an electropherogram labeled with ^{3}H: (a) the autoradiography, 2 weeks exposure; (b) the evaluation of track B of the autoradiography, 6000 dpm ^{3}H; (c)measurements with the old TLC-Berthold scanner, 7 hr per track; (d) the measurement of track B with the new analyzer LB 283 in about 1 hr (4000 sec); (e) VSV protein, ^{3}H-lysin-labeled, ∿ 6000 dpm. The electropherogram and the autoradiography were obtained from Professor J. Kruppa, Institut für Physiologische Chemie Universität Hamburg.

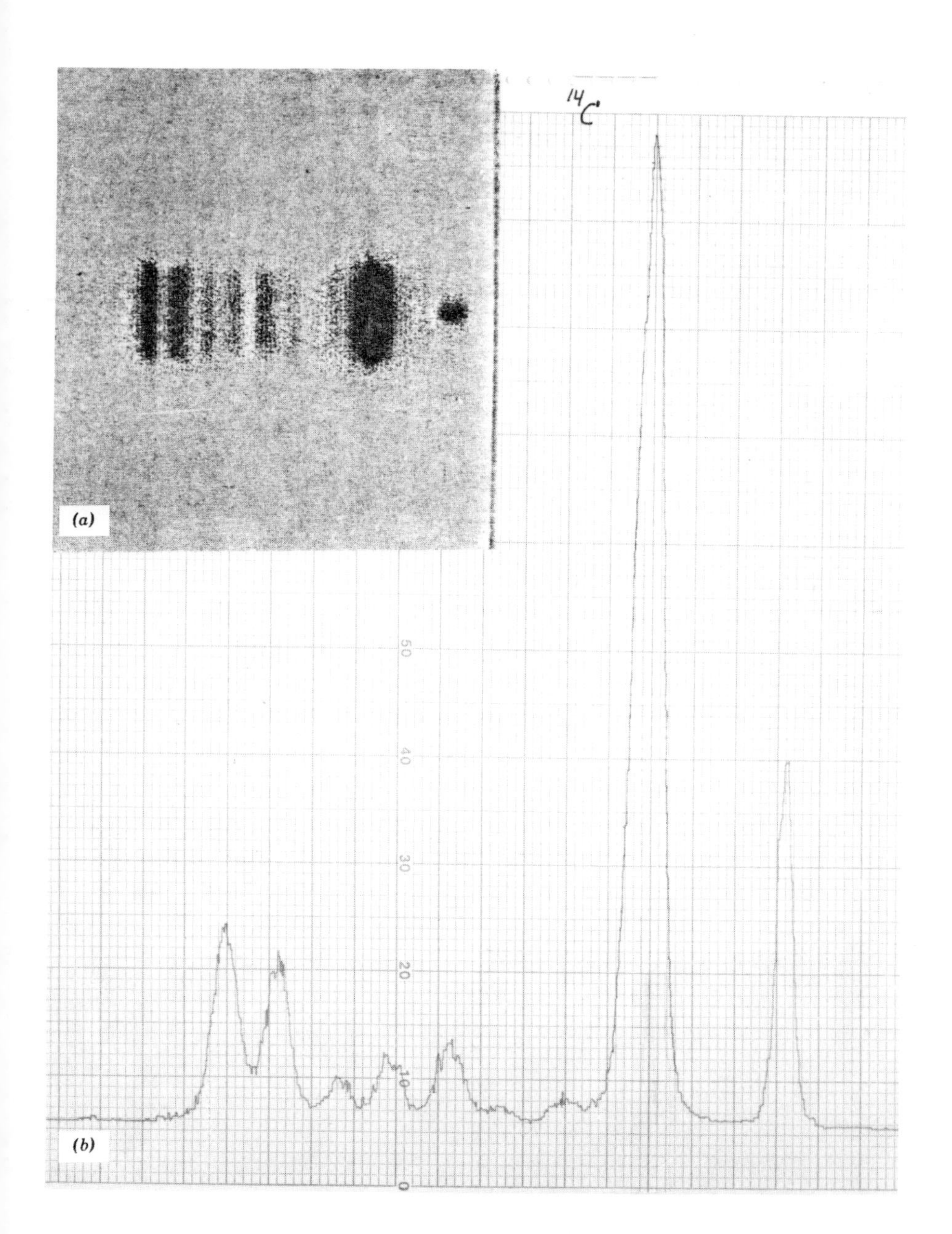

Figure 16. ^{14}C TLC chromatogram, (a) spark chamber photograph, (b) linear analyzer, 2 hr exposure time.

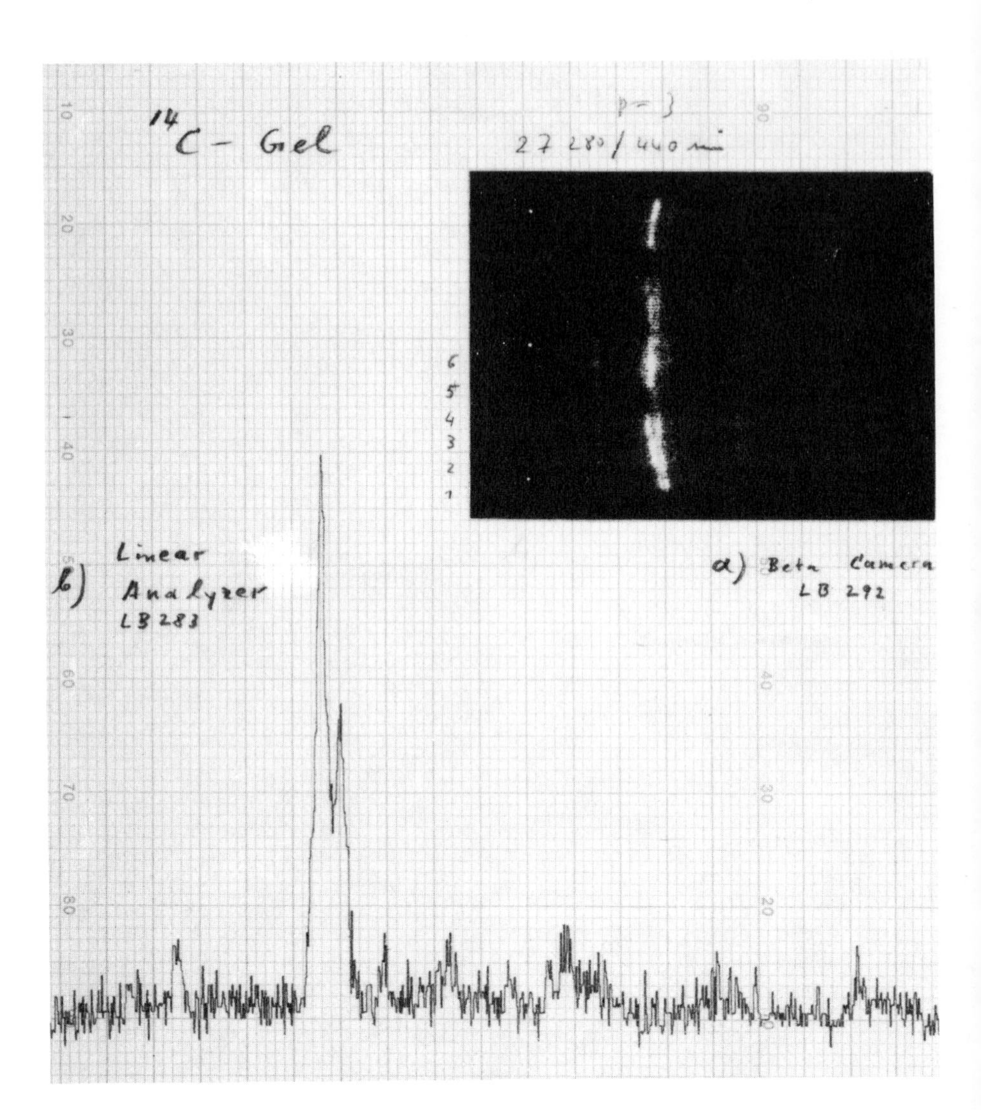

Figure 17. ^{14}C - gel, electropherogram of very low activity, (a) spark chamber picture, 400 min exposure time; (b) linear analyzer, 440 min (track 2); (c) linear analyzer, 3360 min (track 2).

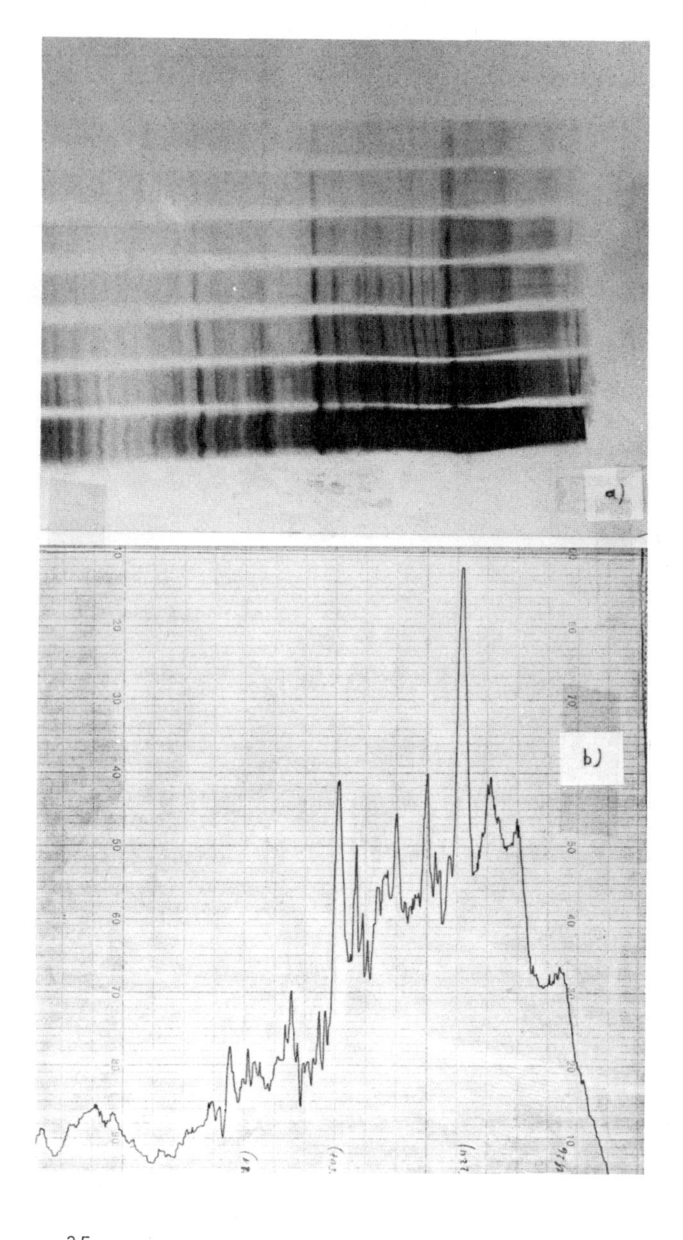

Figure 18. ^{35}S-labeled electropherogram, protein separation, (a) autoradiography, 2 weeks exposure; (b) linear analyzer, 106 min exposure.

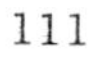

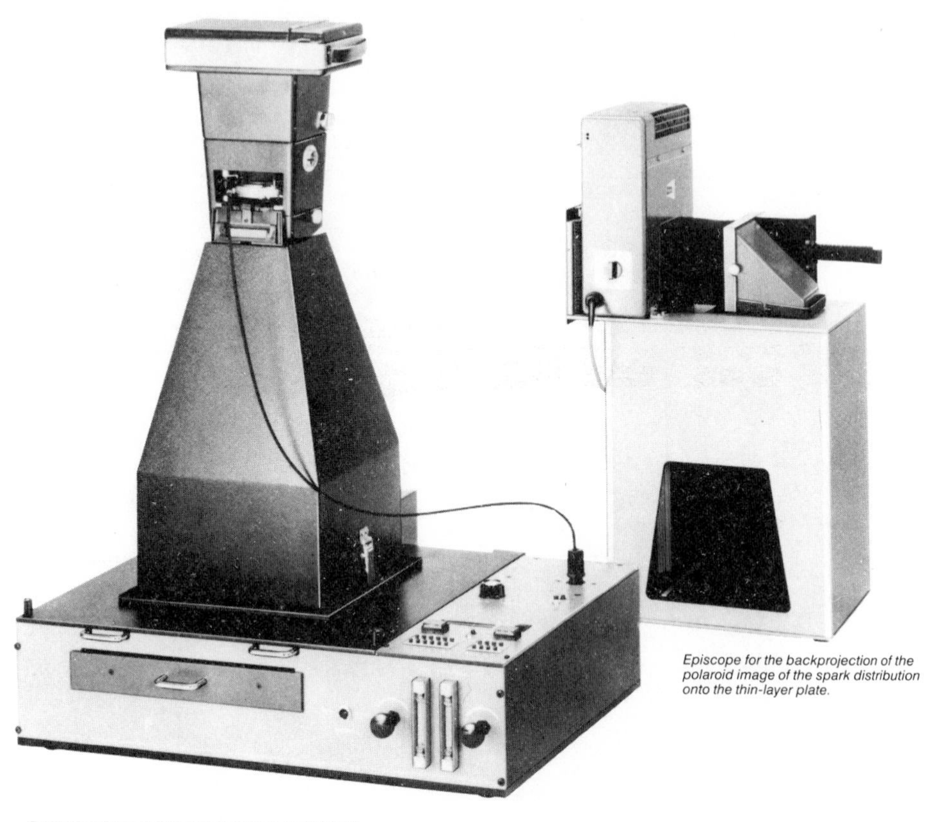

Figure 19. The Beta camera LB 292: (a) spark chamber with high-voltage supply, scaler-timer, and Polaroid camera; (b) Episcope for the backprojection of the polaroid image of the spark distribution onto the thin layer plate.

compounds, can easily be located and removed from the plate for a subsequent quantitative analysis.

Detector

The detector is a self-triggered wire spark chamber of 20 × 20 cm sensitive area (see Figure 20). The chamber cannot be triggered

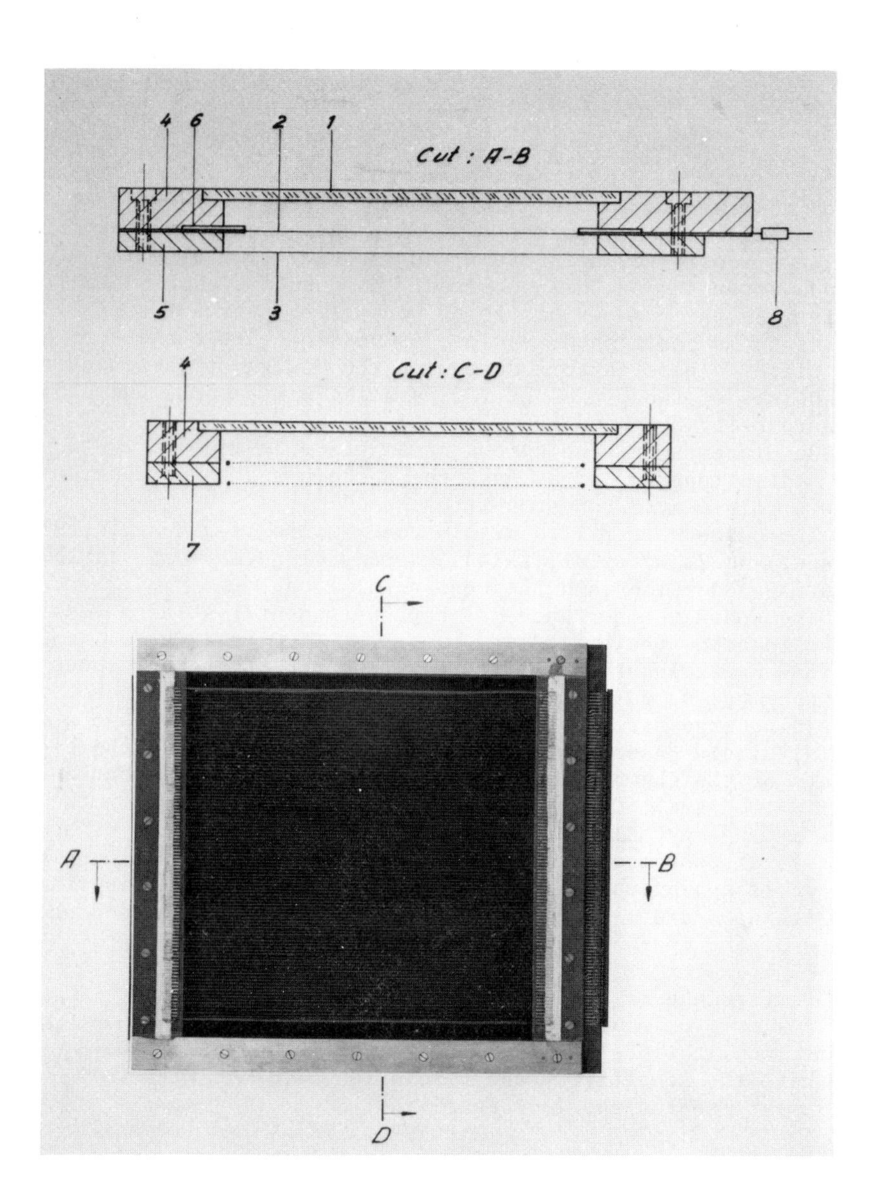

Figure 20. (a) Schematic drawing of the wire spark chamber: 1-conductive glass plate, 2-50-μ gold-plated tungsten wire, 3-200-μ gold-plated tungsten wire, 4-Stesalit frame. (b) Photograph of the wire spark chamber. The chamber consists of 2 wire planes. The anode plane: 100 wires, 50 μ diameter, gold-plated tungsten, 2 mm distance. The cathod plane: 100 wires, 200 μ diameter, gold-plated tungsten, 2 mm distance. The plane distance is 3 mm.

by the ^{3}H and ^{14}C β-particles because of their low energies, 18 keV maximum and 155 keV maximum respectively. Therefore, the high voltage is continuously applied to the anode, and the chamber is always sensitive, apart from the recovery time of a few microseconds after a spark has occurred. To obtain a good spatial resolution, the chamber has to be very flat, over a few millimeters, and has to be positioned as close as possible above the chromatogram plate. These factors determine the design of the chamber. It consists of two planes of 100 parallel wires each, the plane distance being 3 mm and the wires spaced by 2 mm. The cathode plane facing the chromatogram plate is made of 200-μ gold-plated tungsten wires, and the anode plane above consists of 50-μ gold-plated tungsten wires.

The chamber is filled with a gas mixture of 89% argon, 9% methane, and 2% methylal, this mixture giving the most stable operating conditions and the best resolution.

The spark chamber anode is kept permanently at high tension somewhate below the breakdown voltage at 3500 to 4000 V. An ionizing particle traversing or entering the chamber produces enough charge by gas multiplication ($> 10^{8}$) at the anode to start a gas discharge resulting in a visible spark. The voltage drops down within 10^{-8} sec. After the formation of a spark, the high voltage of the chamber is maintained at a low value, about 300 V, for 5×10^{-3} sec. During this time the ions generated in the spark are being collected. This prevents the formation of spurious sparks and spontaneous discharges in the chamber, when the high voltage is again applied. Figure 19 shows a photograph of the complete apparatus, and Figure 20, a schematic drawing and photographs of the spark chamber.

Performance of the Chamber

In most applications ^{3}H- and ^{14}C-labeled compounds are used. We give here the sensitivity and resolution obtained with this instrument for the two isotopes.

1. Background

The sensitivity of the chamber depends, obviously, on the background radiation, which should be kept small. If the chamber is not exposed to a radioactive source, it will still produce sparks generated by the cosmic radiation ($1/cm^{2}$ min) and the radioactive background radiation from the chamber material and the environment. The chamber having a surface of 400 cm^2 would have a background of at least 400/min if it is made sensitive to minimum ionizing particles. This is not necessary, because the β-particles from ^3H and ^{14}C are slow and heavily ionizing. Normal operating conditions amount to about 200/min to 400/min background sparks.

2. Sensitivity

In the operating region, 3800 to 4000 V, the chamber detects about 50% of minimum ionizing particles resulting in a background of $0.5/cm^2$ min to $1/cm^2$ min. With a source of β-particles from ^3H and ^{14}C of 5 mm diameter as detection efficiency the ratio of the measured spark rate S to the measured rate with windowless gas flow GM-counter G is defined. Under these conditions the efficiency S/G = 10 to 20% for ^3H is obtained, and for ^{14}C the efficiency is S/G = 50 to 70%. The detection efficiency of ^3H β-particles is so low because of their low energy, E_{max} = 18 keV, and therefore short range, R_{max} = 4 mm air. In average, a β-particle from the ^3H decay stops within 0.5 mm above the chromatogram plate. It forms electrons by ionization at the plate surface, which are drifted to the anode of the chamber by the electric stray field.

Figure 21 shows the photograph of the spark distribution from ^3H spots in a thin layer plate. The activity of the spots varies from 250 to 10^4 dpm/0.25 cm^2. By self-absorption in the plate only about 1 to 2% of the ^3H-betas are emitted from the surface into the 2π hemisphere, that is, 100 dpm results in 5 to 10 sparks/min, which is about 10 to 20 times the background radiation.

3. Resolution

To determine the spatial resolution a plastic strip of 1 mm width and 0.3 mm thickness is placed above ^3H and ^{14}C spots in thin layer plates. The results (Figure 22) show that in both cases the resolution defined in this way is better than 1 mm. Polaroid positive-negative film type 665, 75 ASA, 20 DIN is used here. For a given radioactivity one has to use the smallest workable aperture of the photocamera to achieve the highest resolution.

Application

The main application of the spark chamber is the detection of

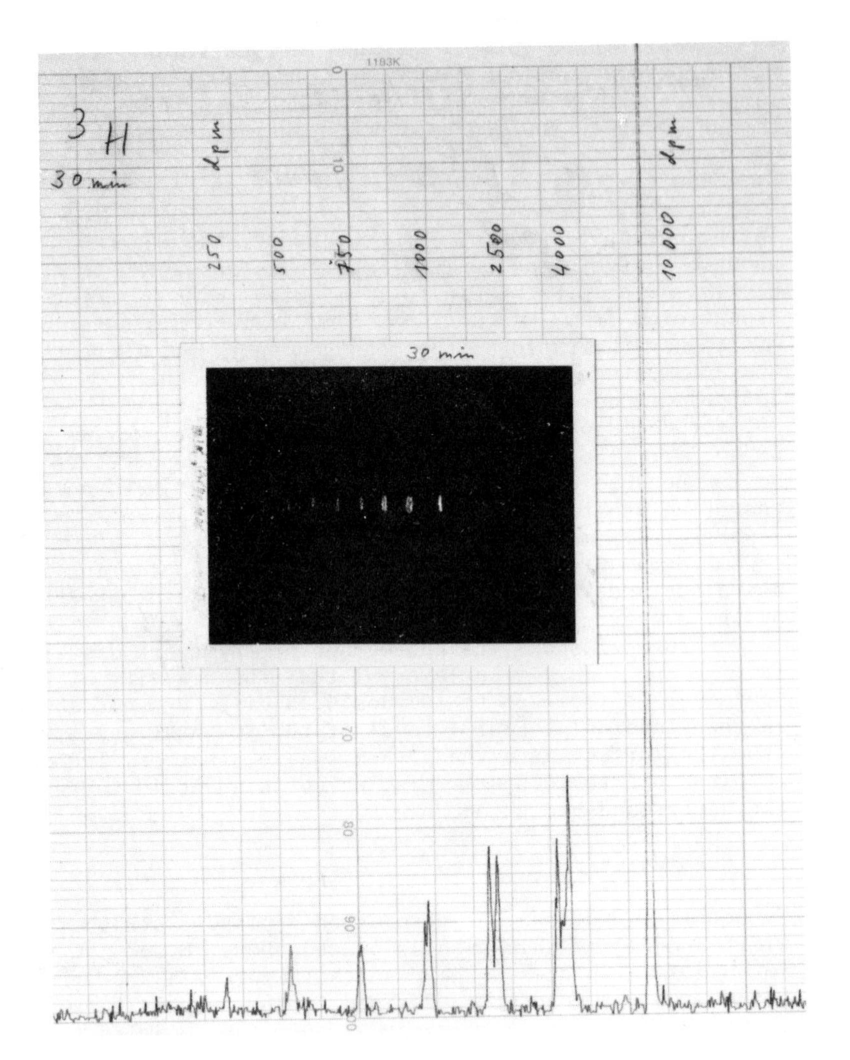

Figure 21. Photograph of the spark distribution from ^{3}H spots in a thin layer plate. The activity of the spots varies from 250 to 10^4 dpm/0.25 cm^2. By self-absorption in the plate only about 1 to 2% of the ^{3}H betas are emitted from the surface into the 2π hemisphere, that is, 100 dpm results in 5 to 10 sparks/min, which is about 10 to 20 times the background radiation.

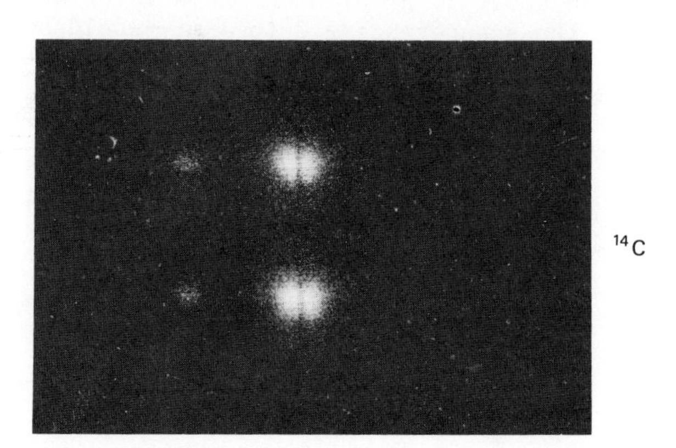

Figure 22. Resolution for ^3H (top) and ^{14}C (bottom). The radioactive spots of ^3H and ^{14}C on the TLC plate are covered with a plastic strip of 1 mm width. The resolution is better than 1 mm. (Polaroid film 665, 75 ASA (20 DIN), aperture 8).

radioactive-labeled compounds in thin layer chromatography plates. The application is also valid for the evaluation of electrophoretic plates and the detection of ^{3}H or ^{14}C or other nuclides, distributed in biological tissues. In Figures 23 to 26 spark chamber photographs of some ^{3}H and ^{14}C chromatograms are shown. For comparison Figures 23a to 23c give the evaluation of the same chromatogram with the spark chamber and a scan with a (Geiger Muller) GM-tube and a measurement with the linear analyzer.

CONCLUSION

The wire spark chamber is an adequate tool applied in radiochromatography. The visual method is simple and of low cost. The relatively small event rate, 10^{-2} to 10^{2}/s is well adapted to the spark chamber repetition rate of about 10^{3}/s.

The great advantage over other methods is the high sensitivity and the high resolution. In about 20 minutes 250 dpm ^{3}H and 30 dpm ^{14}C can be detected. The resolution is better than 1 mm for both, ^{3}H and ^{14}C. Exposure times of several weeks with autoradiography and about 24 hr with the thin layer scanner, the resolution of the scanner being 2 to 4 mm, would be comparable.

The new linear analyzer LB 283 offers the best resolution, the same sensitivity as the spark chamber, and the best performance for one-dimensional--including parallel-developed--chromatograms and electropherograms. It has the great advantage of being able to perform quantitative measurements.

The Beta camera LB 282 is the favored alternative for two-dimensional distributions.

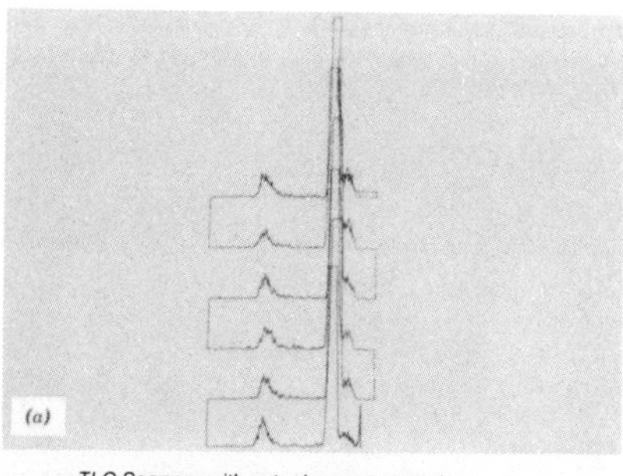

TLC Scanner with autochronous recorder,
counting time 24 h.

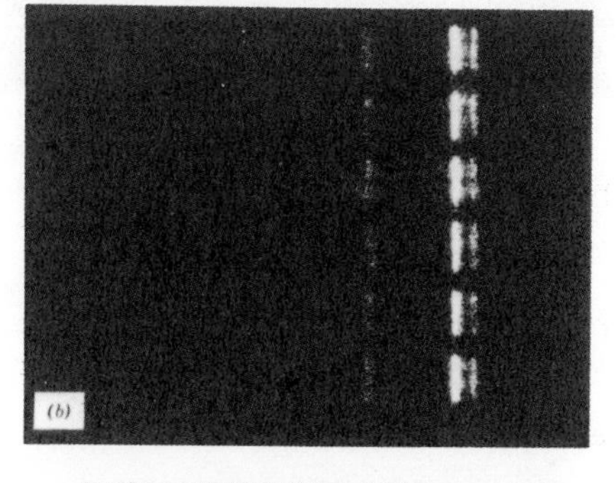

Beta Camera LB 292, 33 000 sparks discharges in 5 min.

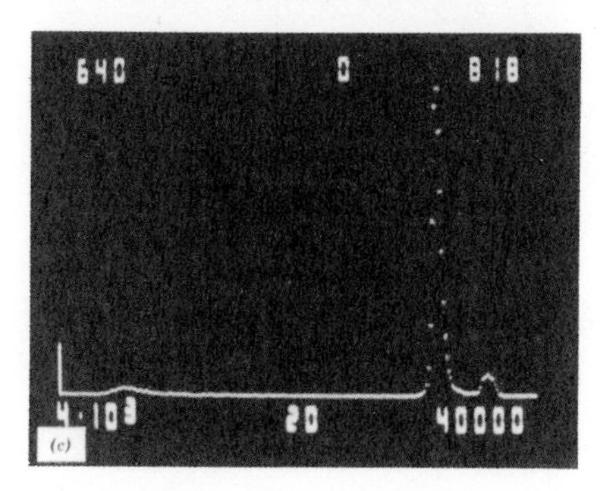

Linear Analyzer LB 282, count of bottom trace,
counting time 10 min.

Figure 23. Comparison of count results for the same ^3H thin layer chromatogram. The Beta camera has the shortest counting time and is the preferred alternative for two-dimensional chromatograms. (a) TLC scanner with autochronous recorder, counting time 24 hr. (b) Beta camera LB 292, 33,000 sparks discharges in 5 min. (c) Linear analyzer LB 282, count of bottom trace, counting time 10 min.

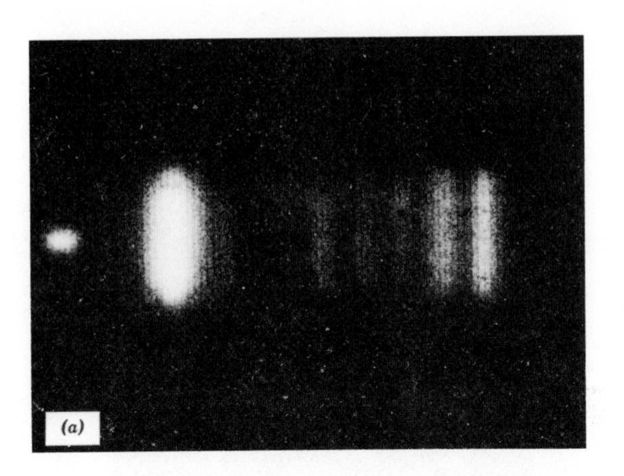

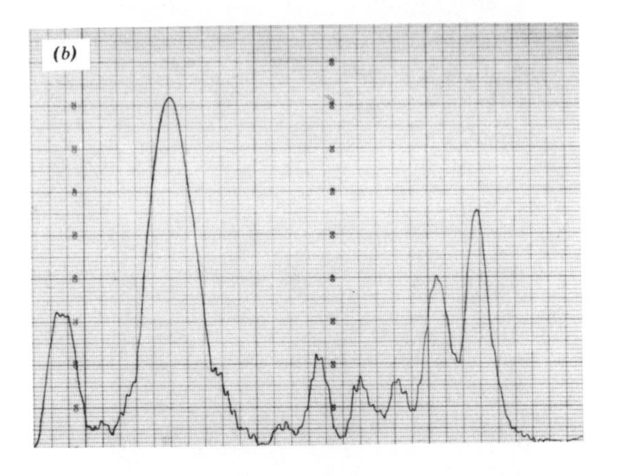

Figure 24. Spark chamber picture of (a) a ^{14}C chromatogram being evaluated with the (b) Berthold densitometer LB 281.

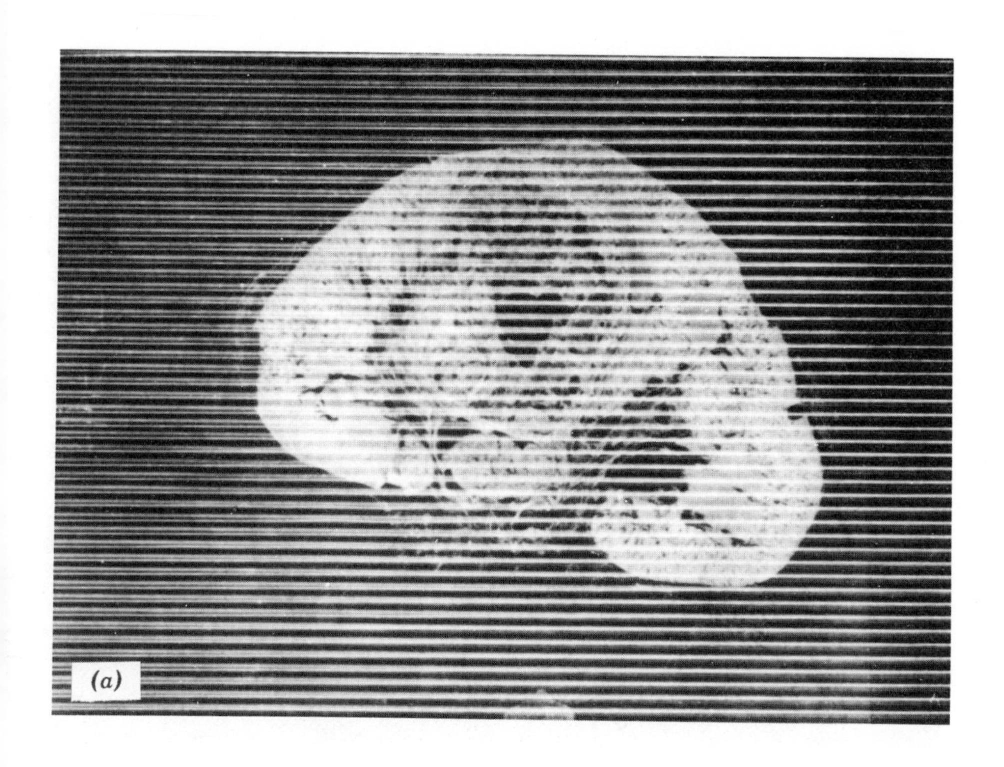

Figure 25. (a) Five-microtom section of a cancerous human kidney placed inside the spark chamber. The ^{3}H labeled microtom section was obatined from Professor Dr. Rabes, Pathologisches Institut der Universität, München.

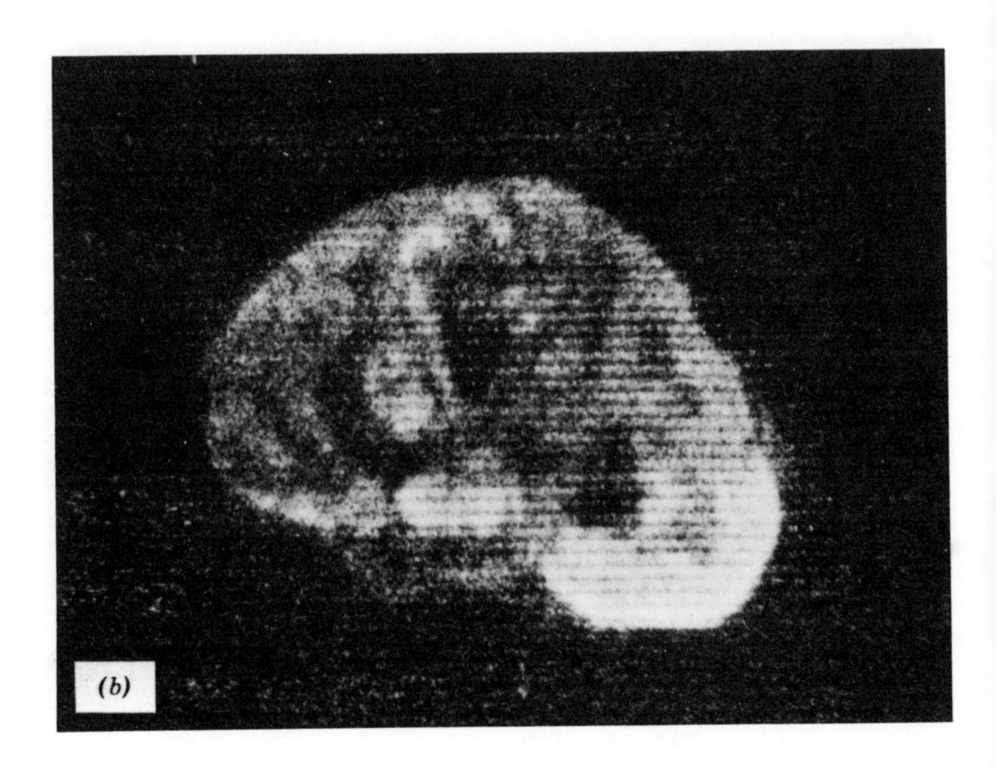

Figure 25. (b) Spark chamber recording of ^3H distribution in the kidney section. The kidney has been flushed with ^3H-labeled thymidin after operation.

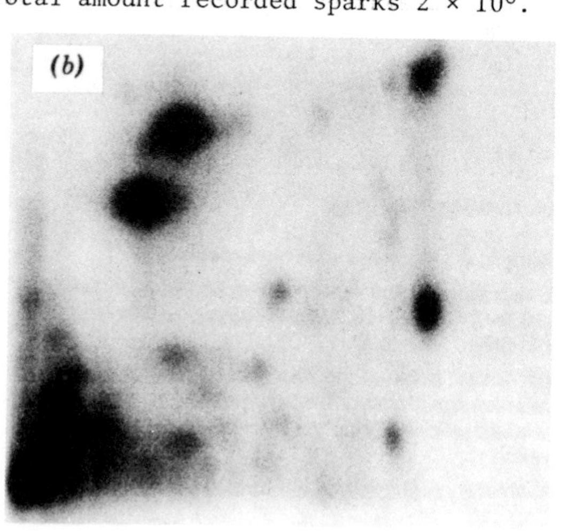

Figure 26. (a) Autoradiogram of a thin layer plate marked with ^{14}C. Exposure time approximately 4 weeks. (b) Same plate as in (a) but recorded with the spark chamber (negative). Exposure time 6 hr, aperture 16. Polaroid film positive-negative 665 (20 DIN). Total amount recorded sparks 2×10^6.

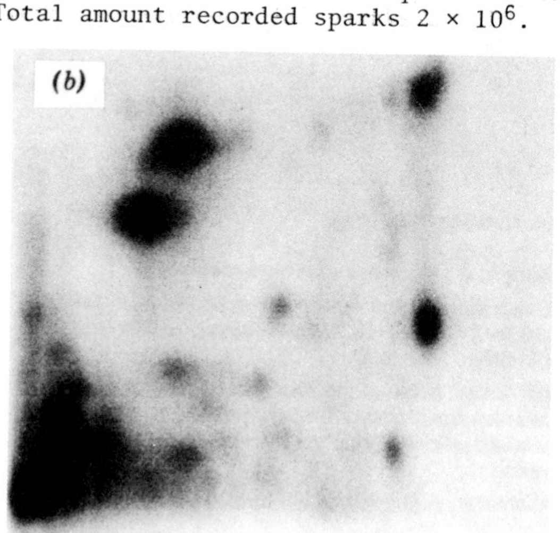

CHAPTER 8

Quantitative Analysis by Imaging Radiation Detection

Seth Shulman

INTRODUCTION

TLC has been shown to yield reproducible, quantitative results in many applications, and its simplicity, low cost, and extensive literature of tested separation techniques make it a very attractive tool. In order to remain competitive with new techniques such as HPLC, however, improvements are needed in the process of translating separations on the plate into useable, quantitative data for the researcher.

For radioisotope-labeled TLC, accurate, quantitative analysis is often a tedious and difficult process. The existing techniques all present problems of one kind or another. Autoradiography is slow, has poor sensitivity, and is difficult to use quantitatively. Scintillation counting has excellent quantitative accuracy and sensitivity, but presents substantial problems in sample preparation and spatial resolution. Conventional radiochromatogram scanners require no sample preparation but are generally slow, have poor sensitivity, and often are not instrumented to provide accurate, quantitative results.

In an effort to make quantitative radiochromatogram analysis simpler, faster, and more competitive with other techniques, several groups[1-4] have investigated the use of imaging proportional counters (IPC). These instruments have characteristics that retain most of the advantages of the previously mentioned analysis techniques while eliminating many of the drawbacks. As in autoradiography and conventional scanners, the IPC can analyze the samples on the plate, requiring no extra sample preparation. Unlike these two other techniques, however, the IPC has very high sensitivity. Its detection efficiency at any position is comparable to that of the scanners, but it views the entire sample at one time, thereby making its overall sensitivity or speed about 100 times as great as conventional scanners.

Some capabilities of the older techniques can still not be matched; especially the resolution available with autoradiography, and the ultimate sensitivity of scintillation counting, where the self-absorption of the thin layer is eliminated by dissolving the sample in scintillation fluid. (These constraints are discussed in more detail below.) However, for most research involving radiochromatograms, the IPC provides an ideal analysis technique that combines high sensitivity, rapidity, and ease of handling to produce accurate, quantitative results.

INSTRUMENTATION

To understand the imaging proportional counter (IPC), it is instructive to first review the workings of a standard proportional counter as shown in Figure 1.

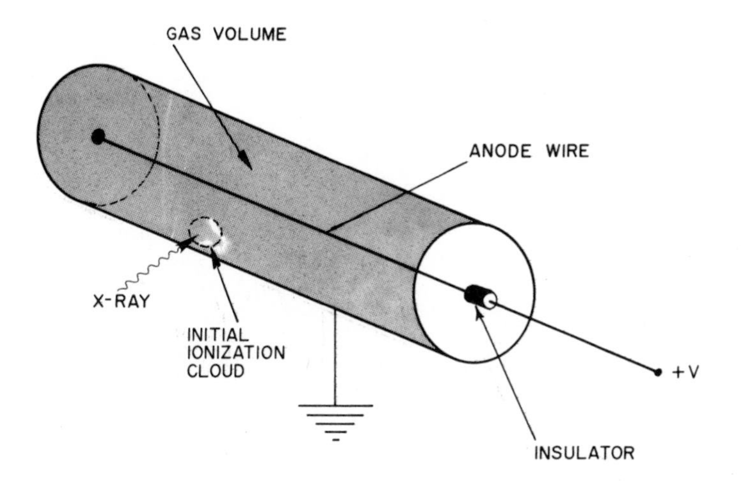

Figure 1. Schematic drawing of a conventional proportional counter.

The counter has a gas volume, usually circular or rectangular in cross section, with a high-voltage anode wire running through it. Radiation, such as X-rays or beta particles that ionize the gas, produces free electrons that are accelerated toward the anode wire. When the electrons enter the very high electric field near the wire surface, they are accelerated to sufficient energy to produce further ionization of the gas. This increase in the number of electrons produces a pulse on the anode wire of sufficient magnitude (only about 1000 electrons are required with modern amplifiers) to detect with electronics.

In some counters the gas volume is completely sealed, and the radiation enters through a window made of plastic, beryllium, or other material. In the case of low-energy alpha- and beta-particle detection, the entrance aperture must be windowless, and gas lost to the surroundings must be continually resupplied to the detector volume. Many gases can be used, but the most common are mixtures of noble gases and hydrocarbons. A mixture of 90% argon and 10% methane (called P-10) is widely used and can be readily obtained from most gas suppliers.

To add imaging capability to the counters, the design must be altered to provide an electronic signal that varies as the position of the incident radiation varies over the sensitive area of the detector. One such scheme for obtaining position information in one dimension is shown in Figure 2. The metal anode of Figure 1 is replaced by a resistive anode made with a carbon coating on a quartz fiber. A preamplifier is attached at both ends instead of only one end. When a pulse of electrons is collected at the resistive anode, it behaves like a current divider. Part of the pulse flows toward each end, with the ratio of the two parts of the pulse determined by the amount of resistance between the original collection point and each of the preamplifiers. The two amplified pulses are then converted to a digital result on a pulse height scale, and the quotient shown is computed to give a numerical result for the position. A digital image can then be built up in a computer memory by adding 1 count to the total stored in the memory location which corresponds to a particular interval along the anode.

The resolution that can be achieved with this (or other) imaging scheme is on the order of 0.2 to 0.5% of the total detector length. For radiochromatograms, the detector length used is 25 cm, and the data are stored in 256 locations in the computor memory, corresponding to 1-mm intervals. However, the main limitation on the resolution in practice is due to the finite depth of the detector and the omnidirectional nature of the

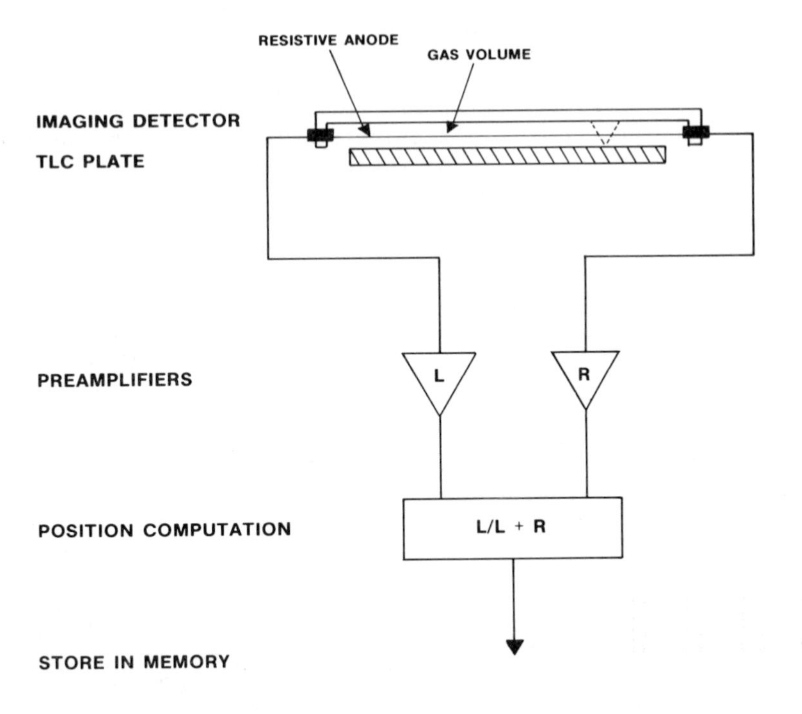

Figure 2. Schematic drawing of a resistive anode imaging proportional counter (IPC), its electronics, and the logical flow of event data.

radiation emanating from the sample. The dashed lines in Figure 2 illustrate two possible paths of betas radiated from the same location on the TLC plate. The betas will produce ionization all along their paths, and the detector will measure the centroid of this ionization. In general, the spread (or defocusing) will be about comparable to the detector depth for high-energy betas and will decrease as the energy and penetrating power of the betas decreases. Thus for ^{32}P, the resolution will be limited to the detector depth of 5 mm, while for ^{14}C it will improve to about 3 mm and for tritium (^{3}H) 1 to 2 mm.

The relative resolution performance for tritium and ^{14}C is shown in Figure 3. Radioactivity was spotted on a TLC plate with

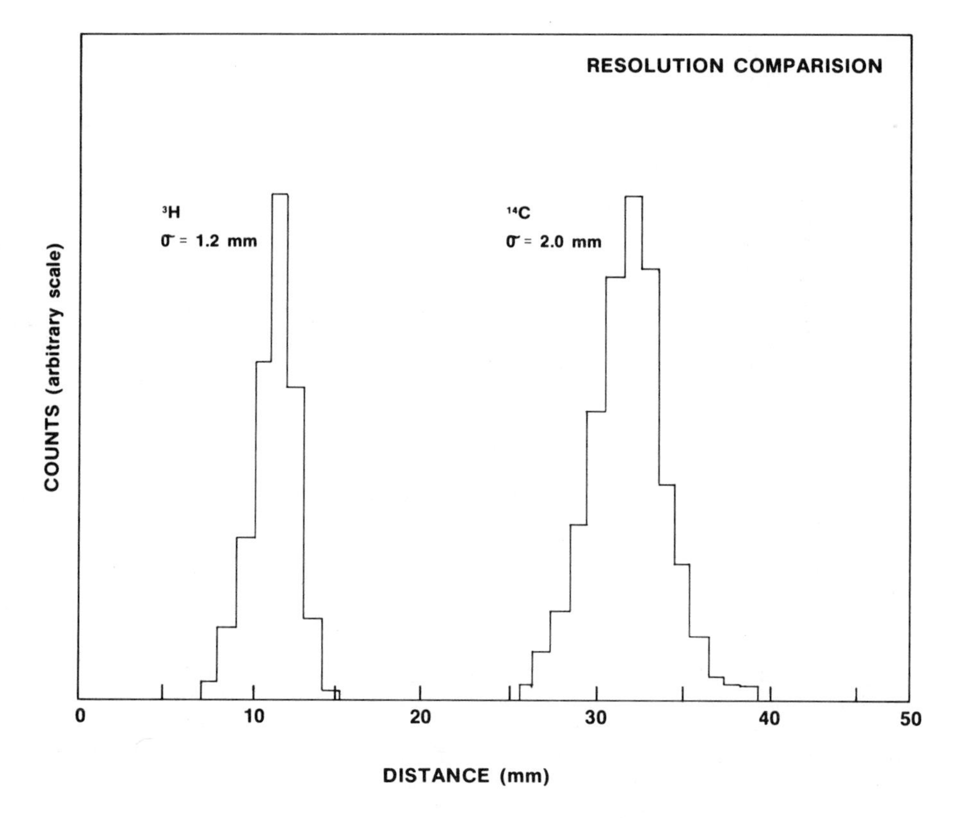

Figure 3. A resolution comparison between tritium and ^{14}C. The spots analyzed had intrinsic diameters of 1 to 2 mm.

a micropipette and then analyzed using an IPC instrument (Bioscan, Inc. BID System 100). The spot sizes are themselves on the order of 1 to 2 mm in diameter, so the actual detector resolution is somewhat better than the gaussian-fit parameters shown.

TLC APPLICATIONS

In applying IPC techniques to TLC analysis, one of the fundamental limits is the self-absorption of betas by the thin layer material itself. The extent of the problem for the three most commonly

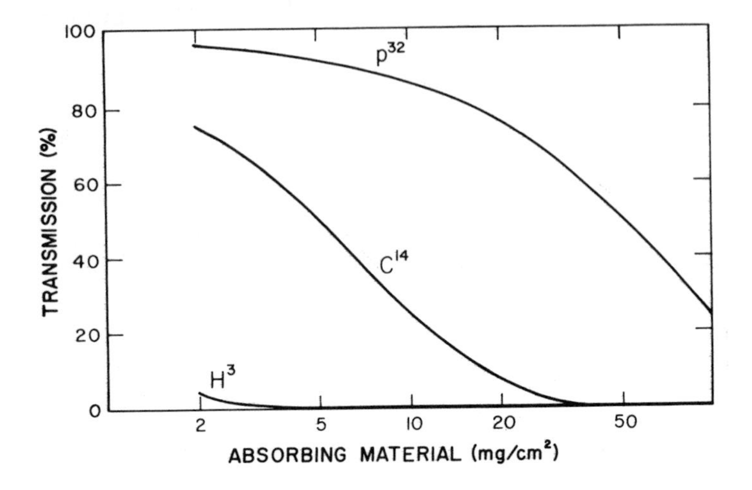

Figure 4. Transmission of betas from various isotopes through intervening matter.

used isotopes is shown in Figure 4. Clearly, the fraction of tritium betas transmitted through a 200-μm layer of density 1 g/cm^3 is negligible, while most ^{32}P betas are transmitted. Actual measurements of counting efficiency show a range that depends on the type of TLC plate used. With soft silica gel plates having no binder, as much as 2 to 3% of the tritium betas will be counted by the IPC, whereas with hard plates having inorganic binders, the counting efficiency can be as low as 0.3 to 0.5%. Efficiencies for ^{14}C are about 10 times those for tritium, whereas for ^{32}P an efficiency of 30% or more (compared to a maximum possible efficiency of 50% for the 2π detector geometry) is obtained. It should be emphasized that the efficiency limitations are a direct result of counting the activity on the plate and are not introduced by the IPC technique. Any other counting technique that can view the radiation only after it has traversed the thin layer would be subject to the same limitation.

In spite of the low efficiency for tritium, the IPC technique is very sensitive because it views all regions of the chromatogram for the entire analysis period. It has an intrinsically low background of about 0.2 to 0.3 cpm/mm and can easily detect a 5-mm-wide tritium spot of 1000 dpm in 10 minutes. The background in 5 mm will be about 10 counts, while the spot, at a counting efficiency of 0.5%, will produce 50 counts.

Figure 5 shows a 30-minute analysis of tritiated dansyl chloride derivatives of catecholamine metabolites. The smallest peaks seen, at the origin and at 45 mm, correspond to 800 and 1200 dpm, respectively. To perform the same analysis by scintillation counting, 5-mm intervals scraped from the plate would require 30 samples and probably a minimum of 60 minutes (2 minutes per sample) of counter time. The scraping and sample preparation might require an additional 30 minutes or more of effort.

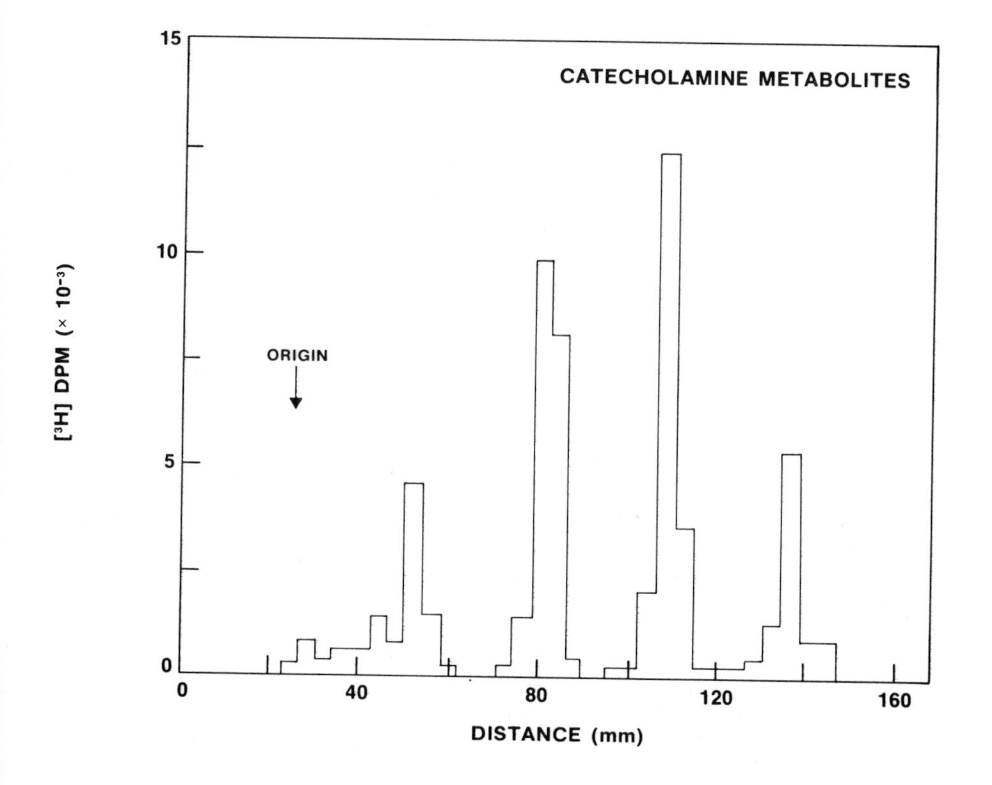

Figure 5. A 30-minute IPC analysis of tritium-labeled catecholamine metabolites.

Since the principal quantitative technique currently in use for radiochromatogram analysis is scintillation counting of material scraped from the TLC plate, it is important to have a careful comparison of the quantitative results from this technique with those from the IPC. The most thorough comparison to date is that published by Baird et al. (3). Samples of carcinogen ([^{3}H] benzo(a)pyrene) metabolite separations were analyzed by the two methods. Figure 6 shows part of the results from that work. The scintillation counting results are shown in Figure 6a and the IPC results in Figure 6b. Two different metabolite extractions are shown analyzed by each method. The filled dots and the open dots in the IPC results correspond to the solid line and dashed line, respectively, of the scintillation counter results. The percentage of each metabolite calculated from each method of analysis agrees to better than the 10 to 15% variation among several samples analyzed by scintillation counting alone. Thus, the quantitative results from the IPC technique are at least as accurate as the experimental reproducibility in this type of experiment. Further tests still need to be carried out to determine the ultimately achievable accuracy.

FUTURE PROSPECTS FOR IPC

Although IPC techniques are now developed to the point where they could handle a large fraction of current TLC radiochromatographic analysis, there are some areas in which improvements are needed to extend the applicability. One such area is that of spatial resolution. For some TLC, HPTLC, and gel electrophoresis, the separation techniques are capable of resolving bands with as little as 1 mm spacing, and it would be very useful to improve the IPC resolution to be comparable. One possible improvement is to reduce the depth of the detector since it determines the resolution for ^{14}C and higher-energy isotopes. A detector with only a 2.5-mm depth has been tested and gave a resolution of 1.4 mm (full width at half-maximum response) for ^{14}C betas (4).

A second approach to improving resolution is through the use of mechanical collimators. These are constructed in such a way as to accept only a limited angular divergence along the TLC lane while accepting almost the full divergence across the lane, thus providing the maximum efficiency for a given spatial resolution. With a collimator, a 5-mm-deep detector can still maintain a resolution of better than 2 mm. The major drawback is a loss of efficiency, which can drop to 10% of the uncollimated efficiency.

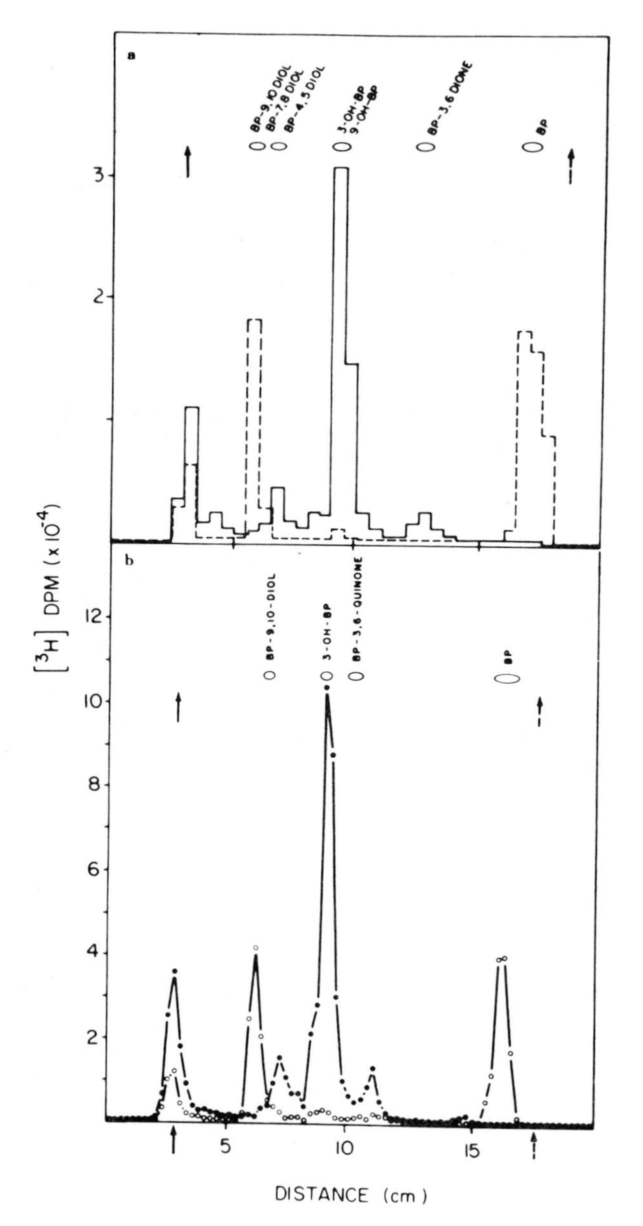

Figure 6. A comparison between (a) scraping and scintillation counting and (b) IPC analysis. The open and filled circles correspond to the dashed and solid lines, respectively. The IPC analysis time was 20 minutes.

Because of the low efficiences, collimators are most useful in applications where the sample activity is high, such as purity analyses of radiochemicals. Figure 7 shows such a radiochromatogram The autoradiograph at the top shows at least two impurity spots to the left of the main product, some activity at the solvent front, and a general background all along the lane. The main peak contains 96% of the activity as determined by scintillation counting. The IPC results without a collimator were obtained in only 10 minutes and gave the same fraction of activity in the main peak. However, with these results it is clearly difficult to decide where the integration boundaries for the main peak should be set. Also the impurity spot just to the left is lost in the wings of the main product, not surprisingly, since it is only 6 mm away and contains less than 2% of the activity.

The bottom panel of Figure 7 shows the same sample analyzed with a collimator. The main peak is much more clearly defined, and both spots to the left are clearly separated. The separation of the solvent front to the right of the main peak is also much clearer. With such collimation techniques, it should be possible to perform quantitative radiochemical purity checks during all phases of synthesis and use. Impurities down to 0.1% can be detected and determined without recourse to autoradiography and scraping and counting.

A second IPC development that may have interesting applications in TLC work is two-dimensional imaging capability. At present 2-D IPCs are in routine use in high-energy physics, X-ray and neutron diffraction studies, and X-ray astronomy. In TLC separations, 2-D samples are produced when different separation techniques are used in perpendicular directions on a plate to resolve compounds that cannot be separated with a single technique. With the exception of a brief note published on some work in the Soviet Union (2), however, 2-D IPC techniques have not been used to analyze these more powerful TLC separations.

To simulate the kind of quantitative data that could be produced, a one-dimensional IPC can be mechanically scanned in the second dimension. An example of this is shown in Figure 8, where histone phosphopeptides have been separated by a combination of chromatography and electrophoresis on cellulose TLC plates. The top panels show autoradiographs of the ^{32}P-labeled sites that were obtained in 1-week exposures. After approximately one month (two ^{32}P half-lives), the plates were analyzed by scanning a 1-D IPC across the plate. Contour maps of the results are shown in the bottom panels. These maps are a graphical representation of quantitative data obtained with about 5 minutes of analysis time

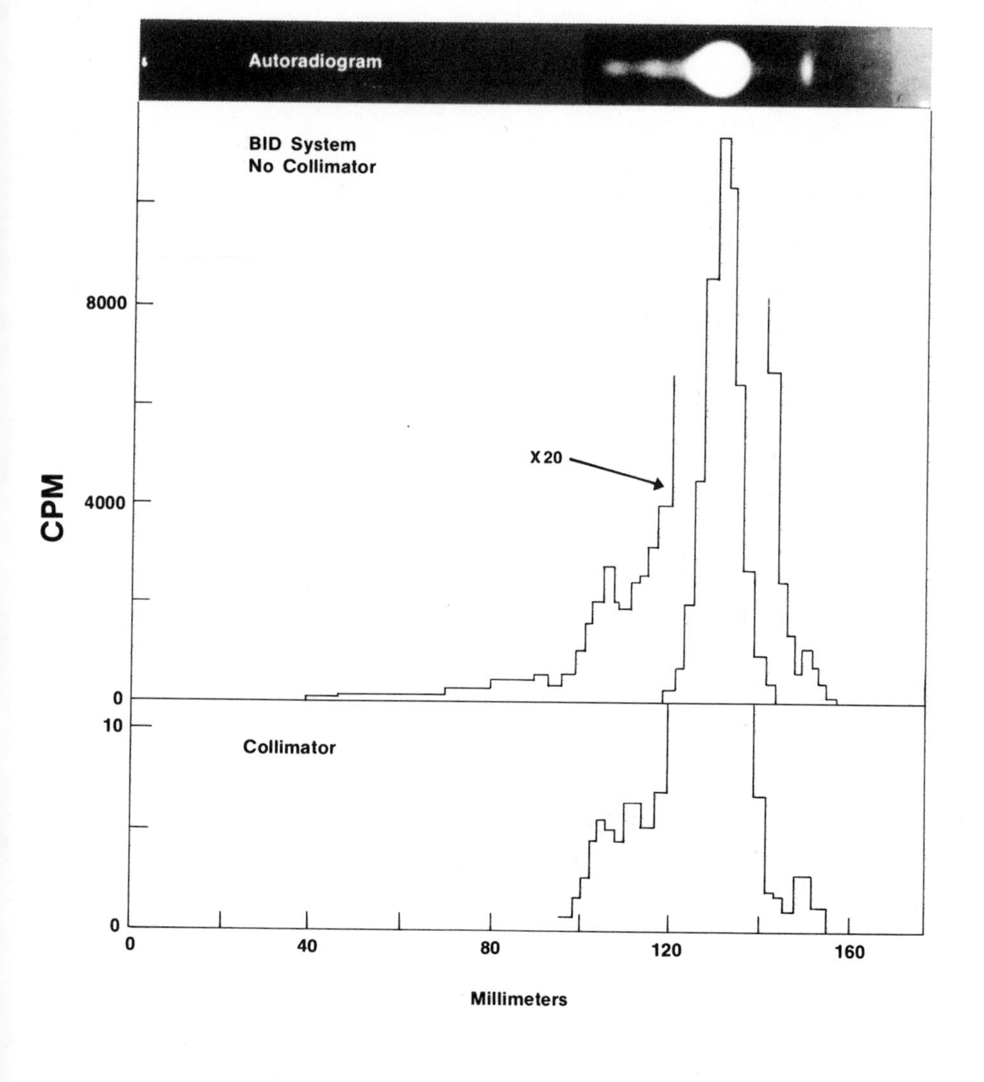

Figure 7. A comparison of autoradiography and IPC analysis. The TLC plate is a purity check of a ^{14}C-labeled compound. The IPC data, both with and without a collimator, are shown below the autoradiogram. The analysis time with no collimator was 10 minutes, while that with the collimator was 20 minutes.

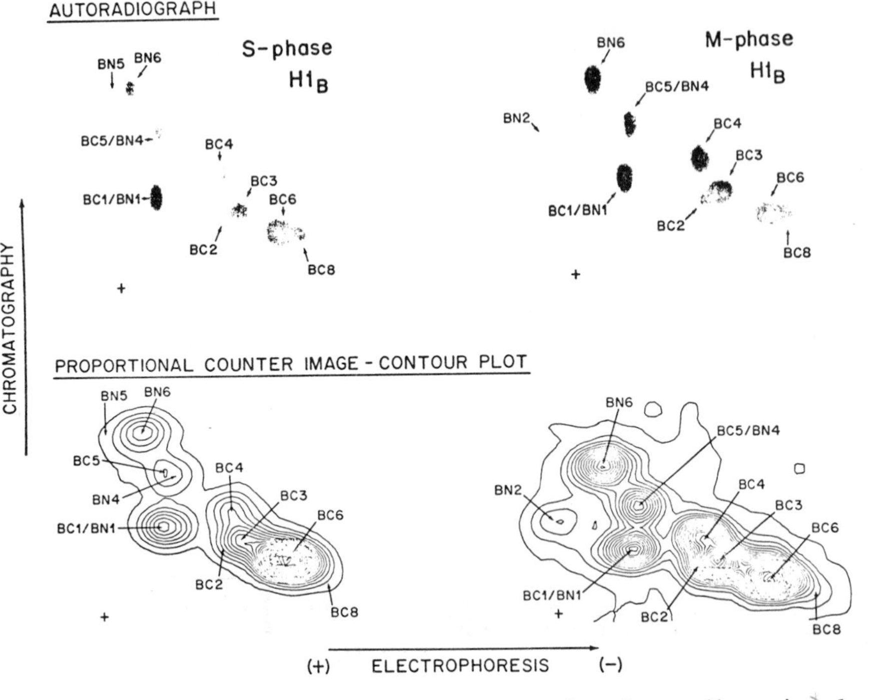

Figure 8. Autoradiography and IPC analysis of two-dimensional TLC separations of phosphopeptides. The IPC data are shown as contour plots to quantitatively indicate the distribution of ^{32}P in two dimensions.

at each 3 × 3 mm area of the 20 × 20 cm plates. On the basis of these data, the total activity in the peaks was computed and a comparison made to determine the phosphorylation at various sites during the cell cycle (5).

The above results were obtained by mechanically scanning a 1-D IPC, but a full 2-D IPC would have the same relative sensitivity advantage that the 1-D IPC has over conventional radiochromatogram scanners. The scan took several hours, but a 2-D IPC would obtain the same information in about 5 minutes. Clearly, powerful quantitative analysis techniques are available for 2-D separations on TLC plates. At the present time such techniques are not in widespread use, perhaps because the IPC capabilities are not widely known or commercially available to the biomedical research community. Once it is recognized that quantitative information can be obtained, the resolving power of TLC will be greatly enhanced by the routine use of 2-D separations.

REFERENCES

1. A. Gabriel and S. Bram, FEBS Lett., 39, 307 (1974).
2. Yu. V. Zanivsky, S. P. Chernenko, A. B. Ivanov, L. B. Kaminir, V. D. Peshekhonov, E. P. Senchenkov, I. A. Tyapkin, and V. N. Kalinin, Nuclear Instruments and Methods, 153, 445 (1978).
3. W. M. Baird, L. Diamond, T. W. Borun, and S. Shulman, Analyt. Biochem., 99, 165 (1979).
4. K. Goulianos, K. K. Smtih, and S. N. White, Analyt. Biochem., 103, 64 (1980).
5. K. Ajiro, T. W. Borun, S. Shulman, G. McFadden, and L. Cohen, Biochemistry, 20, 1454 (1981).

CHAPTER 9

Improved Methods for Detection of Low Energy Beta Emitters in TLC

Bart W. Bartelsman, Shari L. Ohringer, and Dennis L. Fost

INTRODUCTION

Autoradiography is the technique of detecting radiolabeled compounds by film imaging methods. Very simply, decay products strike a film emulsion that contains silver halide, forming a latent image that is visible after photographic processing.

Fluorography is a technique that is slightly more involved. Instead of directly exposing the film to decay products, the film is sensitized by light emitted from an energized (activated) fluor. A fluor is defined as a compound that emits light energy after being struck by radioactive decay products. Many radioactive decay products are quite energetic. The decay products from ^{32}P and ^{125}I are more energetic than the light emitted by the fluor, and it would be nonproductive to detect these isotopes by fluorography. On the other hand, for ^{3}H, ^{14}C, and to an extent ^{35}S, the decay products are low-energy beta particles. These isotopes are suitable for fluorography, since the average energy of the emitted beta particles is so low that a significant percentage of them do not reach the film emulsion.

The concept of film autoradiography has been known since the turn of the century, when it was observed that some ores blackened film emulsions. In the late 1940s, serious research utilizing autoradiography sprang up. The U.S. Atomic Energy Commission funded several projects studying plant metabolisms using ^{14}C and ^{32}P. In one project, labeled metabolites were isolated by paper chromatography, and the chromatograms were exposed to X-ray film at room temperature (1).

In 1965, Drs. Ursula Luthi and P. G. Waser published a paper on the effects of temperature on fluorography of thin layer chromatograms (2). They used finely ground anthracene and silica gel, which were cast directly onto TLC plates. Based on film-darkening studies, it was concluded that anthracene fluoresced more efficiently at colder temperatures than at ambient conditions. However, Randerath (3) showed that the effects of temperature on fluorography are primarily related to film chemistry, and that reduction of temperature during exposure reduces latent image failure. This theory has been supported by others (4,5) and is consistent with accepted photographic chemistry.

Efforts to improve sensitivity by incorporating fluors into thin layer plates led Randerath (3) to try mixtures of ether and PPO to convert emitted β^- particles to light energy. It was found that there was an optimum concentration for maximum enhancement of ^3H signals. In 1975, Laskey and Mills (4) showed that by preexposing X-ray film to a flash of light, the sensitivity of fluorography could be improved. In 1978, Bonner and Stedman (5) published a report describing several compositions of fluor mixtures, among them melted PPO, into which TLC plates are dipped to improve sensitivity. All of the methods have drawbacks ranging from smearing of the isolated compounds to separation of the adsorbent from the support.

In 1977, New England Nuclear Corporation started a group to develop new commercial products for improving fluorographic detection of low-energy β^- emitters for gels, thin layer chromatography plates, and tissue. EN^{3}HANCE™ was introduced in 1979 and was designed for fluorography of gels. A similar product was then developed for use with thin layer plates. This product was introduced in 1980 under the name EN^{3}HANCE™ Spray.

RESULTS AND DISCUSSION

During the development of EN^{3}HANCE™ and EN^{3}HANCE™ Spray, a number of parameters were studied that affect sensitivity of autoradiography and fluorography. The results of these studies are presented below.

Experimental

All of the sensitivity tests were conducted as follows. Serial dilutions of ^3H-leucine or ^{14}C-leucine were spotted onto Eastman silica gel chromatogram sheets. The sheets were sprayed with

EN^3HANCE Spray where required and exposed to X-ray film for a period of time at a particular temperature. After exposures, all films were processed at 20°C for 4 minutes in Kodak KLX liquid X-ray developer, washed 2 minutes in running water, fixed 4 minutes in Kodak Rapid Fix, and rinsed in running water for 10 minutes. After air drying, the optical densities of the exposures were measured on a Gelman ACD-18 automatic computing densitometer.

Materials

Thin layer chromatography plates were Eastman Chromatogram Sheets 13179 silica gel; ^3H-leucine and ^{14}C-leucine were from New England Nuclear. The densitometer used for quantitating film densities was a Gelman ACD-18 automatic computing densitometer. X-Ray film was supplied by Eastman Kodak and LKB Corporation. Darkroom supplies, Kodak Liquid X-ray Developer, and Kodak Rapid Fixer were supplied by Eastman Kodak.

One of the most important aspects of fluorography is the film that is used for detecting the labeled compound. Because fluorography is a measurement of light emission, the spectral sensitivity of the film and the light emission of the fluor must match. Additionally, the film must be as sensitive as possible to allow the most rapid detection. Of all the films tested, Kodak X-OMAT R film is the film of choice for fluorography, with Kodak X-OMAT SB being somewhat slower. LKB Ultrofilm ^3H has applications that will be discussed.

Figure 1 shows how Kodak X-OMAT R film compares to Kodak X-OMAT SB and LKB Ultrofilm ^3H. Spotted chromatograms were sprayed with EN^3HANCE^{TM} Spray and exposed at dry ice temperature (-76°C) for 24 hr.

Kodak has introduced a new film, Kodak X-OMAT AR, which uses an emulsion identical to the R film, but the base is changed from blue to gray. This film shows identical response figures to X-OMAT R. Temperature plays another important role in affecting sensitivity. Lowering the exposure temperature reduces latent image fading. Figure 2 shows how temperature plays a role in improving sensitivity. During the photographic process, a quanta of energy hits a silver halide crystal. This sets up a charge in the crystal lattice. With time, this charge fades away, and the crystal reverts to its original state. This reversible process, latent image fading, is reduced by cold temperatures. By slowing down the reverse reaction over a given period of time, more halide crystals become charged, and when the film is developed a stable image is formed.

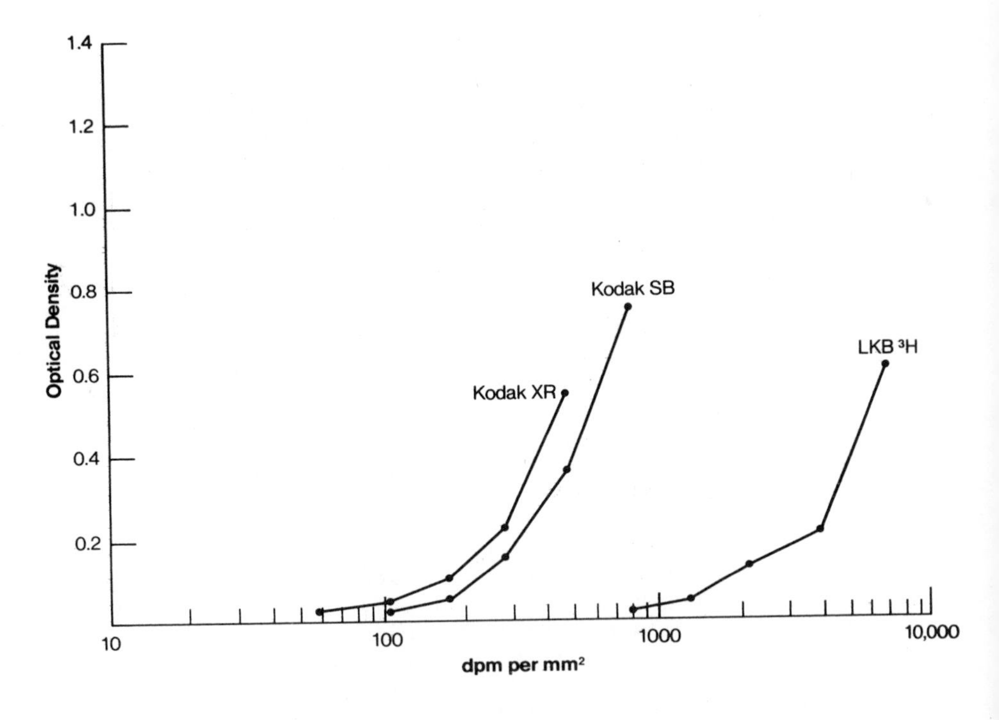

Figure 1. Comparison of Kodak X-OMAT R film, Kodak X-OMAT SB film, and LKB Ultrofilm ^3H.

Figure 3 compares the sensitivity of fluorography to autoradiography of tritium. TLC plates were identically spotted and one of the plates was treated with EN^{3}HANCE™ Spray. The other plate was left untreated. Both plates were exposed to Kodak X-OMAT R film at -76°C for 24 hr, developed, and the optical densities plotted. The mean free path of a ^3H β^- particle is 1.8 µm in air. Absorption losses occur in the TLC plate, at the TLC-air interface, and in the gelatin protective coat that is present on most X-ray films. During fluorography, the β^- particle is no longer required to travel the relatively long distance to a silver halide crystal, but only to the nearest fluor molecule, where the energy is converted to light. The light does not suffer as much attenuation as the electron, and hence the silver halide crystal is exposed more efficiently. A 100- to 250-fold enhancement is shown.

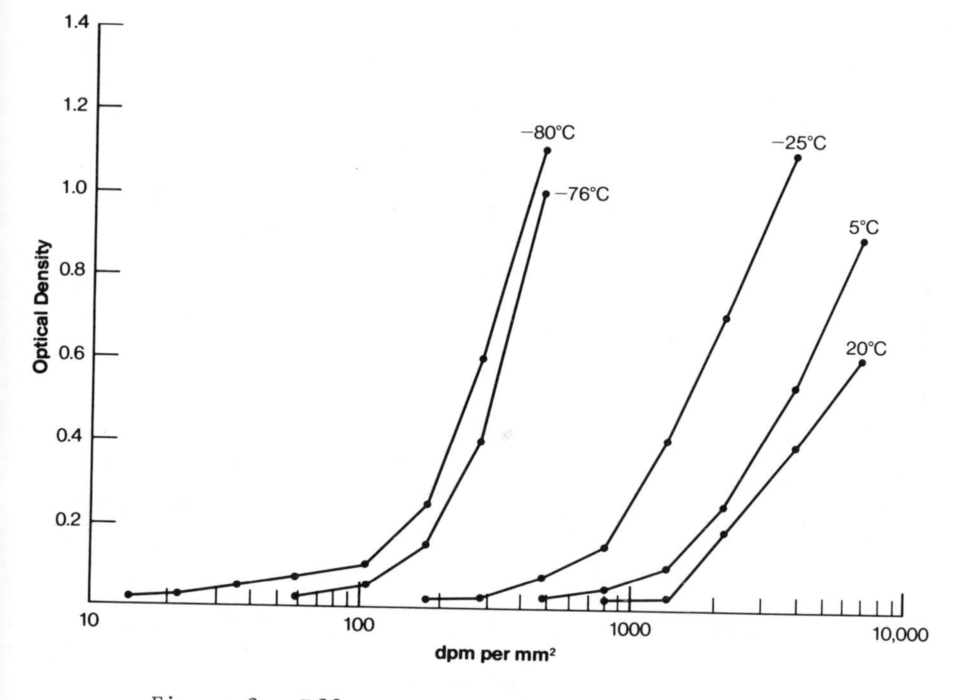

Figure 2. Effect of temperature on sensitivity.

In Figures 4 and 5, optical density (OD) versus disintegrations per minute (dpm), enhanced and unenhanced, for several exposure times is plotted. With these graphs one can predict exposure times of X-OMAT R X-ray film to obtain a desired optical density when the amount of radioactivity on the substrate is known. By staying within the theoretical considerations of latent image formation, it should be possible to improve the detection limit by partially exposing the silver halide crystals prior to exposing them to the sample. Preflashing film gives a means for accomplishing this (see Figure 6). Identical TLC plates were sprayed with EN^{3}HANCE^{TM} Spray and exposed to flashed and unflashed film. The film was flashed using a Canon Canolite D two-battery camera strobe that was modified to allow remote triggering by soldering a momentary contact switch between the terminals on the shoe. Then a Kodak No. 21 Wratten filter and one Whatman No. 2 filter paper were taped over the lens of the strobe. The strobe was mounted 90 cm above a sheet of film in a darkroom. The

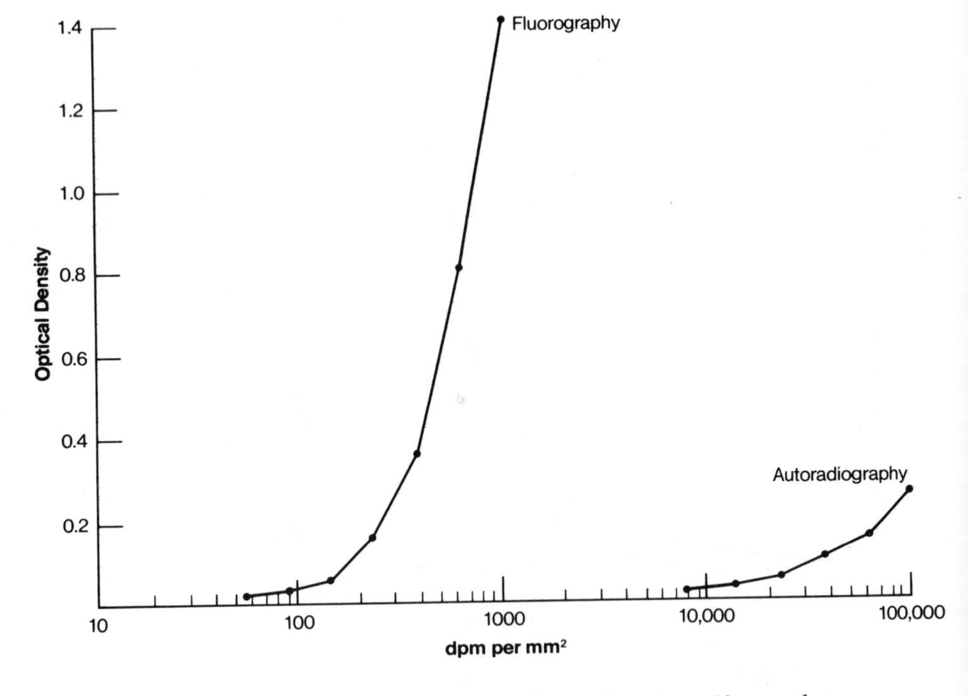

Figure 3. Comparison of fluorography and autoradiography.

flashed side of the film was then placed in contact with the spotted TLC plate and exposed at -76°C for 24 hr. Preflashing increases the background fog levels of the film, depending on the intensity of the flash. In our experiments, background was increased from 0.4 to 0.8 OD on the film. The results were similar to those reported by Laskey and Mills (4), and we would recommend that, if the background is not objectionable, this is an excellent means of improving sensitivity.

In some instances, fluorographic imaging may not be possible. In these cases autoradiographic sensitivity may be improved through the use of LKB Ultrofilm ^{3}H. This film has no protective gelatin coat. By removing this coat, the silver halide crystals can be brought closer to the emission source, thus reducing absorption losses. When used in fluorography, it was found that this film is less sensitive than Kodak X-OMAT R film (Figure 1). Additionally, no further improvement is evident if the LKB film

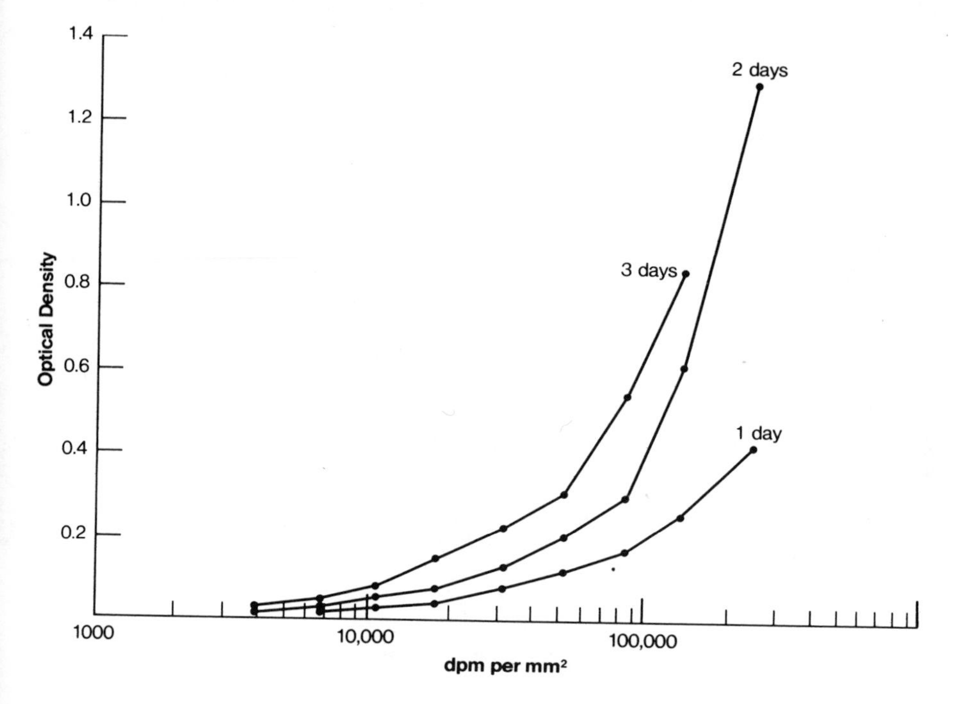

Figure 4. Time effect on sensitivity, unenhanced.

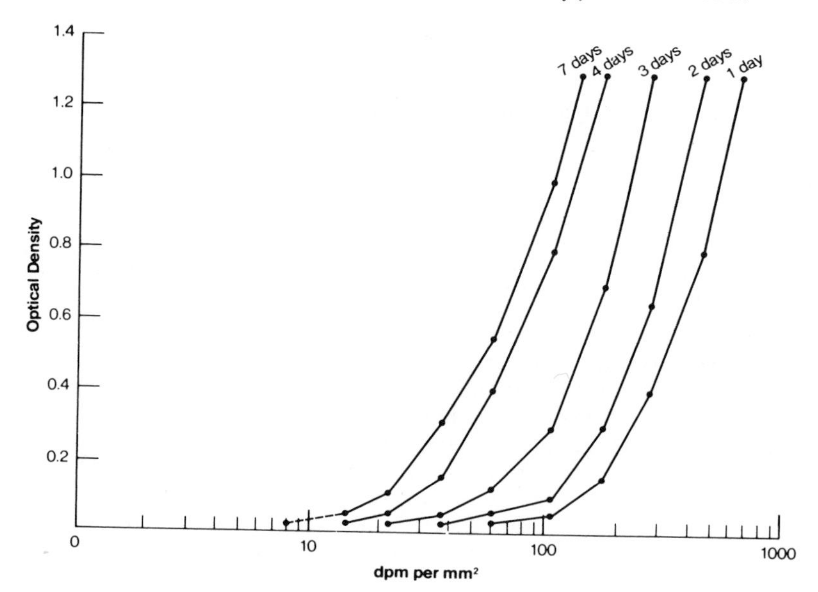

Figure 5. Time effect on sensitivity, enhanced.

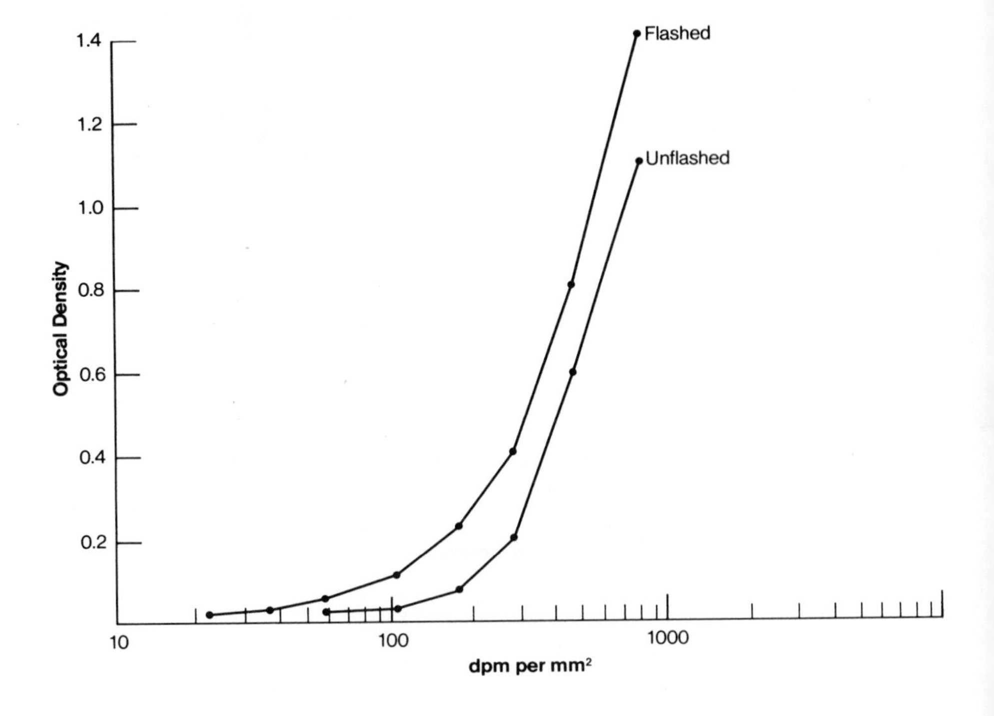

Figure 6. Comparison of flashed and unflashed film.

is fluorographed as shown in Figure 7.

Carbon-14 isotopes are also good candidates for fluorography. The emission energy is somewhat higher, and silver grain exposure occurs more readily, but attenuation from distance and gelatin layers on film still play an important role. In Figure 8, the difference between fluorographed and autoradiographed TLC plates, spotted with ^{14}C-leucine, is shown.

Another useful technique for improving the sensitivity is to ensure that the TLC plate is in very close contact with the film. This can be accomplished by using tension clips to hold the film and TLC plates together. In this configuration the distance that light has to travel to expose the film is minimized.

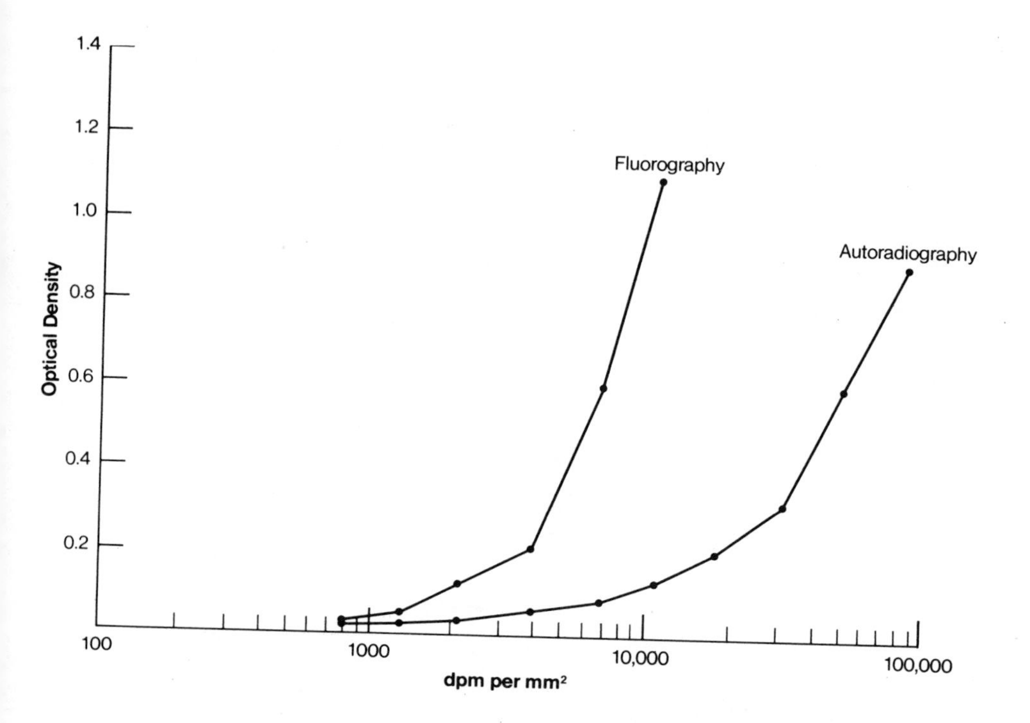

Figure 7. Film sensitivity of LKB ^3H film.

REFERENCES

1. A. A. Benson, J. A. Bassham, M. Calvin, T. C. Goodale, V. A. Haas, and S. Stepka, J. Am. Chem. Soc., 72, 1710 (1950).
2. U. Luthi and P. G. Waser, Nature, 205, 1190 (1965).
3. K. Randerath, Anal. Biochem., 34, 188 (1970).
4. R. A. Laskey and A. D. Mills, Eur. J. Biochem., 56, 335 (1975).
5. W. A. Bonner and T. D. Stedman, Anal. Biochem., 89, 247 (1978).

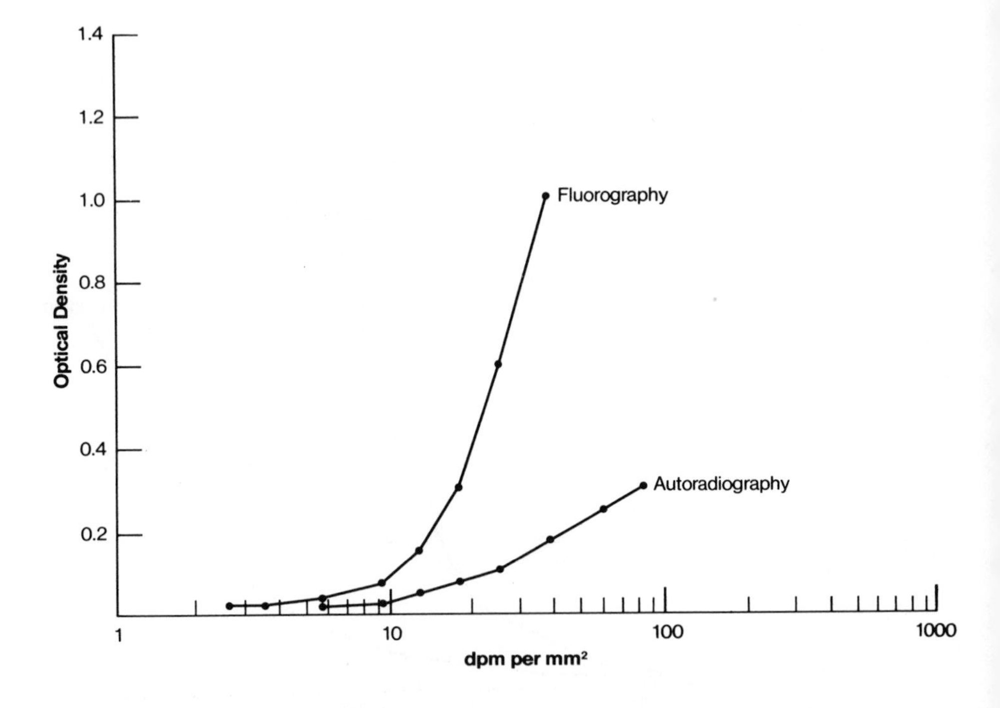

Figure 8. Comparison of fluorographed and autoradiographed TLC plates spotted with ^{14}C-leucine.

CHAPTER 10

Effect of Scanner Resolution on System Performance

Richard Apotheker

INTRODUCTION

Historically, the resolving power of an optical scanning instrument has been considered to be a rather minor parameter and generally classed as information that is more significant to the manufacturer than the user. This assumption becomes less and less valid as thin layer (and other) separation approaches the theoretical ideal, that is, greater density distributions, nondiffused boundaries, higher scan rates, and minimum development distances. These characteristics can (and often do) combine to cause <u>any</u> scanning system to operate in an unpredictably nonlinear mode. A technique was required that would objectively report the absolute resolution characteristics of the system apart from the considerations of the thin layer parameters.

THEORY

In the context of optical scanning, resolution can be defined as a measure of the instrument's ability to accurately perceive the zone between adjacent components, or, conversely, the minimum size component of the separation that can be accurately perceived. The former is usually observed as a "return to baseline" between peaks, whereas the latter is, unfortunately, seldom observed directly, its effect being noted simply as "error."

There are two major mechanisms involved:

1. Optical resolution: the combined "field of view" of the optics and the response characteristics within the field.
2. System response: the total response-time characteristics of all system components.

These mechanisms can interact to a large degree dependent upon the dynamic optical properties of the scan. A major factor affecting the degree of interaction is the type of optical change encountered, which may be roughly grouped as either step or slope changes induced by:

1. Edge density gradients.
2. Edge shape/angle variables.

Although the significance of the above factors is visibly critical in the determination of resolution, it should be noted that since the effects are continuously active, optical gradients occurring in the core of any given zone are also subject to the same error functions.

In considering the dynamics of edge detection, the adjacent-equal-density-squares approach was selected as a test pattern for the following reasons:

1. No edge density gradients.
2. No shape/angle variables.
3. Accurate and repeatable step changes.

The "perfect" optical step change would then be used as the baseline for system performance. The resulting response thus becomes an indicator of the type and magnitude of resolution limiting factors. This report describes how this was carried out.

EXPERIMENTAL

Test Pattern

A test pattern was created (Figure 1) consisting of 22 rectangles of equal length (7 mm) with steadily decreasing widths (1.25 to 0.20 mm in 0.05-mm increments). Ordinary photographic film was

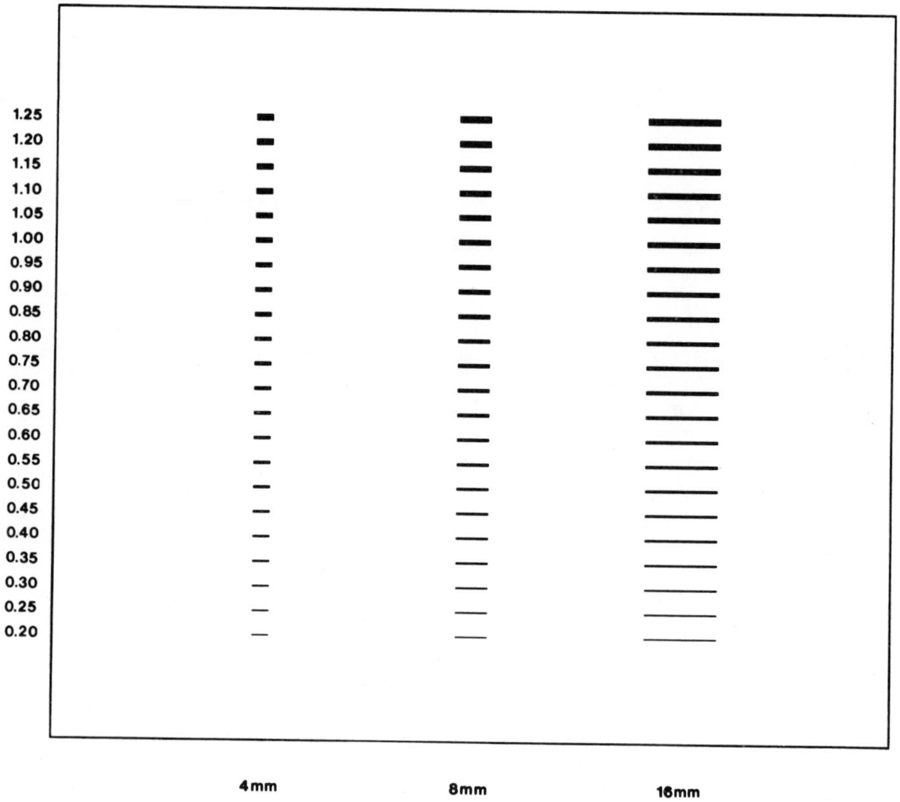

1.25
1.20
1.15
1.10
1.05
1.00
0.95
0.90
0.85
0.80
0.75
0.70
0.65
0.60
0.55
0.50
0.45
0.40
0.35
0.30
0.25
0.20

4mm 8mm 16mm

Figure 1. The resolution test pattern used for scanner evaluation.

chosen as the medium, with the rectangles being totally opaque. Dimensional accuracies were held to ±2%, yielding area tolerances of ±4%. This approach effectively removed linearity of opto/electronic response as an output variable.

Scan Conditions

All scans were made on a Kontes model 800 scanner in the single-beam transmission mode using an 8-mm head. The various degrees of resolution were produced by changing the head-to-media distance. All settings remained unchanged except where noted. A Hewlett Packard 3390A reporting integrator was the output indicator, and, although its own characteristics can create large variables, the net effects were summed as a constant by only changing the chart speed.

Results of Scans

The expected results were a series of flat-topped peaks of constant amplitude and steadily decreasing width. Any progressive decrease in amplitude would thus indicate a loss in true resolution.

Figure 2 shows six scans of the test pattern, at the same scan rates, with various head-to-media distances. The lower right scan was done with the smallest distance as evidenced by the longest straight-line portion of amplitude response. The double arrows on each scan indicate the approximate points at which loss of 100% resolution has occurred. It should be noted that any percentage of resolution could have been chosen, but succeeding area reports would have become more complex.

Figure 3 shows four scans, at the same head-to-media distance, operated at scan rates of 2, 5, 10, and 30 cm/min, respectively. As indicated on the parameter listings, the chart speed was adjusted accordingly to provide equally sized recordings. The center scale shows the widths of the bars on the test pattern at various points.

Scan 1 shows a significant loss of amplitude beyond the peak identified by the dot. This corresponds to the last bar resolved to 100%. Scan 2, at a higher scan rate, shows identical characteristics. Both scan rates resolve at 0.35 mm. From this it can be stated that any further decrease in scan rate will not substantially alter the resolution. Thus, the optical resolution limit has been indicated.

Scan 3, at 10 cm/min, shows a marked loss of resolution which may be defined as system resolution. The effective system resolution, at 10 cm/min, is thus 0.55 mm. Scan 4, at 30 cm/min, shows further evidence of this effect. These results may then be restated. The scanner under test, with it's companion integrator, at the head-to-media distance specified, will produce total resolution of 350-μm phenomena provided the scan rate does not exceed 5 cm/min.

Figure 4 shows three scans at head-to-media distances of 0.1, 1.5, and 5.0 mm. The resulting peaks were reported as height and area, the area was normalized via external standardization with the calibration being one dimension of the absolute area, that is, 1.25, 1.20, 1.15, . . . The amount/area was thus the predicted response factor. The three sets of peak heights data were plotted in Figure 5. The data at bar widths of 0.85

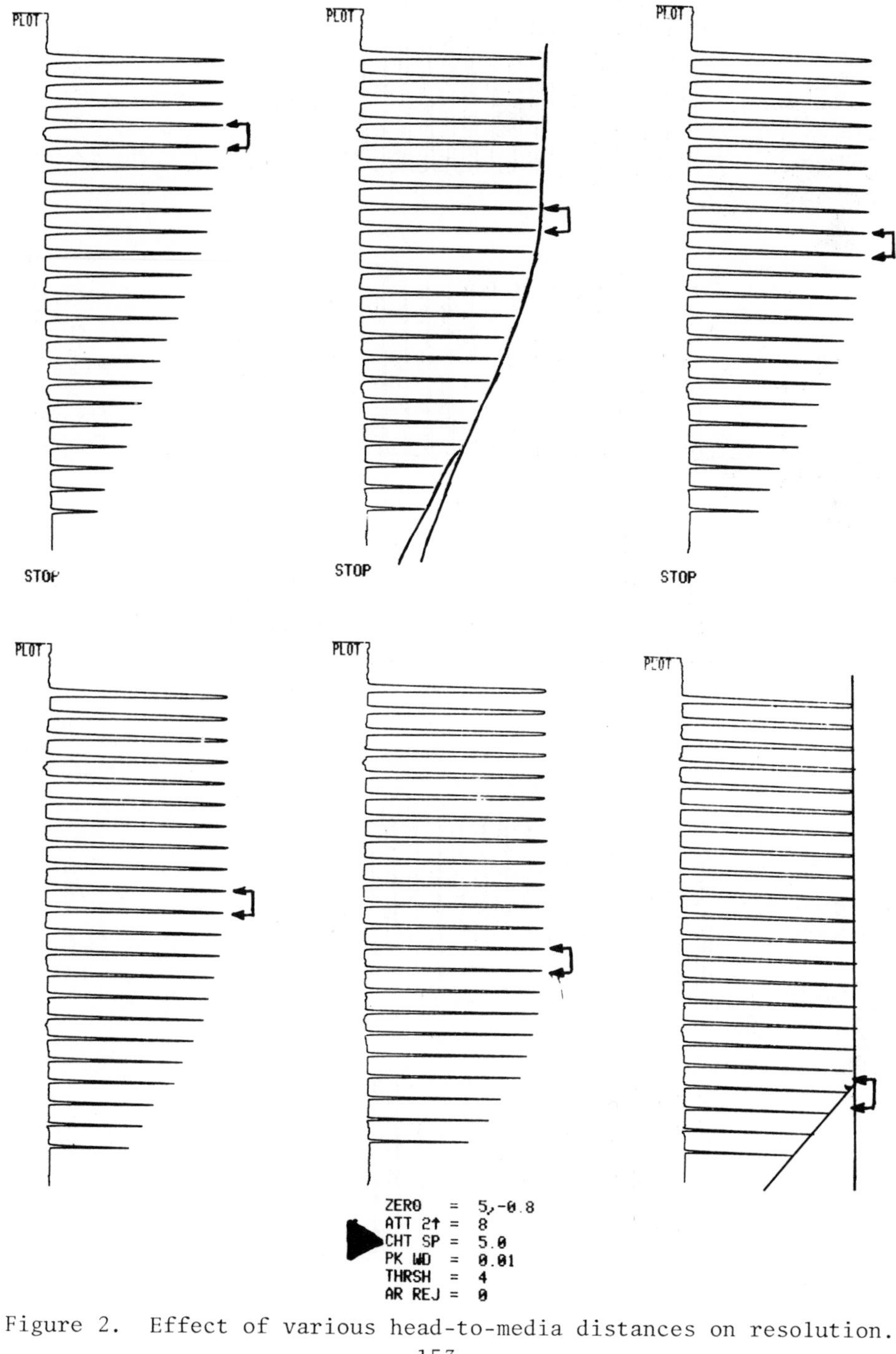

Figure 2. Effect of various head-to-media distances on resolution.

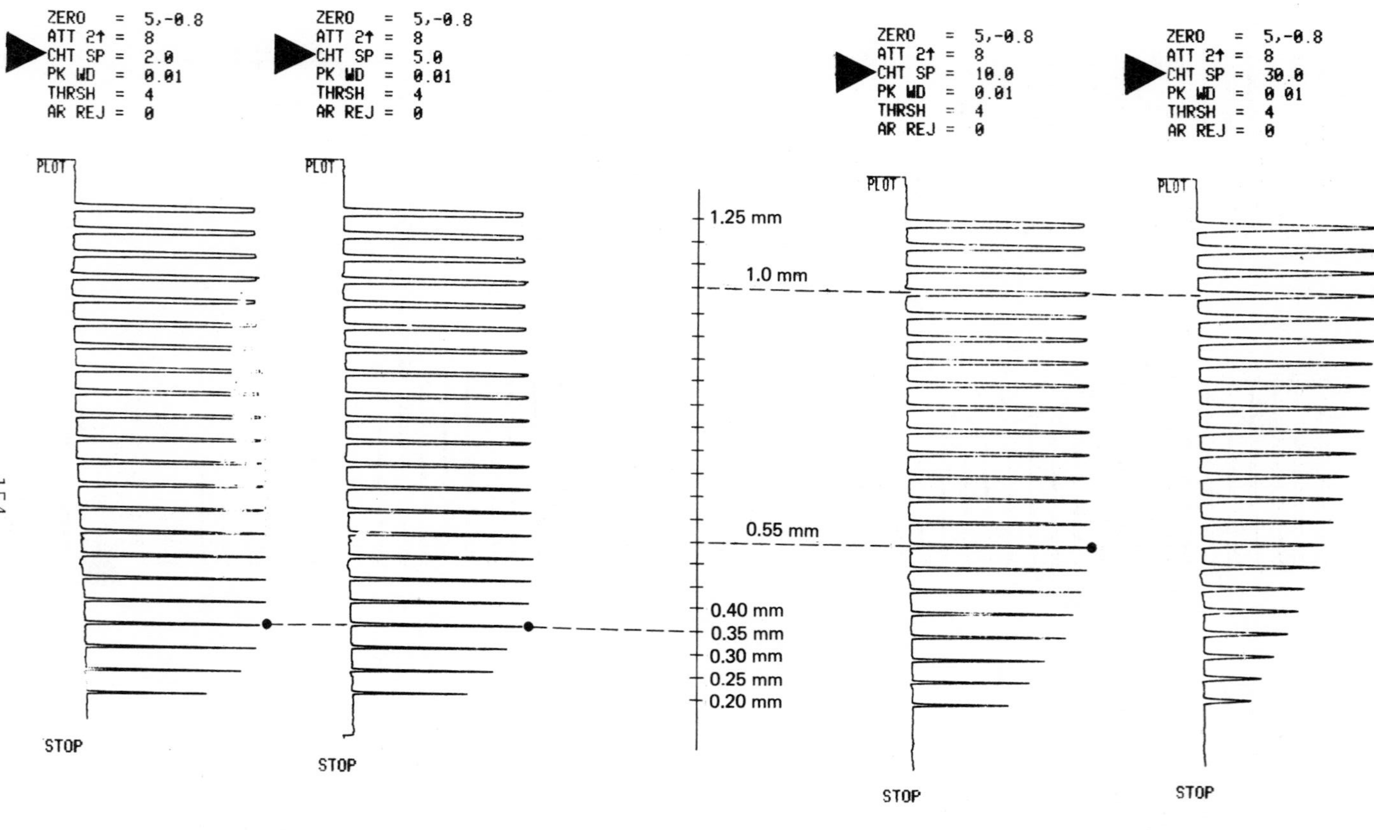

Figure 3. Effect of scan speed on resolution. (a) Scan 1, 2 cm/min. (b) Scan 2, 5 cm/min. (c) Scan 3, 10 cm/min. (d) Scan 4, 30 cm/min.

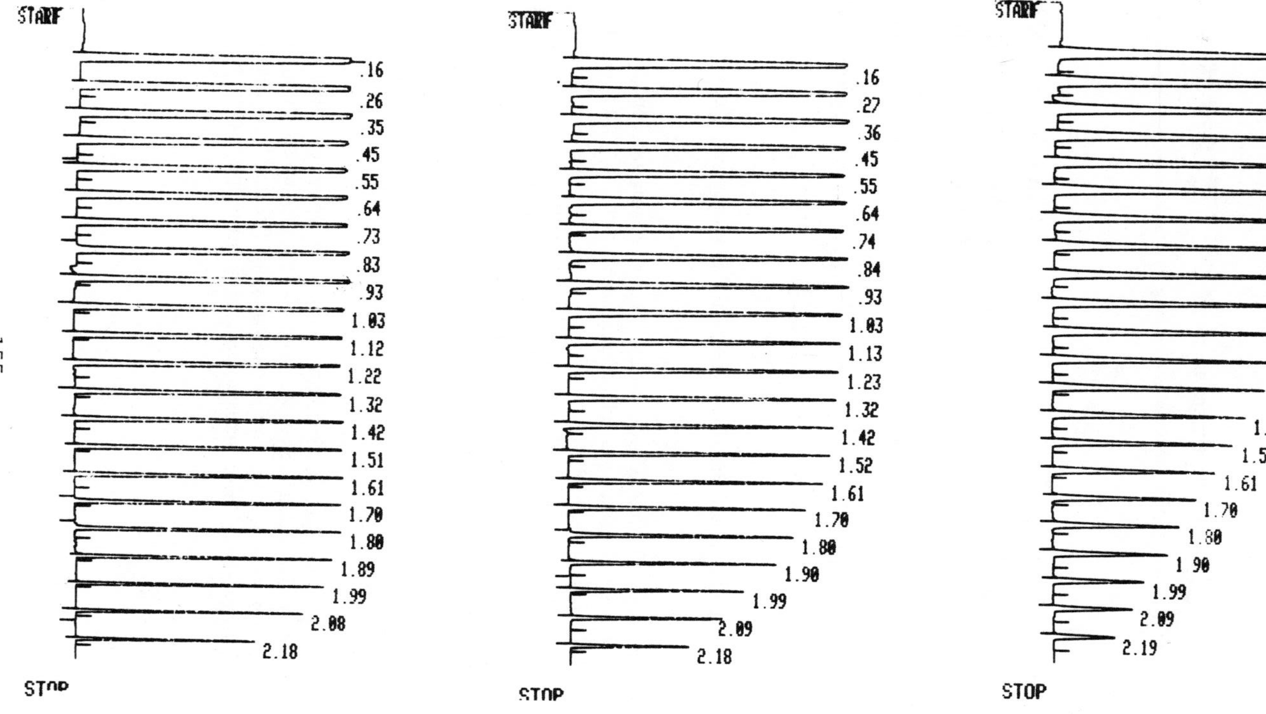

Figure 4. Effect of hea-to-media distance on resolution.

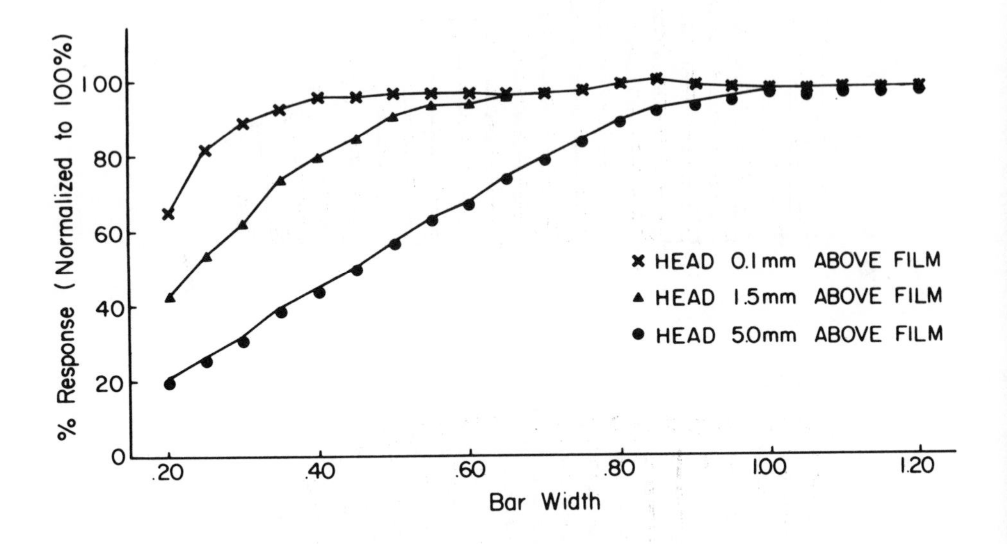

Figure 5. Effect of head-to-media distance on resolution.

and 0.80 mm were disregarded when close examination of the test pattern at these locations revealed aberrations in the film. When area/amount deviations were plotted in Figure 6, the influence of optical resolution on area calculation accuracy became apparent. Using the 0.35-mm bar width as a reference (from Figure 5 as first loss of 100% resolution), the errors beyond ±4% **tolerances** were noted as:

Scan ● = 2%
Scan × = 9%
Scan ▲ = 16%

Realizing that height deviation (loss of optical resolution) was a major factor in area errors, area/amounts were replotted in Figure 7 using the height normalization factor to unmask the remaining variable. The errors at the same point were:

Scan ● = 0%
Scan × = 2%
Scan ▲ = 90%

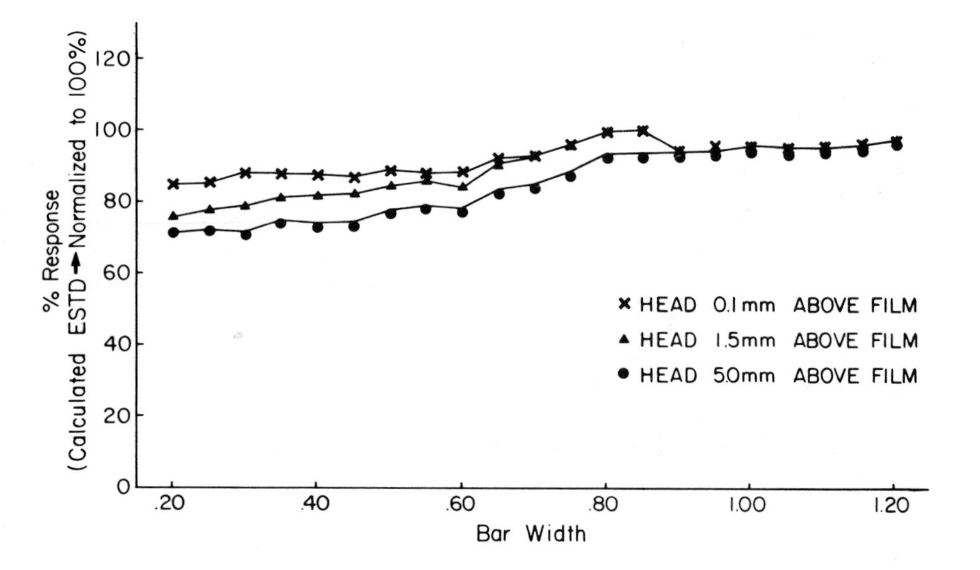

Figure 6. Effect of optical resolution on accuracy of area calculation.

The improvement in the first two scans and dramatic loss of accuracy in the third point to a clear conflict between the integrity of the pattern, accuracy of the scanner, and computational ability of the integrator.

DISCUSSION

With the results indicated above, the practitioner of TLC may well question the validity of the test and the meaning of its results. The answer is uncomplicated: The optical scanning of a TLC plate requires data in the form of density per unit area. An opaque rectangle of known dimensions reduces the problem to its simplest terms.

The meaning of the results can best be described in terms of why and where. The limits of resolution in optical scanning instrumentation exist because of the enormity of the data present and the compromises in instrument design required. With the question of who must compromise (and where, the user faces the real problem of knowing where the true working limit is, not

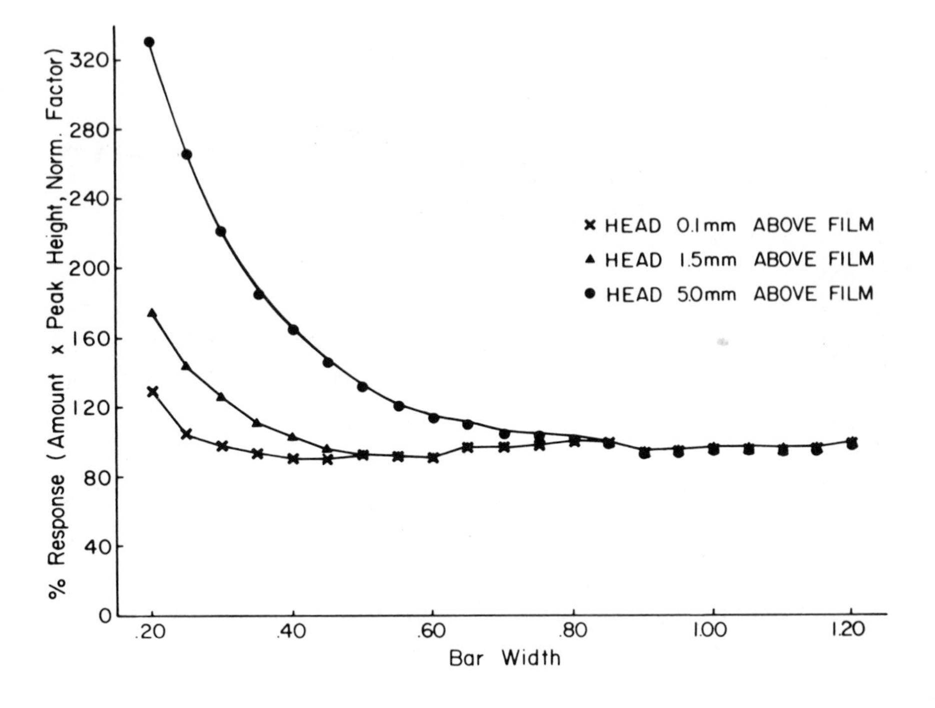

Figure 7. Effect of height deviations on area.

why it is.

It is not the intent of this test to explain the why but to define the working limits of resolution within which the user may expect to obtain accurate and reproducible results.

CONCLUSION

It has been shown that resolution can have a major impact on the accuracy of results. It can also be seen that the test pattern devised represents a "worst case" test of resolution. Although these conditions are only infrequently encountered, it is valid to conclude that the limits derived from these conditions represent the absolute limits of performance of the system under test. Operating beyond the extreme of any given system resolution limit point may not always produce nonlinear response, but if the dynamic optical characteristics of the scan area approaches or exceeds the resolution limit point, errors will certainly occur. It is the variability of this type of distortion that can often be the origin of nonreproducible or inaccurate results.

The technique described represents a quick and accurate method for evaluating the performance of all optical scanning systems using the same criteria. This approach to evaluation can be easily implemented and can also define the interaction of system components (scanner, linearizer, integrator, recorder, etc.) as they affect resolution.

CHAPTER 11

Design and Features of a New TLC Scanner

John D. Johnson

INTRODUCTION

An extremely versatile scanner, the Quick Scan R&D, has been developed utilizing the basic mechanics and electronics of the time-proven Quick Scan Jr. series of instruments produced by Helena Laboratories. The Quick Scan Jr. TLC (see Figure 1) is an instrument that will handle a full 20 × 20 cm TLC plate; its primary limitation in TLC work is that it is a transmittance-only instrument. The Quick Scan R&D (Figure 2) was developed to fill the needs of various researchers in electrophoresis and TLC laboratories.

The front panel controls (Figure 3) of this instrument present information relative to the features of the instrument. The detector voltage control is a 10-turn calibrated control that regulates the photomultiplier voltage between -300 and -1300 V DC. The analog gain control allows chart amplification of the analog signal with the range of 1 to 20 times. The zero control allows for adjustment of the background level, as well as allowing relative measurements to be made. The scan speed is continuously variable from less than 1 mm/sec to more than 15 mm/sec. The function switch allows for signal amplification in a linear or logarithmic mode, while the mode switch presents either increasing or decreasing signals as up-scale deflection for both the analog and integral signals. The pen position controls provide for positioning of the traces without changing the data being presented.

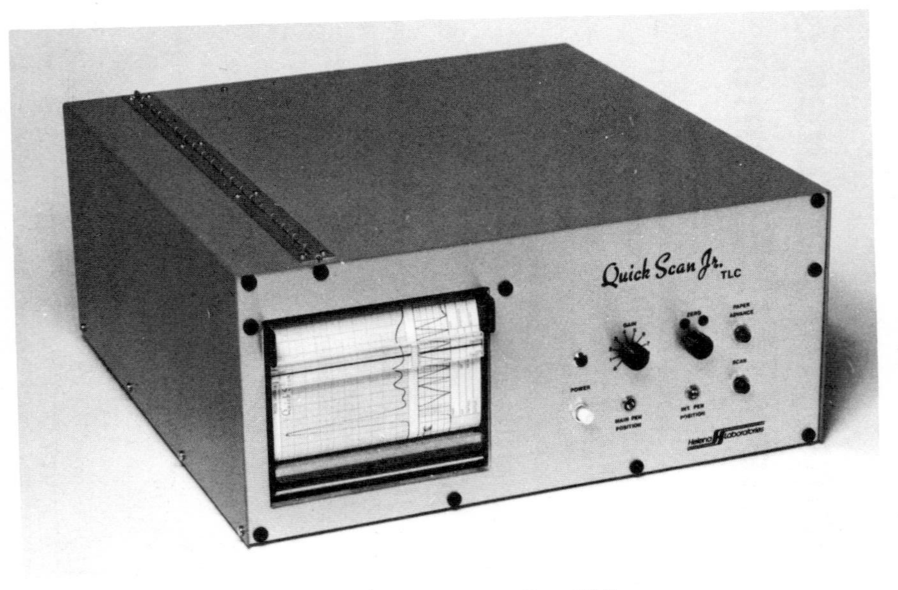

Figure 1. Quick Scan Jr. TLC.

Figure 2. Quick Scan R&D.

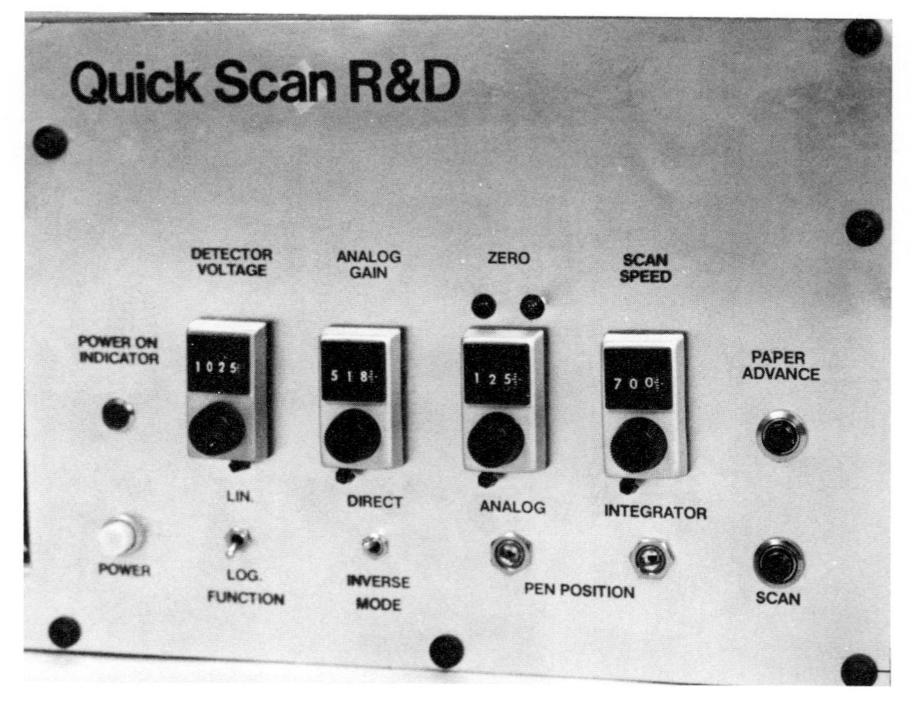

Figure 3. Front panel controls of Quick Scan R&D.

The scan switch brings the integrator pen to baseline and initiates the scanning process.

Optics versatility is achieved through the use of various plug-in modules with multiple configurations possible. The first configuration is transmittance with 1.5 multiplication of the slit plate, also referred to as standard resolution (Figure 4). In this configuration, the light from the lamp is focused on the slit plate, the image of the slit is transferred by a projection lens system via a front surface mirror through an interference filter, and then through the sample to the photomultiplier, which is mounted on the cover of the instrument.

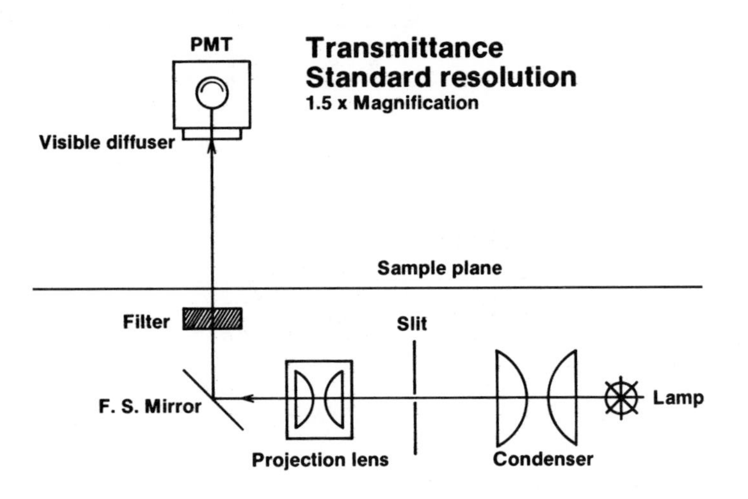

Figure 4. Optics configuration of Quick Scan R&D with magnification of 1.5×.

In the high-resolution optical configuration (Figure 5), the projection lens is moved to a location that transfers the slit image with reduced magnification. Figure 6 tabulates the slit plate and images routinely supplied with the instrument. The limits on the slit plate are such that any image up to 1 mm × 1 cm may be projected with the standard resolution optics.

Visible reflectance (Figure 7) is achieved by moving the detector to a lower position and imaging the illuminated area onto the photomultiplier with another set of lenses. Fluorescence

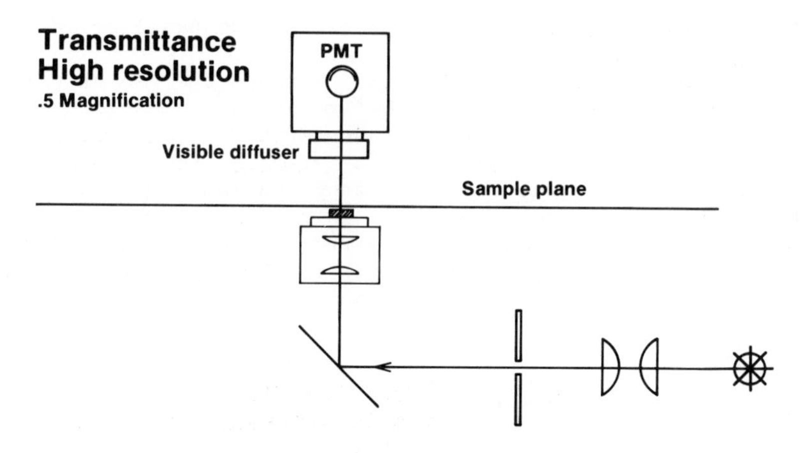

Figure 5. Optics configuration of Quick Scan R&D with magnification of 0.5×.

Slit size, mm

Actual	x .5	x 1.5
2 x 0.2	1 x 0.1	3 x 0.3
3 x 0.3	1.5 x 0.15	4.5 x 0.45
4 x 0.4	2 x 0.2	6 x 0.6

Figure 6. Comparison of slit size for configurations shown in Figures 4 and 5.

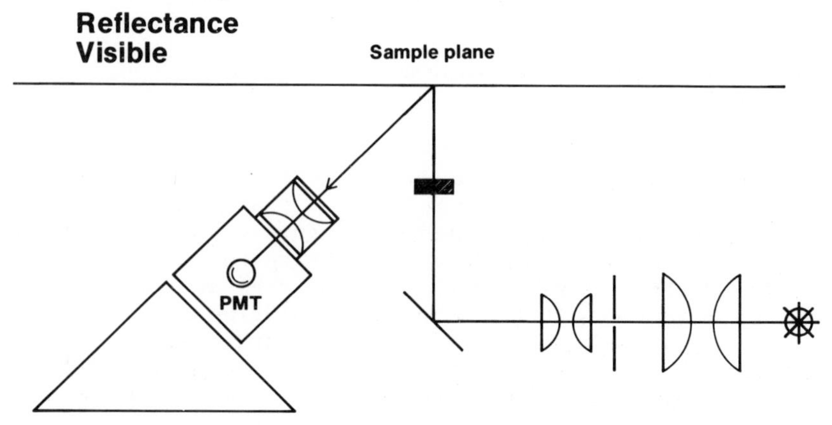

Figure 7. Configuration for visible reflectance.

utilizing 366-nm excitation (Figure 8) is achieved by using a mercury lamp with a phosphor convertor and cobalt blue filter. The fluorescence is viewed through slit optics, which incorporate a UV blocking filter. Reflectance or fluorescence utilizes 280- or 254-nm sources (Figure 9) and is achieved with slit optics on the source and focusing optics on the detector. UV transmittance (Figure 10) utilizes the same source as reflectance, but places the detector on the opposite side of the sample.

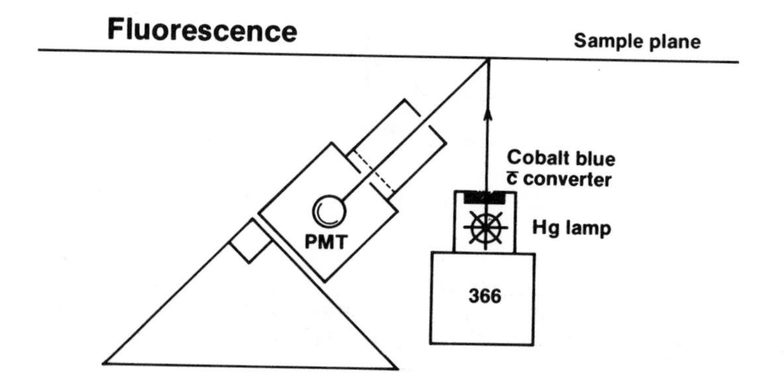

Figure 8. Configuration for fluorescence.

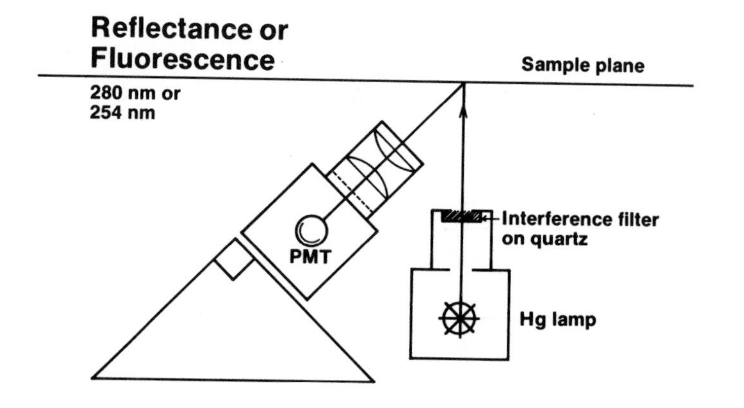

Figure 9. Configuration for reflectance or fluorescence.

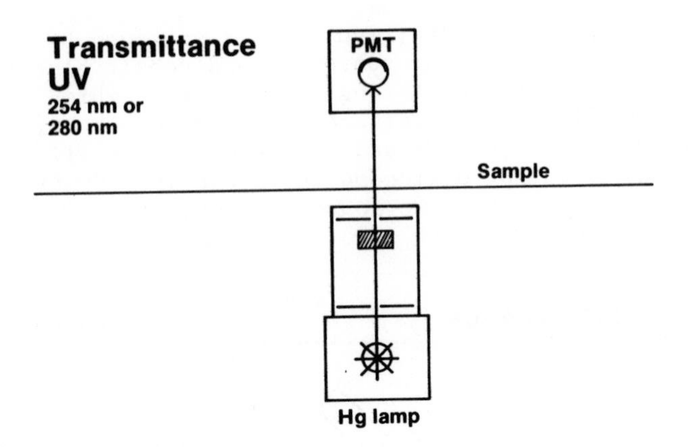

Figure 10. Configuration for UV transmittance.

By selection and manipulation of the many operating parameters of the Quick Scan R&D, a large number of sample types can be made to yield a great deal of useful data.

CHAPTER 12

Automated Sample Preparation for TLC: Extraction and Concentration of Drugs from Physiological Fluids

Linda M. St. Onge, Eugenia Dolar, Edwin M. Richardson, and Maryann Anglim

INTRODUCTION

The Du Pont PREP™ I Automated Sample Processor has been used to extract anticonvulsants (1, 2, 3) and benzodiazepines (4) from serum. The drugs were quantitated by high-performance liquid chromatography (HPLC) or gas chromatography (GC). In this work the PREP™ I was used to extract a variety of acidic and basic drugs from human urine followed by a qualitative analysis by thin layer chromatography. A wide variety of drugs were identified by TLC using standard development procedures and sequential spraying techniques.

MATERIALS AND METHODS

Apparatus

The PREP I (Clinical Systems Division, E. I. du Pont de Nemours and Company, Inc., Wilmington, Delaware) was used to prepare the extracts. It utilizes centrifugal force to process samples through extraction cartridges. Up to 12 samples may be processed simultaneously, and wet or dry extracts may be prepared according to program selection. Solvents are automatically dispensed at the appropriate time in the programmed extraction.

The Type W extraction cartridge (Du Pont) consists of a cap, an extraction column, an effluent cup, and a recovery cup (Figure 1). The resin is polystyrene divinylbenzene, which is hydrophobic and lipophilic. The cartridge was modified for use with urine by adding approximately 1 g of Du Pont glass microbeads to the reservior of the extraction column prior to the addition of the urine and buffer. The glass beads serve as a filter to trap particulates and amorphous materials that are naturally present in urine.

Silica Gel G coated on glass plates (250 μm thickness, 20 × 20 cm) was used for TLC (Analtech, Inc., Newark, Delaware).

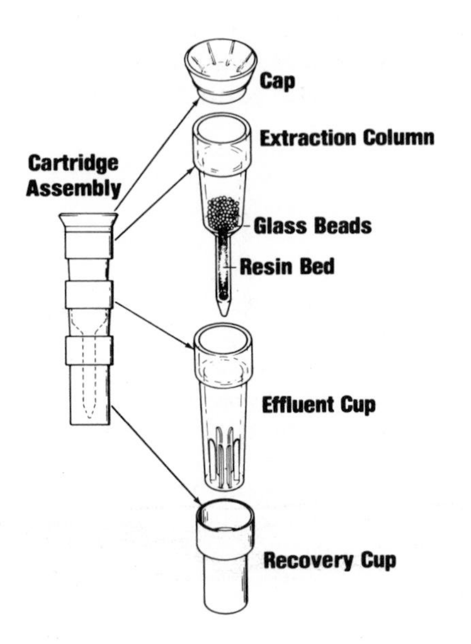

Figure 1. Type W extraction cartridge with glass microbeads added to the reservoir.

Reagents for Extraction

An ammonium chloride buffer was used to make the urine basic prior to extraction. It was prepared by adding 40 g of NH_4Cl to 1 liter of purified water and adjusting to pH 9.3 with NH_4OH.

Purified water was used to wash each column prior to elution with organic solvent. This provided cleaner extracts.

A mobile phase of methanol : acetone : methylene chloride (5:1:15 v/v/v) was used as the eluting solvent to recover drugs from the resin. All solvents were spectral grade.

Reagents for Thin Layer Chromatography

Each extract was reconstituted with 30 μl methanol : methylene chloride (1:1 v/v).

Two mobile phases were employed (5). Mobile phase A consisted of ethylacetate : methanol : ammonium hydroxide (170:20:10 v/v/v). Solvent B consisted of ethyl acetate : methanol (194:6 v/v).

The following spray reagents were used:

Ninhydrin	0.1 g/100 ml isopropanol.
Diphenylcarbazone	0.1 g/100 ml chloroform.
Mercuric sulfate	0.25 g in 10% sulfuric acid.
Iodoplatinate	2 g potassium iodide in a few drops of 1N HCl. Add 230 mg chloroplatinic acid, and dilute to 100 ml with 1N HCl.

Procedure

One gram of glass microbeads was transferred to the reservoir of each Type W extraction column. Then 3 ml of urine and 1 ml of ammonium chloride buffer were added to the reservoir of each column. The cartridge components were separated and loaded into the rotor of the PREP I. The rotor is divided into 12 pie-shaped compartments. Each compartment has an inner rotor position and two outer rotor positions. The extraction columns are placed in the inner rotor position, and the effluent and recovery cups are placed in their respective outer rotor positions. Then 15 ml of water was added to solvent reservoir 1, and 10 ml of eluting solvent (methanol : acetone : methylene chloride) was added to solvent reservoir 2. Program 1 was selected, and after depressing START the programmed extraction was accomplished automatically.

The operating sequence of the PREP I for this extraction is

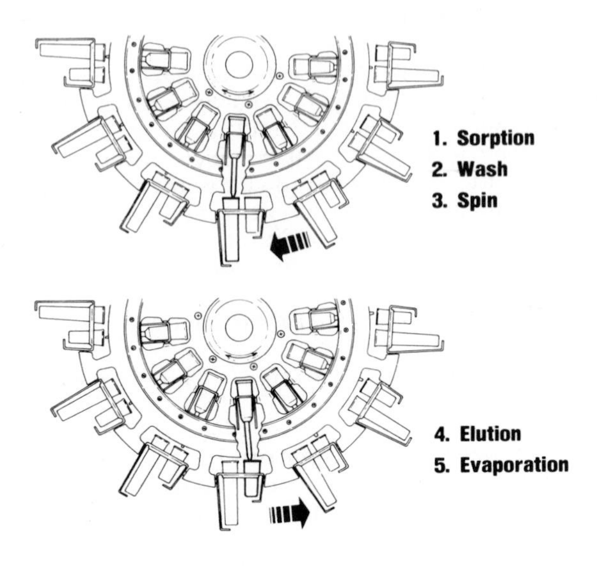

Figure 2. Operating sequence of the PREP I.

shown in Figure 2. The drugs are sorbed to the resin as the sample passes through the resin bed. The water wash is dispensed to remove unsorbed and water-soluble components. The effluent from the extraction column is collected in the effluent cup as the rotor spins in a clockwise direction. A high-speed spin is used to remove excess water from the resin bed before the rotor slows and reverses direction. The drugs are desorbed from the resin as the eluting solvent passes through the resin bed, and the eluate is collected in the recovery cup. The rotor then slows, and jets of warm air evaporate the solvent, leaving a dry residue in the recovery cup. Programmed extraction is completed and the dry extracts are presented in 20 minutes.

Each extract in the recovery cup was reconstituted. The entire extract was spotted on the TLC plate. The plate was developed first in developing solvent A. The plate was removed and allowed to air dry and then developed in developing solvent B. Again, the plate was removed and allowed to air dry.

The following sequence was used to detect the drugs (5). The plate was observed after each step for fluorescence or color reaction.

1. Heat plate at 75°C for 10 minutes.
2. Spray with ninhydrin.
3. Irradiate with UV light for 2 minutes.
4. Spray with diphenylcarbazone.
5. Spray with mercuric sulfate.
6. View under UV light.
7. Spray with iodoplatinate.

RESULTS AND DISCUSSION

Urines of 118 patients were extracted and analyzed using this procedure. Table I shows the drugs that were extracted and detected. A split sample correlation was run with this procedure against Jetube* (formerly manufactured by Manhattan Instruments, Santa Monica, California) extraction and the same TLC procedure. The PREP results correlated exactly with the Jetube results. The identification of all drugs was confirmed by GC, EMIT, or other TLC procedures.

TABLE I. DRUGS IDENTIFIED WITH PREP™ I TLC PROCEDURE

Barbiturates	Chlorpromazine
Morphine	Perphenazine
Codeine	Meprobamate
Cocaine	Hydroxyzine
Benzoylecgonine	Chlordiazepoxide
Methadone	Diazepam
Propoxyphene metabolite	Flurazepam
Meperidine	Glutethimide
Pentazocine	Phenylpropanolamine
Oxycodone	Chlorpheniramine
Phencyclidine	Methapyrilene
Quinine	Amitryptyline
Amphetamine	Nortryptyline
Nicotine	Caffeine

The PREP procedure has the advantage of requiring only 3 ml of urine. A single basic extraction extracted the basic drugs

* Now known as Extralute, manufactured by Analytichem International, Harbor City, California.

as well as the acidic drugs. Small volumes of organic solvent were utilized. The simultaneous and automatic preparation of 12 extracts in the PREP I followed by simultaneous analysis of these extracts on a single TLC plate is a procedure that is ideal for toxicology screening.

REFERENCES

1. L. M. St. Onge, E. Dolar, M. Anglim, and C. J. Least, Jr., Clin. Chem., 25, 1373 (1979).
2. R. C. Williams and J. L. Viola, J. Chromatogr., 185, 505 (1979).
3. P. Koteel and R. H. Gadsden, Clin. Chem., 26, 1053 (1980).
4. R. C. Williams, Fresenius Z. Anal. Chem., 301, 116 (1980).
5. Procedures presented at Workshop of Lab Analysis of Drugs of Abuse, June 3-7, 1974, at the Center for Disease Control, Atlanta, Georgia.

CHAPTER 13

Effect of Humidity on TLC of Phospholipids

Tom R. Watkins

INTRODUCTION

In 1925 Gorter and Grendel postulated that lipids may function at the cell-water interface (1). Within the next decade Davson and Danielli had described the character of the cellular membrane as a bilayer of lipid molecules (2). Since their initial observations about the bilayer nature of the cellular membrane, other investigators have been grappling with the issue of membrane chemical composition. Singer has described the membrane structure as a lamellar lipid matrix with interspersed proteins (3). Investigators now know that biological membranes comprise phospholipid, cholesterol, glycerides, sphingolipid, and protein, among other constituents.

Considerable differences occur in membrane chemical composition between biological species, and also within individuals of a species. Cellular compartments delineated by membranes often contain lipid molecules associated with genesis of the organelle, as is the case with mitochondria, which contain cardiolipin, which is thought to substantiate a line of development from bacteria (4). Such unique features of membrane composition occur among species of both animals and bacteria. In fact, species lipid membrane composition has been used as a taxonomic tool by biologists (5). Within a cell, a lipid fingerprint may be used to identify particular fractions, because of the heterogeneous distribution of some lipid classes (6). The biochemist studying physiological function within the cell ultimately hopes to assign functional properties to specific structural components in the membrane.

In this instance, we wish to study the relation of intestinal mucosa phospholipids and the ability of the mucosal cell to transport dietary fat. This topic has been studied in considerable detail from perspectives other than cellular lipid structure-function relations (7, 8). In one such study designed to reveal the role of dietary fatty acids in cholesterol absorption and metabolism in the Mongolian gerbil (Meriones spp.), Hegsted and co-workers noted that impaired fat absorption in the gut and large changes in serum cholesterol occurred together. If highly saturated fatty acids were fed in the form of coconut oil, along with dietary cholesterol (0.2% w/w), cholesterol in the serum remained very low. On the other hand, if unsaturated fats such as olive oil or safflower oil were fed, serum cholesterol levels doubled. Furthermore, intestinal lipids increased from 5 to 15% of gut wet weight (9). Before explaining the effect of dietary fatty acids upon serum cholesterol, the lesion in gut fat absorption had to be explained, for when dietary fat absorption occurred normally, serum cholesterol rose on these diets. Hegsted observed that inositol addition to the synthetic diet alleviated the problèms associated with obstructed dietary fat absorption, followed by a return to normal serum cholesterol levels (9).

One approach to explain the lipodystrophic lesion in the gerbil has been to perform chemical analyses of cellular compartments, measuring both inositides and total mucosal lipid in animals fed synthetic diets supplemented with or deprived of dietary inositol. Such chemical analyses of cellular domains might locate the cellular defect linked with the inositol-limited diet. Dietary inositol deprivation was presumed also to alter tissue phosphatidylinositol levels.

An approach method of analysis had to be chosen that was sufficiently sensitive to detect microgram quantities, was rapid and quantitative, and was selective enough to resolve each of the major phospholipid classes. Methods that had previously been used to measure inositol in tissues had been microbiological growth assays or quantitative stoichiometric assays, which required relatively large masses of sample. For example, Wildier reported that some natural extracts contain a water-soluble factor that supported the growth of certain yeasts, called the "bios factor" (10). This sort of yeast assay might require two or three days from workup until results are obtained, and might lack adequate specificity. Other workers who had more material relied upon standard quantitative procedures with gravimetry. These were somewhat more specific than microbial assays, although they were tedious, requiring more time per assay than the yeast assays (11). The recent development of more rapid, specific, and sensitive quantitative assays has expedited phospholipid research in general, and phosphatidylinositol studies in particular.

The advent of gas liquid chromatography (GLC) has allowed separation and measurement of various low-molecular-weight, thermally stable molecules. The sensitivity of GLC can be very great for such volatile compounds, especially the halogenated hydrocarbons, with detectability in the nanogram range. For highly polar compounds having low vapor pressures, such as polyhydric alcohols, acids, and sterols, the boiling point can often be reduced markedly by derivative formation of ethers or esters, for example. In the case of natural products, such as phosphatidylinositol, prior hydrolysis may be required to liberate a moiety like inositol from other tissue constituents. Inositol occurs in tissues mainly as the phosphate ester in phospholipids. The completeness of formation of the derivative and its subsequent stability must be considered when evaluating the convenience and usefulness of GLC. If derivatives lack thermal stability and would decompose by oxidation, reduction, or polymerization during analysis, this aspect of the analysis should also be reckoned. And, practically speaking, the time invested per sample for extraction, hydrolysis, derivative formation, cleanup, and concentration before gas chromatographic analysis has to be accounted for in terms of budgetary constraints. These tedious preparative steps, particularly hydrolysis, cleanup, and derivative formation, became so time-consuming in the case of inositol by GLC that other methods were considered.

Most tissue samples contain a heterogeneous mixture of lipids, particularly biological membranes. If the phospholipids alone are considered, their functional groups range from alcohols such as glycerol, inositol, and sugars to bases including ethanolamine, choline, and acids, especially serine. Derivative formation reactions might form satisfactory derivatives of some of these functional groups but not others. Therefore, an analysis would yield information about some but not all of the phospholipids. When other methods of analysis were considered, then, those were considered that would reveal information about most or all of the phospholipid classes. The most promising method appeared to be thin layer chromatography.

An activated silica gel surface, such as silica gel G (with gypsum), if it could be used to separate the phosphatidylinositols from other phospholipids, would offer several advantages in analysis alluded to above. First, all the material applied to the sorbent persists after sample components have been separated. Derivative formation can usually be dispensed with. Second, since multiple samples may be separated in one chromatogram, the analysis time per sample decreases greatly. Third, most analyses can be done under mild conditions, such as ambient temperature. Fourth, several detection methods exist that may be used to visualize the zones in the chromatogram, including visual stains and fluorescent dyes. As with other methods of analysis, sample bands on thin layer chromatograms may be evaluated electronically, and the data analyzed with digital storage, integration, and calculation devices.

Thin layer chromatography became the method of choice for analyzing gerbil gut-membrane lipids. In using silica gel as the sorbent to analyze gut lipids with satisfactory chromatograms, three important factors of the system had to be considered: solvent strength, selectivity, and atmospheric humidity. The solvent strength of the mobile phase, that is, its position in the eluotropic series of solvents, has been defined as the adsorption energy per unit of sorbent. Solvent selectivity relates to the ability of a solvent to alter the relative rates of migration of different sample molecules under conditions of fixed solvent strength. Atmospheric humidity indirectly affects the hydration state of the adsorbent surface, the affinity of the surface for sample molecules, and hence its activity and ultimate resolving power. A brief description of how each of these factors was determined follows.

First, optimal mobile-phase solvent strength must be determined for an appropriate solvent. The solvent must be such that the sample will readily dissolve in it, and also great enough that the sample will migrate at a reasonable rate. If the solvent strength were too low, no sample molecules would migrate away from the origin; if too great, all of them would migrate toward the solvent front, precluding satisfactory resolution in either case. In order to separate the phosphatidylinositols from lecithins, cephalins, sphingomyelins, phosphatidylserines, and other phospholipids on silica gel, a mobile phase with relatively high solvent strength had to be chosen, because of the high affinity of silica gel for these compounds. Two solvent systems were chosen: chloroform, methanol, and water, with either ammonia or glacial acetic acid. Typical solvent systems contained (volume percent basis): chloroform, 62; methanol, 19; ammonia or acetic acid, 15; and water, approximately 4. Experimentally, the mobile-phase strength was increased by addition of methanol at first, and later base or acid, stepwise, until the R_f values (ratio of distance migrated by sample divided by distance migrated by solvent front) varied from just greater than zero to just less than one (data not shown). Both of the mobile phases mentioned nearly separated the major classes of phospholipids, although wide band diffusion of sample on the chromatogram and the acrid odor of ammonia weighted the final choice in favor of acetic acid. A mixture of phospholipids applied to the plate was well-separated initially, but the phosphatidylinositol-phosphatidylserine pair did not resolve. This sort of pairing has been noted by other workers (12). Hence, solvent selectivity had to be modified in order to resolve this pair.

Under given conditions of solvent strength, the selectivity was adjusted to effect complete resolution of the phospholipids. Without complete separation of sample components, quantitation would be impossible, and the experiment could not be done. When further additions of acetic acid resulted in no improvement of resolution, the proportion of water in the system was increased. With added water in the system, phosphatidylinositol and phosphatidylserine separated completely. However, the quality of day-to-day separation of this pair varied with the weather. Notes about atmospheric humidity in the laboratory during sample migration in the mobile phase, when the plates were uniformly washed, hydrated, and stored under fixed humidity conditions, suggested that the mobile-phase water had to be adjusted finely in accordance with the prevailing atmospheric conditions. The methods and results of the experiments to elucidate the role of atmospheric humidity follow.

The influence of atmospheric humidity upon the separation of 13 polar lipid classes including phosphatidylinositol, phosphatidylserine, and phosphatidylethanolamine has been accomplished in one dimension with an elution time of about 45 minutes on silica gel G. Resolution varied with atmospheric humidity for a given mobile phase, and could be adjusted with water content of the mobile phase. Touchstone et al. (13) have recently reported separation of most phospholipids with a mobile phase containing 8% water that is affected little by humidity, using silica gel layers with an organic binder to strengthen the layers.

A typical analysis of samples containing most of these phospholipids has entailed two-dimensional thin layer chromatography (4, 5). Such analyses require twice as much time, besides limiting the number of samples that can be applied per plate. Other investigators have reported one-dimensional separation with limited resolution, that is, with less than 50% of baseline response between peaks (16). The one-dimensional method presented has also been used to analyze lipids from animal tissues other than gut (amniotic fluid), plant tissue (tomato seed), and bacterial membranes.

MATERIALS AND METHODS

Thin layer plates, silica gel GF, 20 × 20 cm, were obtained from Analtech, Inc. (Newark, Delaware). These were washed by development to the top of the plate with chloroform : methanol (v/v), to remove impurities. After air drying, activation for 30 minutes at 70°C was done to remove traces of solvent before sample application.

Reference compounds were applied with microcapillaries (Drummond, Broomall, Pennsylvania). Chromatograms were developed in tanks lined with No. 1 Whatman filter paper (Whatman, Clifton, New Jersey) in which mobile phase was equilibrated for 10 minutes before the plates were inserted. All solvents were doubled-distilled in glass. Relative humidity was measured with an Airguide meter (Chicago, Illinois).

Chromatograms were dried in air, and traces of solvent were removed by placing plates in an oven at 190°C for 3 to 5 minutes. As soon as the plates were cooled to ambient temperature, they were sprayed with 3% cupric acetate in 8% phosphoric acid until the gel was completely wetted (17). Charring was done by heating in an oven at 190°C for 8 minutes.

To transform chromatogram spot information into usable data, the relation of spot photodensity to mass of reference compound has to be known. This can be expressed in the form of a calibration curve as illustrated in Figure 1 for phosphatidylinositol. With such a plot available for each sample compound of interest, the slope of the line can be measured to express chromatogram spot area per unit of mass, for example, square millimeters per microgram. In subsequent analyses, the amount of sample applied to the silica layer will be such that it falls within the limits of the calibration curves. Typical limits of detectability for these phospholipid assays have been 0.1 to 5.0 μg for a given compound with the cupric acetate stain. The chromatograms in these studies were scanned on a Kontes Densitometer, K495000 (Kontes, Vineland, New Jersey) in the transmission mode.

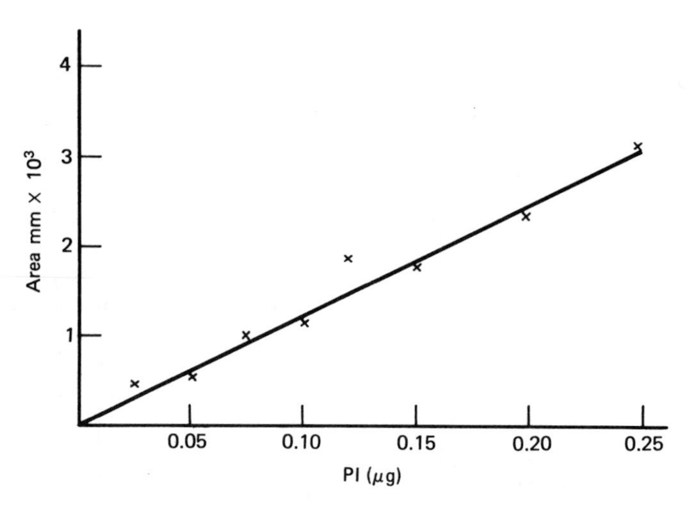

Figure 1. Calibration curve for phosphatidylinositol, with mass on the abscissa and area of photodensity trace on the ordinate.

Reference compounds were obtained from Avanti Biochemicals (Birmingham, Alabama), Biorganics (Newark, Delaware), or Sigma Chemical Company (St. Louis, Missouri). They were: lysolecithin, LL; lecithin, L; sphingomyelin, S; phosphatidylinositol, PI; phosphatidylglyerol, PG; dimethylphosphatidylethanolamine, PEMM; phosphatidylethanolamine, PEA; monomethylphosphatidylethanolamine, PEM; phosphatidylethanol, PE; phosphatidylserine, PS; and cardiolipin (diphosphatidylglyerrol), CL. These were dissolved in 5% methanol in chloroform; 10% ethanol in hexane; or, benzene, to give a concentration of 200 to 500 ng/μl. One to five microliters was applied at the starting point of each lane of the layer in a band 0.6 cm wide.

RESULTS AND DISCUSSION

The chromatogram in Figure 2 shows the separation obtained with the Analtech silca gel GF with mobile phase A. The role of water was demonstrated by developing the chromatograms in mobile phase B, which contained 12% less water at the same relative humidity (RH), Table 1. Mobile phase A resolved all 13 compounds, whereas B did not. Note the PI and L pair in mobile phase B. At increasing humidities, 75 and 89% RH, the second mobile phase, B, failed to

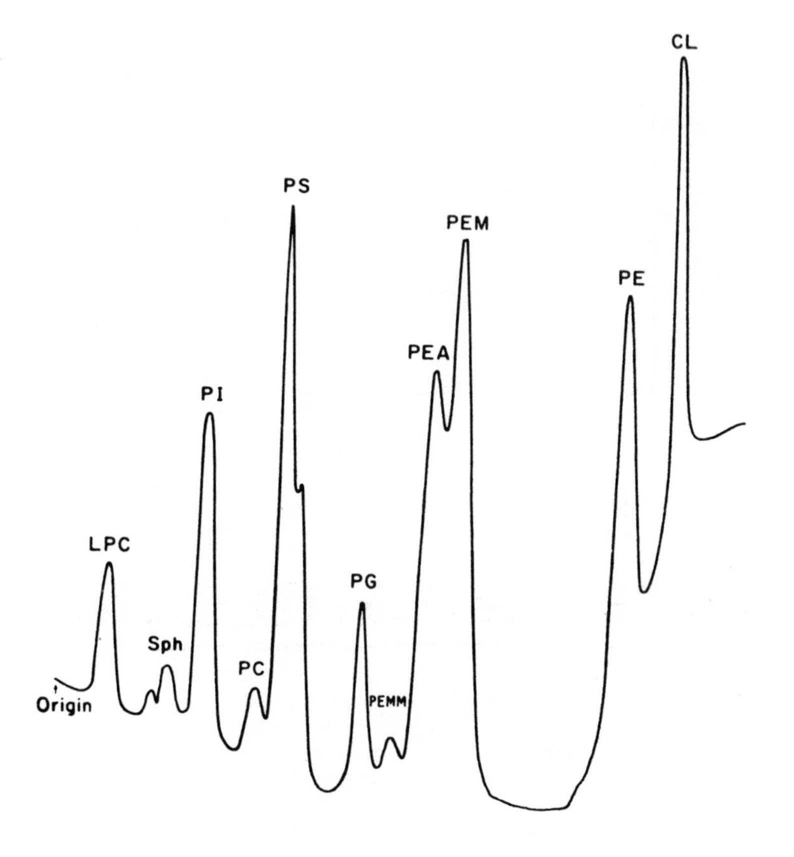

Figure 2. Scan of a chromatogram of the polar lipids designated (see text for key to the abbreviations). The lipids were charred as described in the text.

separate all sample compounds, although standard polar lipids, such as S, PI, PS, and L, were well-resolved; or PEMM, PEA, and PE were resolved, but the others were not. Only under controlled conditions of mobile phase relative to ambient humidity was complete resolution achieved.

Humidity alters the mobility of various compounds selectively (Table I). For example, when the RH was 64%, LL had an R_f (×100) of 5, whereas when the RH was 75%, the R_f was 6, which ratio is expressed as 1.2 in Table I. If the RH were 89%, the R_f was 9, giving the ratio of 1.8. The mobility of S was also largely affected by humidity. In mobile phase B, if the RH were 75%, the ratio of experiments 3/2 was 1.1, but with a RH of 89%, the ratio of experiments 4/2 was 2.1. The RH had less effect on other compounds, such as the phosphatidylethanolamines. If the RH were 75%, the ratio was 1.1; if 89%, 1.3. Humidity did not affect the R_f value of cardiolipin.

From the data in Table I it is seen that the mobilities of S, L, LL, and to a smaller degree, PEMM, PEA, and PEM, increased with relative humidity, based upon the indicated R_f values, other conditions held constant. The mobility of CL, PI, and PG was little affected by ambient relative humidity. It appears that molecules possessing hydroxyl groups (or functional groups that bind water with similar affinity) bind with large affinity to the gel, hence their mobilities varied little with changes in RH (R 4/2 values* 1.0 to 1.1); whereas those without hydroxyl groups but having quaternary or other substituted amine groups migrated faster with increasing RH (R 4/2 values 1.3 to 1.4). Those without amino or hydroxyl groups migrated fastest relatively as the RH increased (R 4/2 values 1.7 to 1.8). LL has a hydroxyl group not readily available to interact directly with the gel surface water, hence it has interaction strength similar to the amine group, and greater humidity speeds its migration rate considerably.

From these data one can appreciate the marked effect of the hydration state of the silica gel upon surface performance in the separation of phospholipids by thin layer chromatography with silica gel GF.

The method described has been satisfactory for studies of intestinal lipids, as originally proposed. No derivative formation has been required, no other chemical modification such as hydrolysis has been needed, and the sample analysis time was reduced several fold over the GLC method referred to above. Furthermore, the GLC system gave information about other lipid sugars than inositol, but none about other phospholipids. So the greatest advantage of this method, perhaps, for studying phosphatidylinositol, in addition to the reduced time of analysis per sample, is the revelation of information about all major classes of phospholipid besides the phosphatidylinositols. Contextual information of this sort about the biological membrane

---

*For R 4/2 values, see Table I.

TABLE I. POLAR LIPID MOBILITY[a] AS A FUNCTION OF RELATIVE HUMIDITY

Compound[e]	Experiment Mobile Phase[b] Relative Humidity Ratio of R_f's	1 A 64	2 B 64	3 B 75	4 B 89	$R_{3/2}$[c]	$R_{4/2}$[d]
LL		9	5	6	9	1.2	1.8
S		23		10	20		
		26	12	13	24	1.1	2.1
PI		31	20	19	21	1.0	1.1
L		48	20	25	35	1.2	1.8
PS		53	29	31	36	1.1	1.2
		58	32	34	40		
PG		66	45	44	47	1.0	1.1
PEMM		83	48	53	66	1.1	1.4
PEA		88	56	61	71	1.1	1.3
PEM		92	58	63	72	1.1	1.3

PE		80	80	77	1.0	1.0
CL	96	88	91	84	1.0	1.0

[a]Mobility expressed as $R_f \times 100$.

[b]Mobile phases, v/v:

	Chloroform	Methanol	Acetic acid	Water
A	60	14	13	2.4
B	60	14	13	2.1

[c]Ratic of Experiment 3 to Experiment 2.

[d]Ratio of Experiment 4 to Experiment 2.

[e]See text for abbreviations.

facilitates interpreting cellular events under specified dietary conditions, as presented here for the gerbil lipodystrophy that occurs during dietary inositol deprivation.

ACKNOWLEDGMENTS

The author wishes to thank Dr. J. C. Touchstone and Dr. S. S. Levin for helpful discussions and the use of the photodensitometer, and Ms. Susan Thomas for technical assistance.

REFERENCES

1. E. Gorter and R. Grendel, J. Exp. Med., 41, 439 (1925).
2. H.A. Davson and J. F. Danielli, J. Cell. Comp. Phys., 5, 495 (1935).
3. S. J. Singer and G. C. Nicolson, Science, 175, 720 (1972).
4. G. B. Robinson, in Biological Membranes, D. S. Parsons (Ed.), Oxford University Press, Oxford, 1975.
5. M. I. Gurr and A. T. James, Lipid Biochemistry, Cornell University Press, Ithaca, 1971.
6. G. Rouser, G. J. Nelson, S. Fleischer, and G. Simon in Biological Membranes: Physical Fact and Function, D. Chapman (Ed.), Academic Press, London, 1968.
7. J. Johnson, Handbook of Physiology, Volume III, American Physiological Society, New York, 1965.
8. T. G. Redgrave, J. Exp. Biol. Med., 49, 209 (1971).
9. D. M. Hegsted, K. C. Hayes, A. Gallagher, and H. Hanford, J. Nutr., 103, 302 (1973).
10. E. Wildier, La Cellule, 18, 313 (1901).
11. E. V. Eastcott, J. Phys. Chem., 32, 1094 (1928).
12. V. Skipski, R. F. Peterson, and M. Barclay, J. Lipid Res., 3, 467 (1962).
13. J. C. Touchstone, J. C. Chen, and K. M. Beaver, Lipids, 15, 61 (1980).
14. G. S. Getz, S. Jakovic, J. Heywood, J. Frank, and M. Rabinowitz Biochim. Biophys. Acta, 218, 441 (1970).
15. O. Reinkonen and A. Luukonen in Lipid Chromatographic Analysis, Volume I, 2nd ed., G. V. Marinetti (Ed.), Marcel Dekker, New York, 1976.
16. T. A. Clayton, T. A. MacMurray and N. R. Morrison, J. Chromatog 47, 277 (1970).
17. M. E. Fewster, B. J. Burns, and J. E. Meads, J. Chromatogr., 43, 120 (1969).

CHAPTER 14

Detection Methods for Lipids on TLC

Joel Bitman and David L. Wood

INTRODUCTION

Direct determination of lipid classes on thin layer plates requires (1) the conversion of the separated colorless lipid spots into colored or fluorescent products and (2) quantitative determination of these products by instrumental analysis. Reagents that react with specific chemical groups such as phosphorus, glycerol, or ester linkages present in some of the lipids may be used but are not applicable to all classes of lipids. Consequently, reagents that react generally with organic compounds have been used to visualize the lipids.

Oxidative reagents such as sulfuric, perchloric, chromic and phosphoric acids have been widely employed for destructive charring to produce brown or black charred areas. Upon heating with the oxidizing agent, cleavage, oxidation, and dehydration reactions occur in the carbon-hydrogen chain of the lipids to yield deposits of elemental carbon or something approaching elemental carbon (1). These charring techniques have been widely used to visualize lipids separated by TLC, followed by densitometric scanning for quantitation. Two of the major problems frequently encountered in the determination of the lipids in situ are (1) differences in charring yield of different lipids and (2) destruction of lipids during TLC.

This investigation was therefore undertaken to accomplish the following objectives:

1. Review existing charring procedures for lipids.
2. Develop an improved technique for charring lipids on thin layer plates.
3. Develop a procedure that would avoid oxidative destruction and hydrolysis of lipids during thin layer chromatography on silica gel.

CHARRING TECHNIQUES

An ideal charring method should quantitatively and completely convert all the component lipid carbon to a carbon residue. Although there is a wide variation in structure and molecular weight of the lipids, when the phospholipids are excluded, there is a fairly close correspondence between the percentage of carbon in the components of the various lipid classes (Table I). Although molecular weight varies from 282 in the free fatty acid, oleic acid, to 885 in the triglyceride, triolein, carbon content only varies from 76 to 84% in any of the simple lipid classes.

In practice, under nonideal conditions, it has been difficult to control reaction conditions to produce complete and quantitative carbonization. For quantitation, however, it is not necessary to achieve complete conversion to carbon, but it is essential that the reaction be reproducible and that there is a constant mathematical relationship between the amount of charring (carbon) and the amount of lipid. Quantitation based upon the yield of carbon could then be indicative of the amount of lipid present in each of the lipid classes.

Several problems are encountered in applying the destructive oxidative charring technique to lipid spots on a thin layer plate. The wide variation in structure of the lipids is an obstacle to achieving completeness of reaction. Some lipids more readily undergo carbonization than others, resulting in differences in charring. Differences in the yield of carbon may arise from:

1. Differences in structure among lipid classes.
2. Differences in structure of compounds within a lipid class.
3. Evaporation of lipid from the thin layer plate before it can be converted to carbon.
4. Oxidation of some of the carbon to carbon dioxide and evaporative loss of the carbon dioxide.

Differences between unsaturated and saturated lipids in the amount of charring have been repeatedly observed (2, 3). Generally,

TABLE I. PERCENTAGE OF CARBON IN LIPID CLASSES

Class	Substance	Molecular Weight	%C
Cholesterol ester	Cholesterol oleate	651	83
Triglyceride	Triolein	885	77
Diglyeride	Glyceryl diolein	621	76
Cholesterol	Cholesterol	387	84
Free fatty acid	Oleic acid	282	76
Monoglyceride	Methyl linoleate	357	78
Phospholipid	Dilinoleyl phosphatidyl choline	782	68

the unsaturated lipids give greater yields of charring. This has been explained by the tendency of the unsaturated lipids to undergo oxidation more rapidly than saturated lipids. The double bond provides a ready means of attack for the oxidative reagent. Privett and Blank (2) demonstrated that triolein gave yields of carbon that were about 2× that given by tripalmitin when sprayed with 50% aqueous H_2SO_4 and charred by heating on a hot plate. Privett and Blank (4, 5) explained this difference as a balance between two processes: evaporation and oxidation. Saturated glycerides give less intense spots than unsaturated glycerides because, being more resistant to oxidation, more evaporation can occur in these compounds prior to their oxidation to carbon. When heated on a hot plate with temperatures on the metal surface up to 360°C, considerable evaporation may occur.

Privett and Blank (4) modified their original conditions to limit the loss by evaporation. They used a stronger oxidizing reagent, a saturated solution of potassium dichromate in 70% sulfuric acid, to complete the oxidation as quickly as possible. They also charred below 200°C in a forced draft oven to avoid direct evaporation that might occur prior to oxidation to carbon, and to avoid drastic conditions that would convert the sample to carbon dioxide with consequent loss of carbon (5, 6). Under these conditions tripalmitin and triolein gave very similar peak areas. The efficiency of the oxidative charring procedure was further evaluated by comparison of the yield of carbon generated from various lipids (6). Most of the lipid classes gave differences in carbon yield, but these differences were not great. Using triolein as the standard for comparison, saturated and unsaturated triglycerides gave equal responses of about 100%, as did fatty acids. Methyl esters of fatty acids gave about 15% greater response while the phospholipids gave about 7 to 10% less response. Cholesterol esters gave slightly higher responses (+5%), but cholesterol yielded 20 to 25% greater charring.

In contrast to these results, in which unsaturated triglycerides yielded greater charred densities than saturated lipids, Barrett et al. (7) found that saturated triglycerides gave a higher response. In their studies, they separated the lipids on silica impregnated with silver nitrate. On silver nitrate, the unsaturated fatty acids form coordination complexes with silver ions. During charring at elevated temperatures with acid, double bond cleavage takes place, and the shorter chain fragments formed may evaporate, thereby lowering the values for the unsaturated triglycerides. Barrett et al. (7) used standards and correction factors in their method to quantitate the lipids.

Chobanov and co-workers (8) in Bulgaria solved the problem of differences in charring between unsaturated and saturated triglycerides in another manner. They also used argentation thin layer chromatography on silica gel. By addition of bromine to the double bonds prior to charring with sulfuryl chloride, they were able to obtain equal amounts of charring from tristearin, triolein, and trilinolein.

Biezenski and co-workers (9) used the conditions of Privett and Blank (2), 50% H_2SO_4 and heating on a hot plate, in a quantitative procedure for lipid classes. Detailed comparisons for individual saturated and unsaturated lipids were not given, but the authors stated that "degree of fatty acid unsaturation appeared to be of little significance."

Downing (1, 10) also used 50% H_2SO_4 and heat to char the lipid classes for quantitative densitometry. By slow heating on the hot plate, Downing (1, 10) found that saturated lipids produced the same yield of carbon as unsaturated lipids. This was achieved by placing the TLC plate on an aluminum slab lying on a cold electric heating plate (20°C). The heating plate was then switched on and, over a 30-minute period, heated to 220°C. Heating was then continued for an additional 10-minute period. Most lipid classes gave similar yields of carbon, but cholesterol gave a much higher yield, apparently owing to the presence of the free hydroxyl function (1).

Fewster et al. (11) demonstrated that neutral lipids and phospholipids could be visualized and quantitated using a copper acetate-phosphoric acid reagent. After spraying the TLC plates, they were heated directly on a hot plate at 180°C for 25 minutes. The relative charring yields of saturated and unsaturated lipids were not given.

COPPER ACETATE CHARRING METHOD

Charring methods that have been used for quantitative densitometry of the lipid classes are presented in Table II. We have used the copper acetate-phosphoric acid reagent of Fewster et al. (11) to visualize the separated lipid classes. Lipids were separated into lipid classes by thin layer chromatography on 20 × 20 cm 250-μm Whatman K5 silica gel plates (Whatman, Inc., Clifton, New Jersey), using the two-stage, one-dimensional procedure developed in our laboratory (12). The mixed standard was separated into the following eight lipid classes, in ascending order from the origin: phospholipids (PL), monoglycerides (MG), free fatty acids (FFA), cholesterol (C), 1,2-diglycerides (1,2-DG), 1,3-diglycerides (1,3-DG), triglycerides (TG), and cholesterol esters (CE). The standard mixture was composed of lipids containing equal amounts of C_{18} monounsaturated (oleic) and polyunsaturated (linoleic) fatty acids in the lipid moiety. The lipids were applied to the TLC plate using the Kontes Chromaflex 12-position automatic spotter (Kontes, Vineland, New Jersey). The spotted TLC plate was subjected to two developments in one dimension to separate the lipids: development 1, chloroform : methanol : acetic acid (98:2:1) to 17.0 cm; and development 2, hexane : ethyl ether : acetic acid (94:6:0.2) to the top of the plate. After drying, the plate was dipped into a solution of 3% cupric acetate in 8% phosphoric acid for 3 seconds, and heated in an oven at 130°C for 30 minutes to char the separated lipid classes. Optical densities of the charred spots were measured with a Shimadzu CS-910 Dual-Wavelength TLC scanner (Shimadzu Scientific Instruments, Inc., Columbia, Maryland). Plates were scanned at 350 nm in either a linear scanning mode using a 1.25 × 10 mm light beam or in a zigzag scanning mode (flying spot) using a 1.25 × 1.25 mm light beam.

The relationship between concentration of the eight lipid classes and densitometric response is shown in Figure 1. These curves show the pattern of response over a wide concentration range from 1 to 35 μg of lipid. There is a linear relationship at low concentrations, but the usual deviations are observed above 10 μg. Cholesterol, cholesterol esters, and free fatty

TABLE II. QUANTITATIVE DENSITOMETRY OF CHARRED LIPIDS ON TLC

Reagent	Heating Conditions	Sensitivity	Lipid Classes	Unsaturated/ Saturated	References
50% H_2SO_4	360°C (Hot plate)	1 μg	Differ	U > S	Privett (2) 1961
$K_2Cr_2O_7$--H_2SO_4	180°C (Oven)	0.5 μg	Equal	U = S	Privett (4) 1962
50% H_3PO_4	340°C (Hot plate)	0.5 μg	Differ	S > U	Barrett (7) 1963
Br_2--SO_2Cl_2	180°	--	Differ	U = S	Chobanov (8) 1976
50% H_2SO_4	20 to 220°C (Hot plate)	0.3 μg	Equal except cholesterol	U = S	Downing (1, 10) 1968
Cupric acetate- -H_3PO_4	180°C (Hot plate)	1 μg	Differ	U > S	Fewster (11) 1969

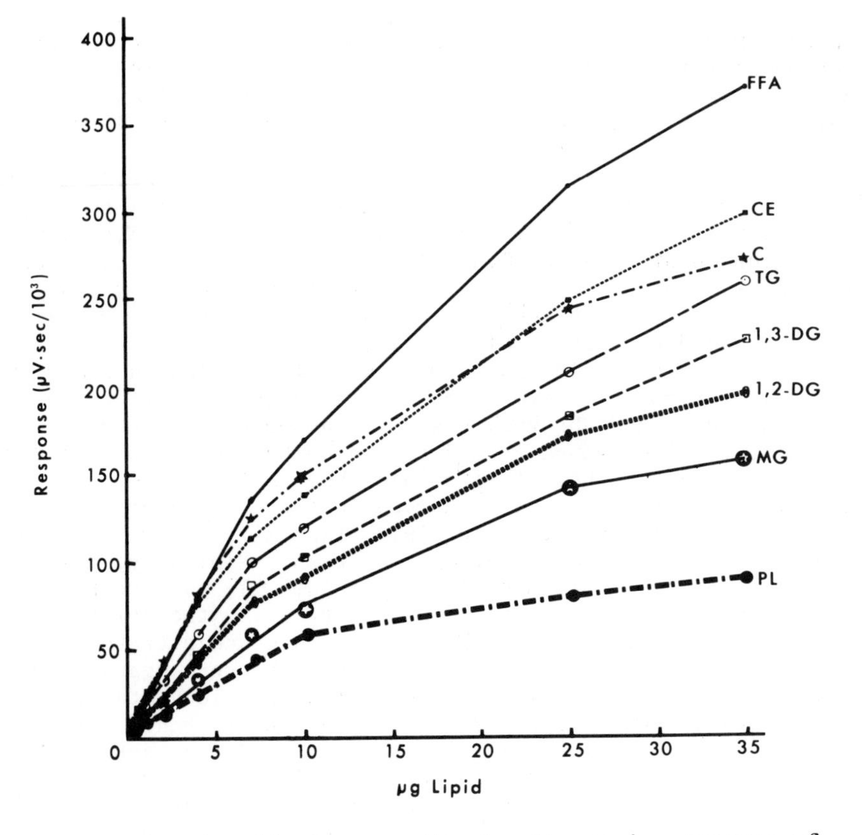

Figure 1. Relationship between the densitometric response of charred lipid and amount of lipid in eight lipid classes.

acids gave the most response. Triglycerides yielded about 70% of this response, and the diglycerides about 60%. The monoglycerides and phospholipids yielded only 30 to 40% of the maximal response.

EVALUATION OF COPPER SALTS AS CHARRING REAGENTS

These differences in charring response led us to consider whether different copper salts might give a more uniform response with the different lipid classes. Eleven copper salts were screened to determine whether a different copper salt might be better than copper acetate for lipid charring. Table III compares the charring yield given by copper acetate with that given by borate, carbonate, chloride, chromate, hydroxide, molybdate, nitrate, permanganate, phosphate, and sulfate salts of copper. An unsaturated lipid class mixture (A), a saturated triglyceride mix (B), and cholesterol (C) were tested with the copper salts to evaluate the intensity of charring. After applying 10 µg of A, B, and C on a micro Whatman K5 silica gel plate, the plate was dipped into a 3% solution of the copper salt in 8% H_3PO_4, and the lipids were charred at 130°C for 50 minutes. The intensity of charring was evaluated on a scale of 1 to 10, 1 being very weak and 10 being very strong. The background was also evaluated since any optical densitometric procedure depends upon the difference between background absorption and the absorption of the sample. Ideally, the background should remain white and exhibit little absorption, while the sample should absorb strongly.

Little difference was observed in the ability of the copper salts to char the unsaturated lipid class mix, column A. All the salts gave a strong charring reaction. Similarly, almost all salts charred cholesterol well (column C), the exception being the chromate. More variation was observed in the ability to char saturated triglycerides. The chromate and nitrate salts gave less response than cupric acetate. The permanganate, phosphate, and sulfate salts gave more response than cupric acetate, and cupric sulfate was clearly the best charring reagent. Except for the chromate and molybdate salts that colored the background green and blue, the background areas of the TLC plate were light in color.

Several of the variables that affect the charring reaction (concentration of copper sulfate, concentration of phosporic acid, and charring temperature and time) were studied in an attempt to optimize the densitometric response. Solutions containing 5 and 10% cupric sulfate gave greater charring than lower concentrations (Table IV). The charring response of the saturated mixture (B), however, was still less than that observed with cholesterol (C) or the lipid class mixture containing 18:1 and 18:2 fatty acids (A). Increasing the temperature at which the plates were charred resulted in an increase in charring (Table V). The unsaturated lipid class mixture (A) was treated with either 5 or 10% cupric sulfate and heated at 180°C instead of 130°C. Charring for 10 minutes at 180°C with 10% $CuSO_4$ in 8% H_3PO_4 appeared to be best, and resulted in maximal intensities. The effect of varying phosphoric acid concentration from 2 to 40% was also studied (Table VI). Concentrations of phosphoric acid above 8% did not

TABLE III. COMPARISON OF COPPER SALTS FOR LIPID CHARRING[a]

Cupric Compound	A[b]	B[c]	C[d]	Background
Acetate	7	3	7	Light
Borate	7	3	7	Light-medium
Carbonate	7	3	7	Light
Chloride	7	3	7	Light
Chromate	7	1	1	Green
Hydroxide	7	3	7	Light
Molybdate	7	3	7	Dark blue
Nitrate	7	2	6	Light
Permanganate	7	4	6	Light-medium
Phosphate	7	4	7	Light-medium
Sulfate	7	5	7	Light-medium

[a] 3% solutions in 8% H_3PO_4. 10 μg of standard lipids A, B, and C was placed on micro Whatman K5 TLC plates. Plates were dipped in the reagent and charred by heating for 50 minutes at 130°C in a vacuum oven. Intensity of charring was evaluated on a scale from 1 to 10, with 1 being very weak and 10 being very strong.

[b] Lipid A: Mixture of eight lipid classes, CE, TG, 1,3-DG, 1,2-DG, C, FFA, MG, PL. The fatty acid moiety consisted of equal amounts of 18:1 and 18:2 acids.

[c] Lipid B: Triglyceride mixture of 30% triolein, 30% tripalmitin, 15% tristearin, 15% trilaurin, and 10% trimyristin.

[d] Lipid C: Cholesterol.

increase charring intensity of the unsaturated lipid standard mixture. We also tested charring with 5% copper sulfate in sulfuric acid. Charring was equal to that obtained with phosphoric, but sulfuric acid gave a much darker background.

The permanganate and phosphate salts had yielded almost as much charring as sulfate in the original screening of 3% copper salts at 130°C (Table III). When these salts were tested at higher concentrations and at the higher 180°C temperature, neither was as good as 10% copper sulfate in 8% H_3PO_4 for 10 minutes at

TABLE IV. EFFECT OF AMOUNT OF $CuSO_4$ ON LIPID CHARRING[a]

% $CuSO_4$ in 8% H_3PO_4	A[b]	B[b]	C[b]
0	5	2	5
1.5	7	5	7
3	7	5	7
5	7	6	7
10	7	6	7

[a]130°C for 45 minutes in oven. Scale 1 to 10: 1 = very weak . . . 10 = very strong.

[b]See notes to Table III.

TABLE V. EFFECT OF AMOUNT OF $CuSO_4$ AND TEMPERATURES ON LIPID CHARRING[a]

% $CuSO_4$ in 8% H_3PO_4	Minutes at 180°C	Charring
5	7	7
5	10	8
10	7	9
10	10	10

[a]Scale 1 to 10: 1 = very weak . . . 10 = very strong. 10 μg of standard lipid class mixture A was placed on micro Whatman K5 TLC plates.

TABLE VI. EFFECT OF CONCENTRATION OF H_3PO_4 ON LIPID CHARRING[a]

5% $CuSO_4$ in % H_3PO_4	Charring
2	3
4	5
8	8
12	8
20	8
40	8

[a] Scale 1 to 10: 1 = very weak . . . 10 = very strong. 10 μg of standard lipid class mixture A was placed on micro Whatman K5 TLC plates.

180°C. Greater charring responses were obtained when lipids were dipped in the (10% $CuSO_4$)--(8% H_3PO_4) reagent than when they were sprayed. Transmission measurements yielded response areas that were 3 times higher than reflectance measurements.

HEATING METHOD--CHARRING

The Whatman K5 silica gel plate has a recommended temperature limitation for charring of 130°C, probably because of the presence of the organic binder. It seemed possible that equal charring of unsaturated and saturated triglycerides might be achieved if the lipids could be heated longer at 180°C or at a higher temperature. The effect of increasing temperature on charring on the Whatman K5 plate is shown in Table VII. The TLC plates were heated on a hot plate. As the temperature was raised, the saturated lipids charred to almost the same extent as the unsaturated lipids, and the ratio of U/S approached 1.0. The background, however, became too dark for quantitative densitometry. Lipid charring on Whatman K5 plates was then compared with charring on Merck silica gel 60 plates. The Merck plates could be heated for periods as long as 38 minutes at temperatures up to 232°C and still be scanned. The backgrounds of Whatman K5 plates heated to 200°C for even 20 minutes were too dark to scan.

TABLE VII. RELATIVE CHARRING OF UNSATURATED AND SATURATED TRIGLYCERIDES

Charring Reagent	Temperature (°C)	Minutes	U/S[a]
3% $Cu(OOCCH_3)_2$--8% H_3PO_4	130	30	10
3% $CuSO_4$--8% H_3PO_4	180	9	1.67
3% $CuSO_4$--8% H_3PO_4	200	2	1.67
3% $CuSO_4$--8% H_3PO_4	200	4	1.33
3% $CuSO_4$--8% H_3PO_4	220	1	1.67
3% $CuSO_4$--8% H_3PO_4	220	1.75	1.25
3% $CuSO_4$--8% H_3PO_4	230	1	1.25
3% $CuSO_4$--8% H_3PO_4	230	1.25	1.20
3% $CuSO_4$--8% H_3PO_4	245	0.40	1.25
3% $CuSO_4$--8% H_3PO_4	245	1.25	1.15

[a]U = standard lipid class mixture A. S = saturated lipid mixture B.

Heating glass thin layer plates on the metal surface of a hot plate did not yield consistently reproducible charring results. This was attributed to variable heating. The metal surface of the hot plate (often 200 to 350°C) affords complete contact between the glass plate and the metal heating surface, thereby providing very efficient heat transfer. However, it is very difficult with this system to provide a definite and reproducible heating period. It is very easy to overshoot heating period conditions both with regard to time and temperature because (1) it is difficult to remove the plate from the heating surface exactly at the end of the heating period and (2) the glass plate continues heating due to absorbed heat even if removed from the metal surface.

Much more reproducible results were obtained by heating from 30 to 180°C under closely controlled conditions. Downing (1, 10) has recommended heating on a cold aluminum slab to 220°C, but we obtained variable results when charring was conducted in this manner. Very reproducible results were achieved by heating in the oven of a gas chromatograph using temperature-programmed heating of 10°C/minute. The TLC plate was placed on a 20 × 20 cm glass plate (0.25 cm thick) that rested on a wire screen in the oven of a Hewlett-Packard model 7620A gas chromatograph. The well-insulated GLC oven provided excellent heat uniformity. At the end of the heating period, the oven was programmed to cool automatically and quickly, thereby avoiding overheating problems.

OXIDATION OF LIPIDS DURING TLC CHROMATOGRAPHY

Two other problems we encountered in TLC of lipids were oxidation and decomposition. It is extremely important to prevent oxidation since the lipids appear to be particularly susceptible to oxidation when spread as a thin film over a surface (13). Lower charring results were obtained, particularly with unsaturated lipids, if care was not taken to avoid exposure to air during spotting and after TLC development (Table VIII). 500 ng of tristearin or triolein was spotted on a thin layer plate. After TLC development, they were either placed in a N_2 cabinet (a vacuum oven that was flushed and filled with N_2), held in a freezer at -20°C, or exposed to room air and temperature. Only keeping the sample in a N_2 atmosphere was effective in preventing oxidation, but placing the sample in a freezer retarded oxidation somewhat. There was a rapid loss of unsaturated lipid in only a few minutes in air.

Many investigators have recommended adding antioxidants to lipids and to solvents used in lipid extraction to prevent oxidation (13, 14). Compounds containing benzene rings plus a hydroxyl group have been found to inhibit oxidation of unsaturated lipids. Three common phenolic antioxidants are butylated hydroxyanisole (BHA), butylated hydroquinone (BHQ), and butylated hydroxytoluene (BHT) (Figure 2). Alkyl substitution in the ortho or para positions greatly enhances antioxidant potency. The addition of the tertiary butyl group in the ortho position is particularly effective. BHT was found to be the least polar of these three antioxidants. After a two-stage development, it would be at or near the solvent front. BHT also did not char well. Five micrograms gave no reaction, whereas BHA gave moderate charring, and BHQ yielded a strong, black residue. Because of this noncharring character and the fact that its high mobility would separate it from the other lipids, BHT was selected as the

TABLE VIII. EFFECT OF TREATMENT AFTER TLC DEVELOPMENT ON LIPID RECOVERY[a]

Treatment	Time	N	Relative Value	
			Triolein	Tristearin
In N_2 cabinet	5 min	6	100	100
In N_2 cabinet	18 hr	6	100	100
In freezer	1.5 hr	6	80	80
In freezer	3 hr	4	70	85
In freezer	18 hr	6	70	60
In air	5 min	4	80	100
In air	100 min	4	33	85

[a]Charred with 10% $CuSO_4$--8% H_3PO_4. 30 to 180°C heating at 10°C/minute.

best antioxidant for use. BHT was added to standards at the level of 0.1% of the lipid and to solvents at the level of 0.005%. Under these conditions there was no loss of lipid if standards were exposed to air for 30 minutes on the TLC plate after sample application.

HYDROLYSIS OF TRIGLYCERIDES ON SILICA GEL

Decomposition of the lipids, primarily hydroylsis of triglycerides during thin layer chromatography, was also encountered. If tristearin or triolein was applied to a silica gel plate in 5 µl of chloroform, the sample developed in hexane : ethyl ether : acetic acid (80:20:1) as a single triglyceride peak (Figure 3). In example A of Figure 3, 500 ng of the lipid was applied using the microcapillary glass pipet Microcap (Drummond Scientific Company, Broomall, Pennsylvania). Sample application and evaporation of the solvent took less than 15 seconds. If the same amount of triglyceride was applied to a silica gel plate in 1 ml of chloroform using the Kontes spotter, a number of peaks were observed. There was clear evidence of six peaks: at the origin, a small monoglyceride/diglyceride peak, an unknown charred spot, free fatty acid, triglyceride, and fatty acid ester peaks. This was

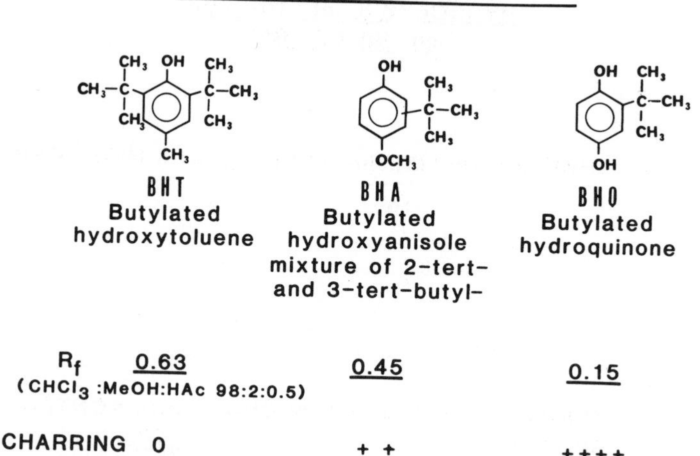

Figure 2. Relative chromatographic mobility and charring of antioxidants.

clear evidence of hydrolysis of the triglyceride that occurred on the TLC plate during sample application (Figure 3B).

Since heat had been used in applying the sample with the Kontes spotter, samples were also applied without heat (experiment 3 in Table IX). Whether chloroform or hexane was used as the solvent, hydrolysis occurred. Even though there was a high flow of nitrogen from each syringe needle of the Kontes spotter, the entire spotting apparatus was placed under a plastic canopy and flushed continuously with nitrogen. Experiment 4 shows that hydrolysis was still observed. Using chloroform that had been purified by passage through an aluminum oxide column to remove peroxides did not prevent hydrolysis (experiment 5). Using a different silica gel plate did not prevent hydrolysis. Experiments 1 to 5 were conducted with Merck silica gel 60 plates. Experiment 6 was performed on the Whatman K5 silica gel plate, and extensive hydrolysis occurred.

The effect of the solvent upon triglyceride hydrolysis was studied by the following series of experiments (Figure 4). Samples of 500 ng of triglyceride were placed on all 12 lanes of a TLC plate by pipetting with a microcap in 5 µl of $CHCl_3$, a procedure that does not cause hydrolysis. Then 1 ml of either

HYDROLYSIS of TRIOLEIN on SILICA GEL

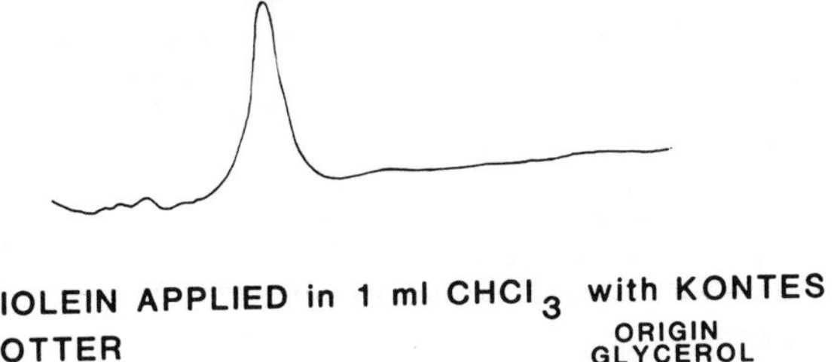

Figure 3. Hydrolysis of triolein on silica gel. Densitometric scan.

hexane or chloroform was pipetted onto the triglyceride, using nitrogen and no heat. Some lanes received no solvent but had N_2 flowing onto the triglyceride for the 30-minute solvent evaporation period. After this, the plate was developed. Only triglyceride could be seen in the lanes that had been exposed to only 5 µl of solvent. In the lanes that had received 1 ml of solvent, hydrolysis occurred. This was true with hexane, chloroform, chloroform purified by redistillation using a fractionating column, chloroform purified by passage through alumina, and with fresher chloroform or hexane from the manufacturers.

The solvents were also tested for their ability to react with the charring reagents. After 1 ml of each solvent was evaporated onto the silica gel plate using the Kontes spotter, the plate was dipped in (10% $CuSO_4$)--(8% H_3PO_4) and charred. Table X shows that the residue from fresher solvents gave less charring than from older solvent. Redistillation reduced the

TABLE IX. HYDROLYSIS OF TRIGLYCERIDES ON SILICA GEL TLC PLATES[a]

Experiment[b]	Lipid	Sample Application	Solvent & Volume	Plate Temperature (°C)	N_2 (1/min)	Application Time (min)	Lipids Observed
1	18:0	M	5 µl $CHCl_3$	25	0	0.25	TG
	18:1	M	5 µl $CHCl_3$	25	0	0.25	TG
2	18:0	KS	1 ml $CHCl_3$	50	24	14-19	Glycerol, MG, DG
	18:1	KS	1 ml $CHCl_3$	50	24	14-19	FFA, TG, FA ester
3	18:1	KS	1 ml $CHCl_3$	25	20	20-35	Glycerol, MG, DG
	18:1	KS	1 ml Hexane	25	20	20-35	FFA, TG, FA ester
4	18:1	KS-N_2	1 ml $CHCl_3$	25	20	20-35	Glycerol, MG, DG FFA, TG, FA ester
5	18:1	KS	1 ml $CHCl_3$- -Al_2O_3	25	20	20-35	Glycerol, MG, DG FFA, TG, FA ester
6	18:1	KS	1 ml $CHCl_3$	25	20	20-25	Glycerol, MG, DG FFA, TG, FA ester

[a]N = 6, M = microcap, KS = Kontes spotter.

[b]Experiment 4: KS was placed under plastic canopy flushed with N_2 during sample application. Experiment 5: $CHCl_3$ purified through Al_2O_3 column. Experiment 1 to 5: TLC plate = Merck silica gel 60. Experiment 6: TLC plate = Whatman K5.

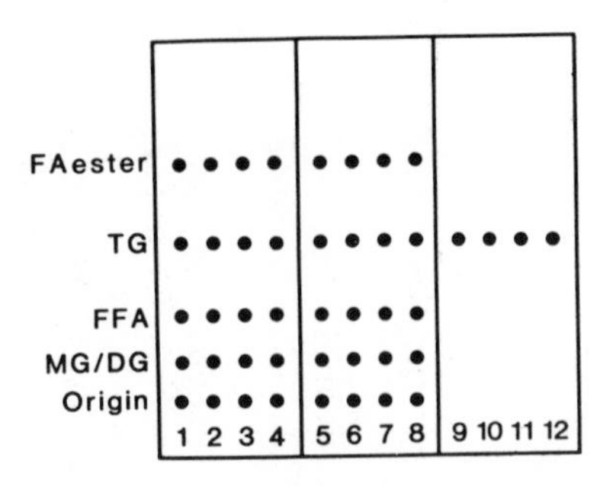

TRI-18:1 on LANES 1-12 in 5μl $CHCl_3$ with MICROPIPET

EXP:	SOLVENT	SOLVENT	SOLVENT
A	$CHCl_3$	Hx	None
B	redist. $CHCl_3$	$CHCl_3$ - Al_2O_3	None
C	fresh $CHCl_3$	fresh Hx	None

Figure 4. Hydrolysis of triolein on silica gel. TLC plate.

TABLE X. CHARRING OF SOLVENT RESIDUE[a]

Solvent	Type	Charring Intensity
$CHCl_3$	Fresh	2.0
Hexane	Fresh	3.3
$CHCl_3$	Old	9.8
$CHCl_3$	Old-redist.	5.5

[a]N = 4, Scale 1 to 10: 1 = very weak, . . . 10 = very strong.

charring somewhat. The amount of charring from 1 ml of the solvent alone was not as great as the charring observed when 500 ng of triglyceride in 1 ml of solvent had been spotted and the plate developed. This suggests that glycerol or another product of triglyceride hydrolysis or decomposition was present in the residue at the origin and is responsible for the greater charring observed.

TABLE XI. POSSIBLE FACTORS CAUSING HYDROLYSIS OF TRIGLYCERIDES ON SILICA GEL

1. Heat
2. Peroxides in solvents
3. Oxygen
4. Nitrogen
5. Light
6. Contaminants in solvents
   Acid
   Alkali
   Metals
   Water
   Alcohol
7. Silica gel
   Metals
   Water
   Acid
   Alkali

Some of the possible factors that could cause hydrolysis of triglycerides on silica gel are shown in Table XI. The first five of these factors were essentially eliminated by previous experiments. Contaminants in the solvents or in the silica gel are more difficult to eliminate. Silica gel itself (Figure 5) is an acid (15), and the possibility cannot be excluded that a silicic acid catalyzed hydrolysis of the triglyceride is occurring. Examination of the analytical data listed by a number of solvent manufacturers demonstrates that a number of components are always present, even in the purest redistilled solvents currently available. Acids, metals, and water are present in both solvent and adsorbent, although they are present in extremely low concentrations.

A component almost always present in chloroform is alcohol, added as a stabilizer to prevent peroxide and phosgene formation. Usually, 1% ethanol is added, and Randerath (16) has pointed out that the elutive action of chloroform is altered by this addition. This fact is generally ignored in most TLC research reports. Indeed, in our own two-stage solvent development system, the first solvent phase consists of chloroform : methanol : acetic acid (98:2:1). Properly, this should be modified to be chloroform :

SILICA GEL

$$HO-Si(OH)_2-OH + HO-Si(OH)_2-OH \longrightarrow (HO)_2Si(O)_2Si(OH)_2 + 2HOH$$

$$SiOH \rightleftharpoons SiO^- + H^+$$

$$Si-O-Si + H_2O \rightleftharpoons 2\,SiOH$$

Figure 5. Silica gel.

methanol : ethanol : acetic acid (97.02:2:0.98:1). The presence of a substance in the fatty acid ester region of the chromatoplate suggests that alcoholysis occurred with the formation of ethyl esters of the hydrolyzed fatty acids. Gordon et al. (17) found that the formation of fatty acid ethyl esters by acid-catalyzed hydrolysis was a rapid reaction difficult to avoid in solvents containing alcohol.

The acid concentration in the solvent is low, but when a relatively large volume (1 ml) is evaporated, 20 µg of acid may be present. This may be 40 times the amount of the lipid (Table XII). This acid may catalyze the hydrolysis of the thin layer of triglyceride spread on the silica gel surface. When 5 µl of solvent is evaporated, 100 ng of acid is delivered to the plate. Conversely, the effect of the large volume of solvent might be to provide a liquid medium in which the hydrolytic reaction could occur. When 5 µl of solvent is spotted, the lipid is in solution on the plate about 15 seconds. When 1 ml of solvent is spotted, the lipid is in and out of solution repeatedly for a period of 15 to 30 minutes.

The susceptibility of lipids to oxidation and other reactions when spread out as a thin film on an active surface has been noted before (13, 14). Pelick et al. (18) have discussed the characteristics of the adsorbents and a number of factors that

TABLE XII. ACID IN SOLVENT[a]

Triglyceride	Sample Application Method	Volume	Acid	Acid/TG
500 ng	Kontes spotter	1 ml	20 μg	40/1
500 ng	Microcap	5 μl	100 ng	1/5

[a]0.002% = 0.002 g/100 ml = 0.02 g/l = 20 mg/l = 20 μg/ml = 20 ng/μl.

could give rise to false results in TLC of lipids. In comparison to silica gel, aluminum oxide is rarely used for chromatography of lipids because it causes hydrolysis of ester linkages and isomerization of double bonds (19, 20). Some alterations of lipids on silica gel have been described. Borgstrom (20) found that 2-monoglycerides were isomerized to 1-monoglycerides on silica gel, and this interconversion was further studied by Privett and Blank (2) and by Thomas et al. (21). The results presented in this report document another alteration of lipids on silica gel, the hydrolysis of triglycerides.

SUMMARY

The effects on lipid charring of using different copper salts were studied. A procedure was developed that used 10% $CuSO_4$ in 8% H_3PO_4 as a charring reagent. Improved reproducibility of charring was obtained by temperature-programmed heating of the thin layer plates from 30 to 180°C at 10°/minute in the oven of a commercial gas-liquid chromatograph. Oxidation of lipids on the thin layer plates could be avoided by (1) the addition of an antioxidant, BHT, to samples and solvents and (2) keeping samples in an atmosphere of nitrogen during spotting and after TLC development. Hydrolysis of triglycerides on silica gel during sample application could be avoided only by applying the lipid to the plate rapidly and in a small volume of solvent.

REFERENCES

1. D. T. Downing, J. Chromatogr., 38, 91 (1968).
2. O. S. Privett and M. L. Blank, J. Lipid Res., 2, 37 (1961).
3. O. S. Privett, M. L. Blank, and W. O. Lundberg, J. Am. Oil Chem. Soc., 38, 312 (1961).
4. O. S. Privett and M. L. Blank, J. Am. Oil Chem. Soc., 39, 520 (1962).
5. O. S. Privett, M. L. Blank, D. W. Codding, and E. C. Nickell, J. Am. Oil Chem. Soc., 42, 381 (1965).
6. O. S. Privett, K. A. Dougherty, and W. L. Erdahl, in Quantitative Thin Layer Chromatography, J. C. Touchstone (Ed.), John Wiley, New York, 1973, pp. 57-78.
7. C. B. Barrett, M. S. J. Dallas, and F. B. Padley, J. Am. Oil Chem. Soc., 40, 580 (1963).
8. D. Chobanov, R. Tarandjiska, and R. Chobanova, J. Am. Oil Chem. Soc., 53, 48 (1976).
9. J. J. Biezenski, W. Pomerance, and J. Goodman, J. Chromatogr., 38, 148 (1968).
10. D. T. Downing, Lipids, in Thin Layer Chromatography, J. C. Touchstone and J. Sherma (Eds.), John Wiley, New York, 1979, Chapter 18, pp. 367-391.
11. M. E. Fewster, B. J. Burns, and J. F. Mead, J. Chromatogr., 43, 120 (1969).
12. J. Bitman, D. L Wood, and J. M. Ruth, unpublished.
13. G. Rouser, G. Kritchevshy, and A. Yamamoto, in Lipid Chromatographic Analysis, Vol. 3, G. V. Marinetti, (Ed.), Marcel Dekker, New York, 1976, pp. 713-776.
14. G. J. Nelson, in Blood Lipids and Lipoproteins: Quantitation, Composition, and Metabolism, G. J. Nelson (Ed.), Wiley-Interscience, New York, 1972, pp. 3-73.
15. H. W. Kohlschutter and K. Unger, in Thin Layer Chromatography, E. Stahl (Ed.), Springer-Verlag, New York, 1969, pp. 7-23.
16. K. Randerath, Thin-Layer Chromatography, Academic Press, New York, 1968, p. 49.
17. S. G. Gordon, F. Philippon, K. S. Borgen, and F. Kern, Jr., Biochim. Biophys. Acta., 218, 366 (1970).
18. N. Pelick, T. L. Wilson, M. E. Miller, and F. M. Angeloni, J. Am. Oil Chem. Soc., 42, 393 (1965).
19. W. Trappe, Biochem Z., 350, 150 (1940).
20. B. Borgstrom, Acta Physiol. Scand., 30, 231 (1954).
21. A. E. Thomas, J. E. Scharoun, and H. Ralston, J. Am. Oil Chem. Soc., 42, 789 (1965).

CHAPTER 15

TLC Technique for Analysis of Plasma Lipids

Howard R. Sloan and Constance S. Seckel

INTRODUCTION

Lipid metabolism, in general, and the metabolism of plasma lipids and lipoproteins, in particular, is an area under intense investigation (1). Among the many disorders of lipid metabolism, the familial hyperlipoproteinemias are of special medical importance (2); they constitute a group of six distinct genetic entities each of which is believed to be inherited by a single-gene mechanism and is characterized by an elevation of the plasma levels of cholesterol or triglyceride or both. In these disorders the elevated plasma lipid levels are associated with elevations in the plasma concentrations of specific lipoproteins. The prevalence of the hyperlipoproteinemias in the general population is not precisely known, but it is not less than 1 to 2%. In addition, there are a large number of hyperlipidemic individuals in the general population who owe their abnormality to a complex interaction of metabolic, environmental, and polygenic factors. The clinical relevance of these disorders is due to the striking incidence of premature coronary vascular disease that occurs in patients with several of the hyperlipoproteinemias (2). Therapy for these conditions is primarily directed at dietary and drug therapy designed to lower plasma lipid levels, particularly the level of plasma low-density lipoprotein cholesterol.

Unfortunately, there is no single test that infallibly separates those individuals who have hyperlipidemia from those who do not. Determining plasma levels of both cholesterol and triglyceride will, however, detect the presence of hyperlipoproteinemia in the vast majority of cases. Methods for quantitation of these lipids in plasma are widely available and have been successfully automated (3, 4). Although these techniques are suitable for the routine evaluation of hyperlipoproteinemia in adults, they have several shortcomings for the evaluation of pediatric patients in general and infants in particular. The most important problem is that the amount of blood required precludes serial evaluation of plasma lipid levels in infants and small children. The automated techniques require as much as 1.0 to 2.0 ml of blood, an amount that constitutes 2.5 to 5.0% of the blood volume of a small premature infant. On the other hand, enzymatic techniques require as little as 50 to 100 µl of blood for the quantification of both cholesterol and triglyceride, but they are relatively expensive to perform and require trained personnel. Only cholesterol and triglyceride may be quantified enzymatically. Neither the automated nor the enzymatic methods permit an estimation of the fraction of cholesterol that is esterified. This paper describes a thin layer spectrodensitometric technique that permits the simultaneous quantification of the plasma levels of cholesterol, cholesteryl ester, triglyceride, and free fatty acids; the method requires as little as 5 µl of plasma. The technique is made possible by the availability of thin layer chromatographic plates with a special preadsorbent area composed of diatomaceous earth that permit the direct application of plasma samples without prior solvent extraction. Development of the chromatographic plate with the appropriate solvent simultaneously extracts the lipids from the lipoproteins, precipitates the proteins, and separates the various lipid classes. In this report, we describe our application of this technique to the simultaneous and rapid analysis of cholesterol, cholesteryl ester, and triglycerides in small samples of plasma.

MATERIALS AND METHODS

A model 2955 Transidyne Scanning Densitometer (Transidyne, Ann Arbor, Michigan) equipped with a Spectro-Physics Minigrator (Santa Clara, California) was employed for scanning. The plates were scanned at 600 nm in the reflectance mode with a 9-mm long and 0.8-mm wide beam. The Minigrator calculated peak areas. For thin layer chromatography, Whatman (Clifton, New Jersey) LK5D plates were used, which have a 250-µm-thick layer and are prescored into 9-mm lanes. All plates were prewashed for at least 48 hr in petroleum ether : ethyl ether : glacial acetic acid (90:10:1), the mobile phase.

Blood was collected in 44.7-μl micropipets containing sodium heparin as an anticoagulant (Clay Adams, Parsippany, New Jersey); 5 minutes of centrifugation in these capillary tubes at 600*g* sediments the erythrocytes. The capillary tube is scored with a file and broken just above the erythrocyte-plasma interface. The plasma is expelled into small glass tubes from which 2- to 10-μl aliquots are readily removed and placed onto the preadsorbent area of the plate, 10 mm from the silica gel interface. The plates are dried thoroughly with a jet of warm air from a hair dryer and developed to a height of 7.5 to 10.0 cm. After development, the plates are again dried thoroughly with the hair dryer and sprayed with the cupric acetate stain as described by Touchstone et al. (5).

Cholesterol, cholesteryl oleate, triolein, and oleic acid standards were obtained from Applied Science Laboratories (State College, Pennsylvania). They were assayed for purity by thin layer chromatography before use. All other chemicals were of reagent grade; the solvents, including the (50 to 110° boiling point) petroleum ether, were not redistilled before use.

In some experiments, methyl oleate, which migrates between cholesteryl ester and triglyceride, was employed as an internal standard at a concentration of 5 mg/ml. In most experiments, plasma samples with known concentrations of cholesterol, cholesteryl ester, and triglyceride (3, 4) were applied to several lanes on each 20 × 20 cm prechanneled plate. The area corresponding to the color density of each lipid spot was compared with the area obtained from known amounts of the same lipid in the control lanes. The ratio of cholesteryl ester to cholesterol in several control plasma samples was determined by (1) chromatographing the plasma samples on the LK5D thin layer plates; (2) cutting three 2 × 20 cm lanes from each plate (one from the middle and one from each end); (3) staining these narrow lanes with iodine; (4) reassembling the plate and locating the position of each lipid in the unstained lanes; (5) scraping the cholesterol and cholesteryl ester bands from each lane; (6) eluting the lipids with chloroform : methanol 1:1; and (7) determining the cholesterol and cholesteryl ester content of each band by the method of Rosenthal et al. (6). The concentration of cholesterol and triglyceride in each sample was also determined by standard techniques (3, 4).

RESULTS AND DISCUSSION

A typical thin layer chromatogram is illustrated in Figure 1. The four major neutral lipid classes, that is, cholesterol, cholesteryl ester, triglycerides, and free fatty acids, are well-separated from each other and from methyl oleate, the internal standard. A light band, moving just ahead of cholesteryl ester, can be observed in each lane. This band is derived from material in the thin layer plate itself that cannot be entirely removed either by repeated or prolonged prewashing of the plate; fortunately the material is separated far enough from cholesteryl esters that it does not interfere with the quantitation of the plasma neutral lipids. The best chromatographic results were obtained if the plates were developed to a height of no more than 10 cm; further development frequently resulted in bands that were bowtie or bullet shaped, irregular in color density, and difficult to quantify spectrodensitometrically.

For photodensitometric studies, several detection methods were investigated, and it was found that anisaldehyde-sulfuric acid and phosphomolybdic acid staining provided too dark a background color to permit facile quantification; in addition, phosphomolybdic acid could not be used to quantify fatty acids and cholesteryl oleate in the range below 1.0 μg. The cupric acetate phosphoric acid stain, on the other hand, provided minimal background coloration and could be used to quantify all of the neutral lipids. For each neutral lipid the area obtained by integrating the photodensitometric tracing is linearly related to the quantity of lipid applied over the range of 0.1 to 5.0 μg.

The values obtained for the cholesterol and triglyceride levels of five plasma samples are presented in Table I and II, respectively. The values for cholesterol represent the sum of the values for cholesterol and cholesteryl ester (expressed as cholesterol) and were derived by (1) determining the distribution of cholesterol between the free and esterified forms (2) calculating the response factor (area to weight ratio) for each form and (3) making the appropriate calculations. The values obtained for cholesterol and triglyceride by the thin layer and conventional chemical methods are compared in Tables I and II. The values obtained by the thin layer method are, in general, somewhat higher than those obtained by the chemical methods. The differences between the two methods are, however, less than one standard deviation from the means of the values obtained by the routine chemical methods. The precision of the triglyceride determination was ±2.5% while that for cholesterol was ±3.2%.

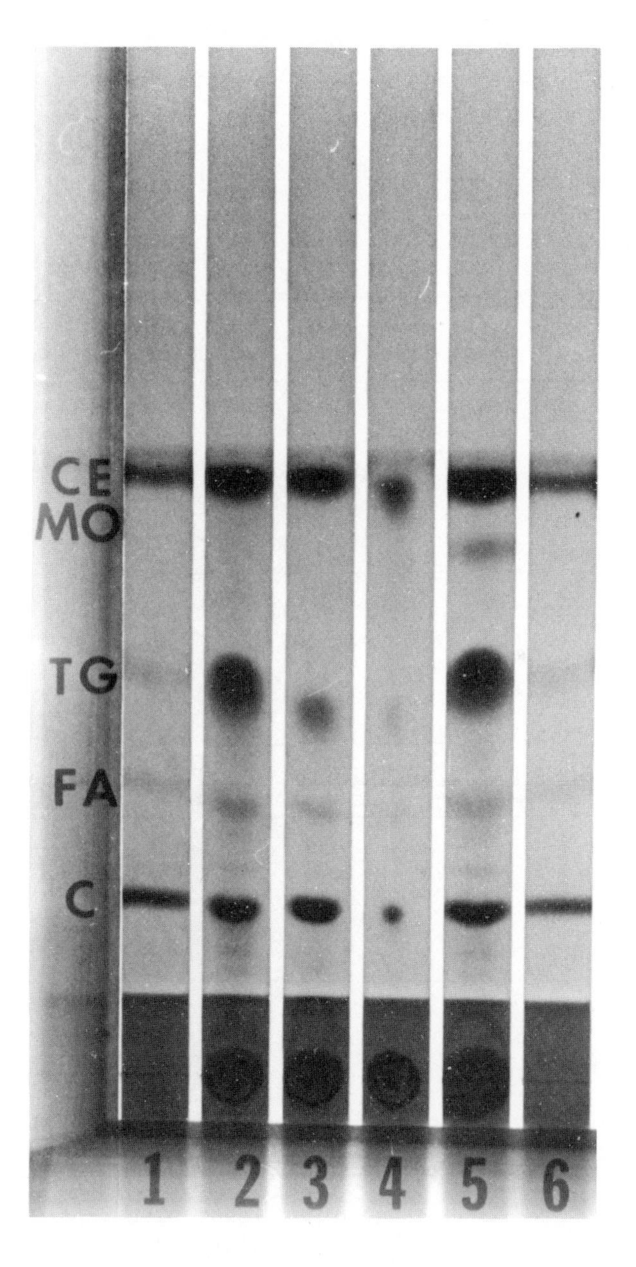

Figure 1. Thin layer chromatographic separation of neutral lipids of plasma. Lane 1: neutral lipid standards dissolved in chloroform; lanes 2 to 6: plasma samples. The solvent used was petroleum ether : ethyl ether : glacial acetic acid 90:10:1. CE - cholesteryl ester; TG - triglyceride; FA - fatty acid; C - cholesterol; MO - monoolein.

TABLE I. CHOLESTEROL CONCENTRATION IN PLASMA

Sample No.	Cholesterol Concentration (mg/dl ± STD Deviation)[a]	
	Routine Method	TLC Method
1	152 ± 8	148 ± 10
2	202 ± 10	209 ± 12
3	243 ± 12	250 ± 16
4	358 ± 18	367 ± 22
5	592 ± 30	610 ± 42

[a]All results are the average of at least five separate analyses. The routine method employed the Technicon Autoanalyzer (3).

TABLE II. TRIGLYCERIDE CONCENTRATION IN PLASMA

Sample No.	Triglyceride Concentration (mg/dl ± STD Deviation)[a]	
	Routine Method	TLC Method
1	87 ± 6	79 ± 8
2	155 ± 10	163 ± 12
3	472 ± 33	494 ± 31
4	704 ± 42	714 ± 48
5	92 ± 6	94 ± 8

[a]All results are the average of at least five separate analyses. The routine method was that of Kessler and Lederer (4).

In the course of these studies, it was observed that there were many errors in technique that could seriously impair the quality of the results obtained. The use of nonanticoagulated sampling tubes makes it very difficult to obtain an adequate sample of serum. If the blood samples are drawn with EDTA as an anticoagulant, the sample may not be stored by freezing. If the sample is frozen before analysis, the lipids migrate as round spots that are very difficult to quantify (Figure 1, lane 4). A similar result occurs if the samples are not adequately dried before being chromatographed. On the other hand, samples with heparin anticoagulation may be successfully analyzed spectrodensitometrically, even after repeated freezing and thawing or after several months of storage at -20°. The application of more than 5 μg of any single neutral lipid to a lane produces an irregularly stained band that cannot be adequately evaluated photodensitometrically (Figure 2, lanes 4 and 5). Therefore, if a sample contains more than 5 μg of a single lipid per microliter, appropriate dilutions are made with a 0.9-gm/dl solution of sodium chloride. The results obtained by adding a known amount of methyl oleate to each sample as an internal standard (Figure 1) were compared with those obtained by spotting samples with known concentrations of cholesterol and triglyceride on the same plate. The results were not significantly different, therefore known samples were used as external standards because it is technically simpler.

In addition to the small sample size required for quantitation, the simplicity and rapidity of the method commend its use in many clinical settings. If a thin layer plate is prespotted with standards of known cholesterol and triglyceride concentrations, a plasma lipid analysis can be readily performed in 90 minutes. Even inexperienced operators, making a visual judgment rather than a spectrophotometric quantitation, may make a reasonable estimate of plasma lipid levels within 45 minutes of obtaining a blood sample. Figure 3 illustrates a practical application of the thin layer chromatographic method. A small premature infant accidentally received, over 60 minutes, an intravenous infusion of a fat emulsion that should have been administered over 14 hr. Five μl of plasma, obtained from five hourly 20-μl blood samples, was applied to lanes 2 through 6. An analysis of the chromatograms revealed that the plasma triglyceride level fell from 1580 to 180 mg/dl within 6 hr. Conventional analytical methods would have required much larger samples of blood, and, at least in this hospital, one to two days to provide the same reassuring information. It seems quite likely that this thin layer method

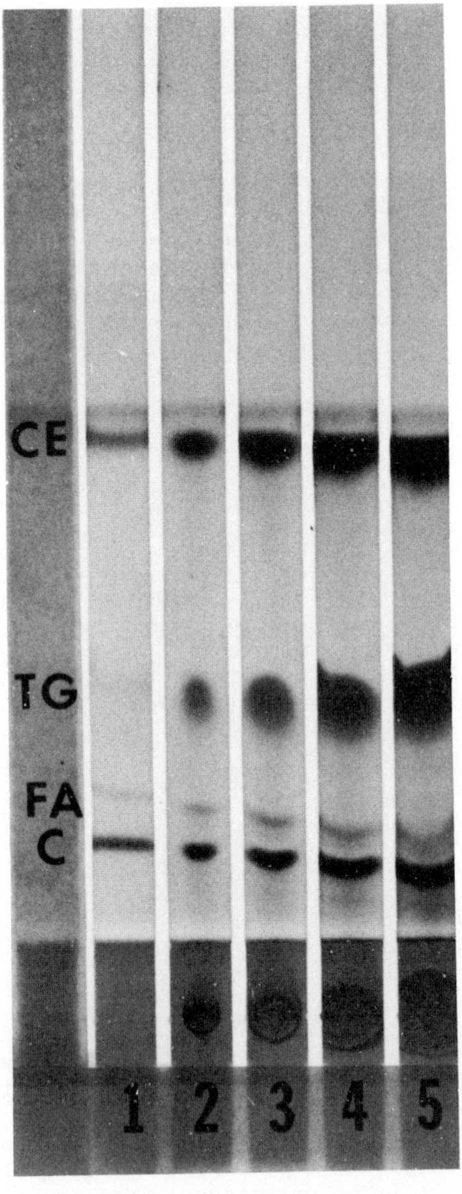

CE

TG

FA

C

1 2 3 4 5

Figure 2. Thin layer chromatographic separation of neutral lipids plasma. Lane 1: neutral lipid standards dissolved in chloroform; lanes 2 to 5: plasma samples. The solvent used and the abbreviation employed are the same as those in Figure 1.

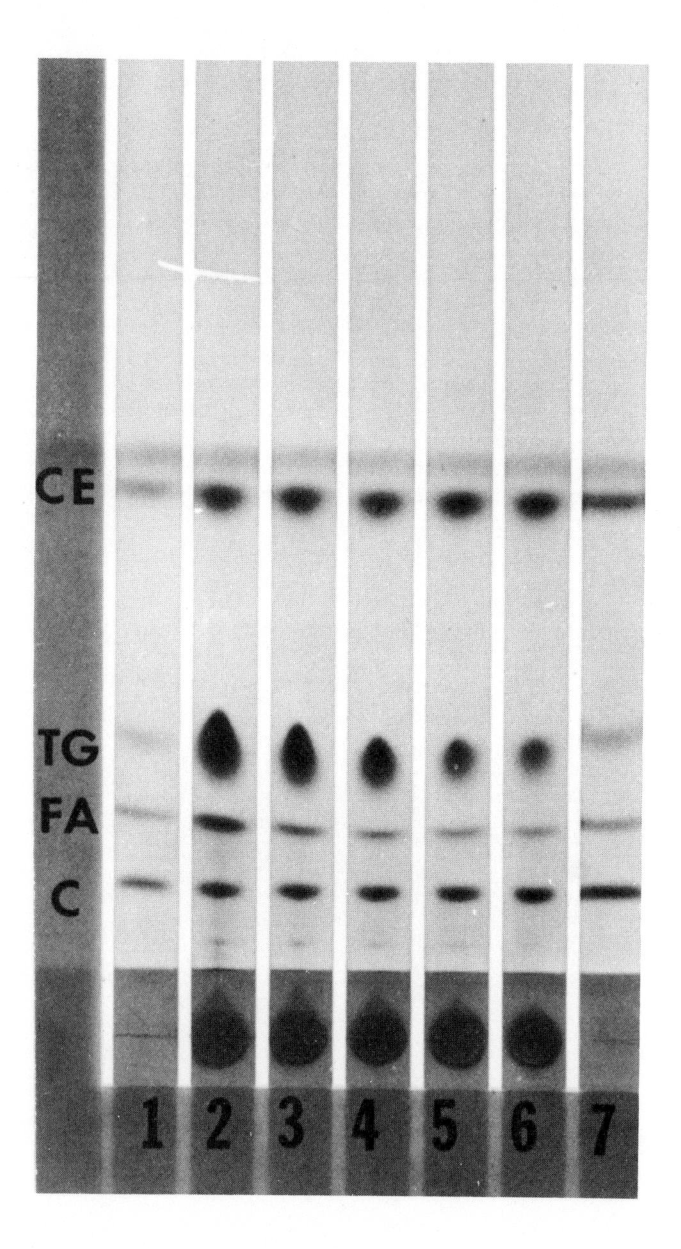

Figure 3. Thin layer chromatographic separation of neutral lipids of plasma. Lanes 1 and 7: neutral lipid standards dissolved in chloroform; lanes 2 to 6: plasma samples obtained at hourly intervals following an intravenous lipid infusion. The solvents used and the abbreviations employed are the same as those in Figure 1.

can be used in a screening program for the hyperlipoproteinemias by employing the same sample of blood obtained for the determination of such routine blood studies as a hematocrit.

REFERENCES

1. P. N. Herbert, A. M. Gotto, and D. S. Fredrickson, Familial Lipoprotein Deficiency, in The Metabolic Basis of Inherited Disease, 4th ed., J. B. Stanbury, J. B. Wyngaarden, and D. S. Fredrickson (Eds.), McGraw-Hill, New York, 1978, p. 544.
2. D. S. Fredrickson, J. L. Goldstein, and M. S. Brown, The Familial Hyperlipoproteinemias, in The Metabolic Basis of Inherited Disease, 4th ed., J. B. Stanbury, J. B. Wyngaarden, and D. S. Fredrickson (Eds.), McGraw-Hill, New York, 1978, p. 604.
3. Technicon Instruments, Total Cholesterol Procedure N24a, in Auto Analyzer Manual, Technicon Corporation, Chauncey, New York, 1964.
4. G. Kessler and H. Lederer, in Automation in Analytical Chemistry, L. T. Skeggs, Jr. (Ed.), Mediad Inc., New York, 1966, p. 341.
5. J. C. Touchstone, M. F. Dobbins, C. Z. Hersch, A. R. Baldino, and D. Kritchevsky, Clin. Chem., 24, 1496 (1978).
6. H. L. Rosenthal, M. L. Pfluke, and S. Buscaglia, J. Lab. Clin. Med., 50, 318 (1957).

CHAPTER 16

Quantitation of Cholesterol in Biological Fluids by TLC with Densitometry

Joseph C. Touchstone, Gerald J. Hansen, Carolyn M. Zelop, and Joseph Sherma

INTRODUCTION

The cholesterol level in blood serum is an important diagnostic aid in various diseases such as diabetes, nephrosis, anemia, hepatitis, hypo- and hyperthyroidism, and hyperlipidemia. The classical clinical analysis for serum cholesterol involves direct treatment of the serum with a reagent composed of ferric chloride dissolved in a glacial acetic acid-sulfuric acid mixture followed by solution spectrophotometry at 540 nm (1). Normal values for serum cholesterol based on this method are reported as (150 to 250 mg)/(100 ml of serum) (2). However, the method is not specific for cholesterol since cholesteryl esters can also react with the color-forming reagent, and their concentration will be included in the value reported for cholesterol.

Thin layer chromatographic separation of cholesterol combined with quantitation by densitometry was reported by Touchstone et al. (3) for determination of cholesterol at nanogram levels in studies of hydrolysis and synthesis of cholesteryl esters by aortic enzymes (3). Standard thin layer chromatography (TLC) and high-performance thin layer chromatography (HPTLC) plates were used, with detection by charring with a cupric acetate-phosphoric acid spray reagent.

TLC with densitometry was also reported by Kupke and Zeugner (4) for the determination of cholesterol in plasma and liver homogenates. Samples (0.5 μl) were applied directly to silica gel 60 HPTLC plates, and cholesterol was scanned after separation and detection as a fluorescent spot with $(NH_4)HCO_3$ reagent. This procedure was complicated by the need to spot the sample over methanol at the origin and to cover the sample with additional methanol, and also a 10-hr heating period to produce fluorescent spots.

Other TLC procedures involved the extraction of cholesterol from serum, chromatography of the extract, and densitometry (5) or scraping and elution of the cholesterol spot followed by solution spectrophotometry (6). In the former in situ determination, both TLC and HPTLC plates were used, and nathoquinone sulfonate reagent was the detection reagent. Methodologies for analysis of cholesterol through 1976 have been reviewed (7).

Thin layer chromatography on silica gel separates sterols from other classes of lipids, so that the values reported would be for free cholesterol in the absence of cholesteryl esters. This should lead to lower concentrations than those found by conventional colorimetry, as indicated by a cholesterol level of 1.38 mmol/l (53.4 mg/100 ml) for pooled plasma (including samples from hyperlipidemic patients) reported by Kupke and Zeugner (4).

This chapter reports the method developed for the densitometric determination of cholesterol in directly applied human saliva using silica gel thin layer plates with a preadsorbent spotting area. Some preliminary values for saliva levels of free cholesterol and comparisons with levels in fingertip blood are also given. Work is now in progress in our laboratories to assess the utility of this method for clinical applications.

MATERIALS AND METHODS

Whatman 20 × 20 cm LK5D precoated silica gel TLC plates with a 3.4-cm preadsorbent strip and nineteen 9-mm-wide scored lanes and Whatman 10 × 10 cm LHPK high-performance thin layer chromatography (HPTLC) plates with a 2.0-cm-wide preadsorbent strip were used. The HPTLC plates were divided into 5-mm-wide lanes using a Fotodyne scoring device and a ruler. Plates were developed with chloroform/methanol (1:1 v/v) and thoroughly air dried before use.

Cholesterol obtained from Supelco, Inc., was prepared in chloroform at a concentration level of 8 ng/μl for use as a standard. Standard and sample solutions were applied by streaking across the preadsorbent portion of each lane (excluding the end two) using a Drummond 10-µl Dialamatic microdispenser. Applications were confined to an area just below the silica gel-preadsorbent interface and a distance of 4 mm above the bottom of the plate. Volumes of 4 to 30 µl (32 to 240 ng) were applied for standards, and 5 µl of diluted saliva and blood serum. Initial zones were dried for 5 minutes with a stream of cool air from a forced air flow dryer.

The mobile phase was chloroform/ethyl acetate (94:6 v/v); the classical mobile phase for lipid separations, petroleum ether/diethyl ether/glacial acetic acid (80:20:1 v/v), was used for comparative purposes. Chromatograms were developed in a Kontes rectangular glass N-chamber containing a filter paper liner and saturated with the mobile phase for 15 minutes before inserting the plate. HPTLC plates were developed in a saturated miniature glass N-chamber designed specifically for 10 × 10 cm plates (Fotodyne). Sufficient mobile phase was added to give a 3-mm-deep solvent pool in each tank.

To ensure that the samples formed into sharp bands at the silica gel-preadsorbent interface, the plate was developed up to this junction, removed from the tank and dried, and then redeveloped until the mobile-phase front had moved 15 cm above the bottom edge of the TLC plate (about 40 minutes) and 8 cm for the HPTLC plate. The plate was then air dried in a fume hood until the mobile phase had completely evaporated (about 3 minutes).

For detection, plates were sprayed evenly and thoroughly (to wetness but not dripping) with a reagent containing 30 g of cupric acetate and 80 ml of phosphoric acid per liter of solution applied with a Kontes sprayer. Heating for 30 minutes at 130° in a chromatography oven produced charred (brown-black) spots on a white background for cholesterol, cholesteryl esters, and other lipids.

Lanes containing cholesterol were scanned (usually immediately after detection) with the Kontes model 800 fiber optics densitometer in the single-beam transmission mode using the 8-mm-long light beam and the "white" phosphor (440-nm emission peak, 300-nm band width) over the shortwave UV light source. An attenuation setting of 16 was typically used. Recorder peak areas were calculated by the formula: height × width at half-height, the half-height line being drawn parallel to the constructed base line. Calibration curves (peak area versus nanogram standard applied) were prepared from standards applied on each plate, and the amounts of cholesterol in samples were interpolated from these curves (Figure 1).

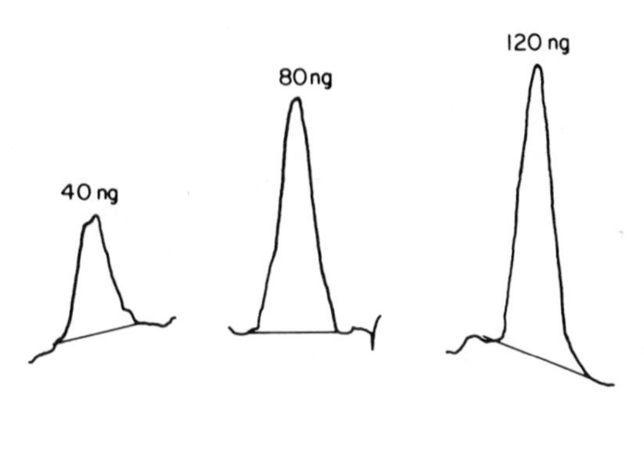

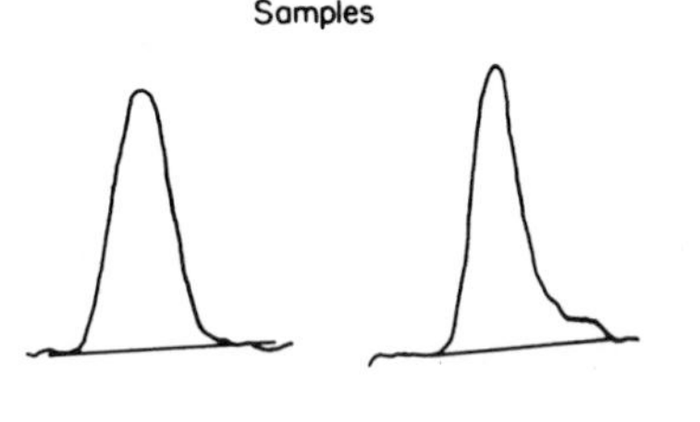

Figure 1. Scans of 40-, 80-, and 120-ng standards and duplicate 5-μl diluted saliva samples spotted on the same plate, with baselines shown. As interpolated from the full calibration curve, the sample peaks represented 84.2 and 87.7 ng of cholesterol, for an average value of 86.0 and a variation of 4%. The attenuation setting of the scanner was ×16, and both scan speed and recorder speed were 2.5 cm/minute. Other conditions for densitometry were as described in the text.

Morning saliva samples were collected upon waking; the mouth was rinsed with water and, after 5 minutes, the sample was expectorated into a clean beaker. One ml was transferred with a volumetric pipet to a small vial and diluted with 1 ml of distilled water. The diluted sample was immediately refrigerated, and was Vortex-mixed before application to the TLC plate. Samples taken at other times of the day were collected and treated in the same manner. A new micropipet was used for each sample, or the pipet was thoroughly rinsed between samples. The dilution factor was included in calculations of cholesterol content of samples.

Blood was drawn from a pricked fingertip into a 1-ml centrifuge tube using a disposable glass Pasteur pipet (approximately four freely flowing drops). The sample was spun in a microcentrifuge at 7000*g* for 2 minutes. Ten μl of the clear, usually colorless or lightly colored, serum layer was removed with a 10-μl Hamilton syringe and was quantitatively diluted with 90 μl of distilled water, which was used to rinse the syringe. The sample was Vortex-mixed prior to application on the layer as described above for saliva. All saliva and blood samples were spotted in duplicate, and results were averaged.

Blood plasma from the blood bank was also analyzed as described; these were diluted 1:20 and 2 μl taken. Rat bile was diluted 1:20 and 1 μl applied to the preadsorbent layer.

RESULTS AND DISCUSSION

Preliminary evaluation of the TLC and HPTLC plated indicated that the former were advantageous for this determination. The commercial availability of plates having prescored lanes with widths that matched the length of the scanning head was very convenient and avoided manual scoring of the HPTLC plates. Calibration curves were linear only to 120 ng on HPTLC plates and then curved downward toward the x axis, while linearity was achieved up to at least 260 ng on the TLC plate. The lowest level for practical quantitation was reduced from 30 to 40 ng for TLC to 10 ng for HPTLC, and visual minimum detectabilities were still lower on both plates. However, the ability to apply larger aliquots of saliva and serum precluded sensitivity as a compelling consideration.

The chloroform/ethyl acetate (94:6 v/v) mobile phase produced an R_f value for cholesterol of approximately 0.40, which is within the ideal range for accurate and precise quantitation by densitometry (8). With this mobile phase, a zone of cholesteryl esters (most likely cholesteryl oleate) was located at R_f 0.76 in chromatograms from most saliva and all serum samples. To ascertain that the cholesterol zone at R_f 0.40 was not mixed with lipids of other classes known to be resolved from it by the classical mobile-phase petroleum ether/diethyl ether/glacial acetic acid (80:20:1 v/v) (in which the R_f of cholesterol is 0.10), analysis of a saliva sample was carried out in parallel on separate plates using both of these mobile phases. Results for cholesterol were identical, confirming the validity of the chloroform/ethyl acetate mobile phase.

The predevelopment of samples up to the preadsorbent-silica

gel layer junction prior to the analytical development was found to produce tighter and more regularly shaped zones that closely matched the standard zones. This resulted in improved precision of quantitation.

The cupric acetate-phosphoric acid reagent was chosen for detection because it provided the greatest contrast between the colored spots and the plate background, as well as the most uniform background. Both sensitivity (signal to noise ratio) and precision were higher with this reagent than with other charring reagents evaluated.

Thorough drying of initial zones on the preadsorbent strip, before and after the predevelopment, was found to be important. However, this should be done with a warm stream of air from a blower-dryer gun rather than with hot air or in a high-temperature oven in order to avoid the possible formation of artifacts or compound decomposition.

Accuracy of the method was evaluated by a standard addition procedure. A saliva sample was analyzed, and a value of 80.0 ng obtained. This represented 80.0 ng/2.5 µl of actual saliva spotted, or 32.0 ng/µl (3.20 mg/100 ml). The remaining sample was spiked with an appropriate volume of cholesterol solution (1 µg/1 µl) to double its concentration, and the analysis was repeated. After correcting the result for the dilution caused by the volume of standard added, the value of cholesterol found was exactly doubled, 6.38 mg/100 ml, indicating essentially 100% recovery of the spike.

To evaluate reproducibility (precision) of the analysis, four replicate samples of 80 ng (10 µl) of cholesterol were applied to a single plate, and the relative standard deviation (coefficient of variation) of the resultant peak areas was 5.65%. When three replicate saliva samples were applied to two separate plates, the RSD values were 3.36 and 3.29%, respectively. Using the Whatman LHP-K high-performance layers, a day-to-day and within-day reproducibility of less than 8% was found. Values for peak areas (square millimeters) for duplicate samples of saliva and blood were always within 10% of each other and usually agreed well within 5%. Figure 1 shows typical scans of standards and separated saliva samples.

Figure 2 shows a typical calibration curve for cholesterol plotted as peak area in square millimeters versus nanograms of cholesterol. The calibration curve is quite reproducible from plate to plate and from day to day, with a usual slope of about 3.0 and intercept of +10, as determined by regression analysis of the data points on an electronic calculator. For maximum accuracy

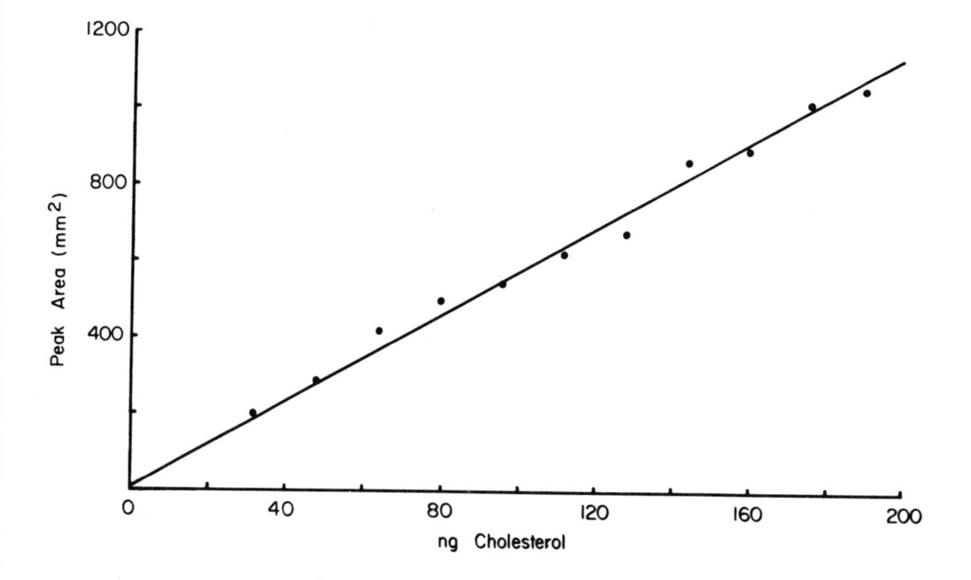

Figure 2. Typical calibration curve for cholesterol on a Whatman LK5D TLC plate resulting from spotting 4 to 27 μl of an 8 ng/μl solution (32 to 192 ng).

and precision, however, standards were always applied on each plate with samples, and an individual calibration curve obtained for each plate.

Table I shows cholesterol levels obtained for two individual subjects and for two pooled saliva samples. The samples from subject 1, a healthy female college student, were collected on 11 separate days over a period of three months. Except for one day, values for morning saliva ranged from a low of 2.28 to 5.70 mg/100 ml, with an average of 3.57. The single very high value of 9.6 mg/100 ml was obtained the day after a 24-hr fast period. The morning saliva values for subject 2, another student, on two different days were in a similar range. The daytime values for the two subjects were much less consistent than the morning values; the midday levels tended to be below the morning values, while the evening levels were either higher or lower than those at midday. No cholesterol at all was detectable for subject 2 in the evening sample on the first day she was monitored. The morning pooled levels, from groups of male and female college students, were somewhat below those for the two individual

TABLE I. CHOLESTEROL LEVELS DETERMINED BY THE PREADSORBENT TLC/DENSITOMETRY PROCEDURE (mg/100 ml)[a]

	Morning Saliva	Saliva Collected before Lunch	Saliva Collected before Dinner	Morning Fingertip Blood
Subject 1				
1	3.20			
2	3.20			
3	5.70			
4	3.48			
5	3.20			
6		1.60	3.80	7.60
7	3.68	3.12	5.44	18.1
8	9.60	2.56	3.12	32.9
9		2.28		26.5
10	3.60			30.0
11	2.56			20.3
Subject 2				
1	3.04	2.72	0	7.60
2	3.36	1.36	1.20	3.04
Pooled Sample (6 Subjects)	2.68			
Pooled Sample (8 Subjects)	3.10			

[a]Average of duplicate spotted samples.

subjects but at the same general level.

Morning blood values for subject 1 ranged from 18.1 to 32.9 mg/100 ml, with the highest level again coming after the fast day. The blood level for subject 2 was significantly below that of subject 1, although their saliva values were very similar. In general, the variation in cholesterol blood levels was greater than in saliva. Table II shows the results obtained with plasma obtained from the blood bank and rat biles.

TABLE II. CHOLESTEROL LEVELS IN BIOLOGICAL SAMPLES (mg/100 ml)

Rat bile	800 to 7400
Seminal fluid	100 to 250
Blood bank plasma	100 to 300

SUMMARY AND CONCLUSIONS

This chapter describes simple methodology for the determination of free cholesterol in human saliva, bile, and fingertip blood using thin layer chromatography with densitometry. No preliminary extraction of the samples is required since direct application to preadsorbent thin layer plates can be made. Sensitivity in the low nanogram range was achieved, and accuracy and precision are adequate for use of the method in clinical screening analysis. Preliminary data and cholesterol levels are reported, but further results for cholesterol and cholesteryl oleate are required before the significance of the numbers on a diagnostic basis is clear. Work is now in progress on collection of further data and comparisons among veinous serum, fingertip serum, and saliva.

ACKNOWLEDGMENT

The authors are indebted to Dr. Bernard Fried of the Lafayette College Biology Department for valuable advice during the course of this research.

REFERENCES

1. A. Zlatkis, B. Zak, and A. J. Boyle, J. Lab. Clin. Med., 44, 486 (1953).
2. G. T. Bender, Chemical Instrumental Analysis: A Laboratory Manual Based on Clinical Chemistry, W. B. Saunders, Philadelphia 1972, p. 17.
3. J. C. Touchstone, M. F. Dobbins IV, C. Z. Hirsch, A. R. Baldino, and D. M. Kritchevsky, Clin. Chem., 24, 1496 (1978); D. Kritchevsky, C. Z. Hirsch, and J. C. Touchstone, Thin Layer Chromatography--Quantitative Environmental and Clinical Applications, J. C. Touchstone and D. Rogers (Eds.), Wiley-Interscience, New York, 1980, Chapter 31, p. 455.
4. I. R. Kupke and S. Zeugner, J. Chromatogr., 146, 261 (1978).
5. R. Wintersteiger and G. Guebitz, Sci. Pharm., 46 (4), 269 (1978).
6. D. Nedeljkovic and R. Topalovic, Arh. Farm., 39 (1), 27 (1980).
7. B. Zak, Clin. Chem., 23 (7), 1201 (1977).
8. J. C. Touchstone, M. F. Schwartz, and S. S. Levin, J. Chromatogr Sci., 15, 528 (1977).

CHAPTER 17

Reverse-Phase TLC for Bile Acids in Rat Bile

Sidney S. Levin and Joseph C. Touchstone

INTRODUCTION

The principal bile acid found in mammals are cholic and chenodeoxycholic acids. A small proportion of the bile acids excreted in the bile is free since they are immediately conjugated with taurine or glycine by the liver and excreted as such. There is considerable variation between species in the nature of bile acid, for example muriocholic acid is specific for rats and mice and is therefore not found in any other species. The bile acid profile, i.e. the proportion and kind of individual bile acid and their taurine or glucuronide conjugates, changes in the presence of various hepatic and gastro intestinal diseases. Profile changes also occur with circulation through the enterohepatic system (2).

Methods for separation of conjugated bile acids are widespread. Most employ a lengthy preparative procedure involving either hydrolysis or derivatization and a complex extraction (3, 4). Hydrolysis produces artifacts that result in unreliable interpretations (5). Large samples are required for extraction to compensate for loss as a result of the "clean-up" procedures involved in the preparative chemistray. The high-performance liquid chromatography (HPLC) methods reported to date (7) give relatively good resolution. However, extensive preparative procedures are required and the number of samples are limited by the length of time necessary to chromatograph a single sample in the HPLC column. Although the separation of the conjugated bile acids on conventional TLC sorbent (silica gel) has been reported (8), resolution has not been complete in single-dimensional systems.

The development of the reverse-phase thin layer chromatogram (RP-TLC) with characteristics similar to the HPLC column permits the efficient separation of the highly polar conjugated bile acids. The advantages of the RP-TLC are (1) as many as 17 samples may be applied and developed on a single chromatogram, (2) the bile can be applied directly without extraction, and (3) the cost of equipment is relatively low. However, since the HPLC binary solvent systems do not always give the anticipated resolution when applied to RP-TLC, a new mobile phase was developed to separate the conjugated bile acids in rat and human bile (9). This new method has been applied to bile from fistula rats.

MATERIALS AND METHODS

Bile is collected directly from the bile duct of male Sprague Dawley rats, weighing 250 to 300 g. The bile duct is cannulated by a modification of the technique first described by Fisher and Vars (10). The cannula is passed through a fistula, and the distal end of the cannula is inserted into an external gall bladder mounted on the back of the animal, just above the fistula. The procedure permits the rat to be housed in a standard metabolic cage and to have complete mobility, without restraint. The animals are able to survive, in an apparently health state, for as long as 120 days, at which time they are sacrificed and liver autopsied. The bile is collected immediately after the cannula is in place and then for varying time intervals throughout the study. If metabolic studies of drugs are to be conducted, for example, phenobarbital, benz(α)pyrene (11), a four- to five-day period of equilibration is allowed before administration of the drug. During this period, the bile is collected for studies of bile acid profile. Following collection, the bile is stored at 4° until analyzed. Aliquots of the bile are diluted 1:1 with water or saline, then 2 to 5 μl of the dilute bile is streaked on 20 × 20 cm Whatman KC_{18} RP-TLC plates. The chromatograms are scored, prior to use, into 1-cm lanes using a Schoeffel model #SDA320 scoring device. The samples are applied with Drummond micropetts that fill end to end. Authentic standards for each of the bile acids listed in Table I are applied on each plate in lanes adjacent to the bile samples. The streak is applied as a

TABLE I. REVERSE-PHASE TLC FOR BILE ACIDS IN RAT BILE[a]

R_f's For Conjugated Bile Acids

Bile Acid	R_f
Tauromuricholic	0.52
Taurocholic	0.38
Glycocholic	0.31
Taurochenodeoxycholic	0.26
Taurodeoxycholic	0.23
Glycochenodeoxycholic	0.20
Glycodeoxycholic	0.17
Glycolithocholic Acid	0.10

[a] LKC_{18} Whatman layer. Mobil phase: Ethanol/0.3M calcium chloride-DMSO (50:50:4).

series of small spots placed linearly across the lane rather than as a single large circular spot. The standards on each plate are checked against the calibration curve to correct for plate-to-plate variation. A stream of warm air is directed over the spotting area while applying the sample so as to concentrate and dry the streak.

The chromatogram is predeveloped to further concentrate and to distribute the sample across the lane. This procedure results in a more concentrated, better resolved chromatogram and may also precipitate interfering protein. To predevelop, the TLC is placed in an open glass trough containing $CHCl_3$:MeOH, 1:1, the predeveloping mobile phase. The solvent is allowed to raise on the plate until the solvent front is just past the origin. The chromatogram is then removed from the trough and allowed to dry in a stream of air at ambient temperature until all traces of solvent have evaporated from the sorbent. This can easily be checked by absence of solvent smell and warm, rather than cold, feel in the back of the glass plate.

The chromatogram is then placed in an unlined developing tank and developed in a mobile phase of 0.3M $CaCl_2$ (pH 3):EtOH:

DMSO, 50:50:4. $CaCl_3$ solution is used instead of water in this highly polar system to prevent the separation of the sorbent from the plate. Development time for a distance of approximately 17 cm is about 2.5 hr; however, since every lane may be used, many samples may be developed simultaneously.

The developed chromatogram is removed from the tank and allowed to dry completely in a stream of warmed air (about 20 minutes). The plate is then placed in a 175° drying oven for 3 minutes to drive off all traces of DMSO. This can be checked by the absence of a garliclike smell. This step may be repeated if DMSO is still present in the plate. When the plate has cooled, the chromatogram is visualized by spraying lightly with 10% H_2SO_4 in ethanol and heating at 180° for 3 minutes. The procedure results in highly fluorescent spots against a white background that can be viewed at 360 nm. Heating longer or at a higher temperature produces a charred spot and dark background. The chromatogram is then scanned at 360 nm with a Schoeffel No. 3000 densitometer at a scan speed of 5 cm/min. An A-74 Corning filter is placed between the plate and the photomultiplier tube to eliminate all stray light below 400 nm. A Hewlett Packard 3385A recording integrater is used for quantitation, the recording speed synchronized with the densitometer in order to provide a 1:1 linear recording of the chromatographic plate.

RESULTS

Bile samples developed on the same plate with authentic standards have similar R_f values. A list of the bile acids studied and their R_f values can be seen in Table I. The recordings of densitometer scans of authentic conjugated bile acid standards, a human bile and a rat bile, are illustrated in Figure 1. The location of the peaks in the human bile corresponds to these found in the standard scan. The major areas in the human bile are taurocholic (TC), glycocholic, taurodeoxycholic (TDC), and glycodeoxycholic. Although the separation of taurochenodeoxycholic (TCDC), glycocholic (GC), glycochenodeoxycholic (GCDC), and glycodeoxycholic (GDC) is difficult, in HPLC as well as TLC, it has been accomplished with good resolution on RP-TLC in this system. The identity of the spots in the bile samples can be verified by comparing not only the R_f values but also the characteristic fluorescent color emitted at 360 nm with each of the bile acid standards.

The rat bile, which has a different profile, has three

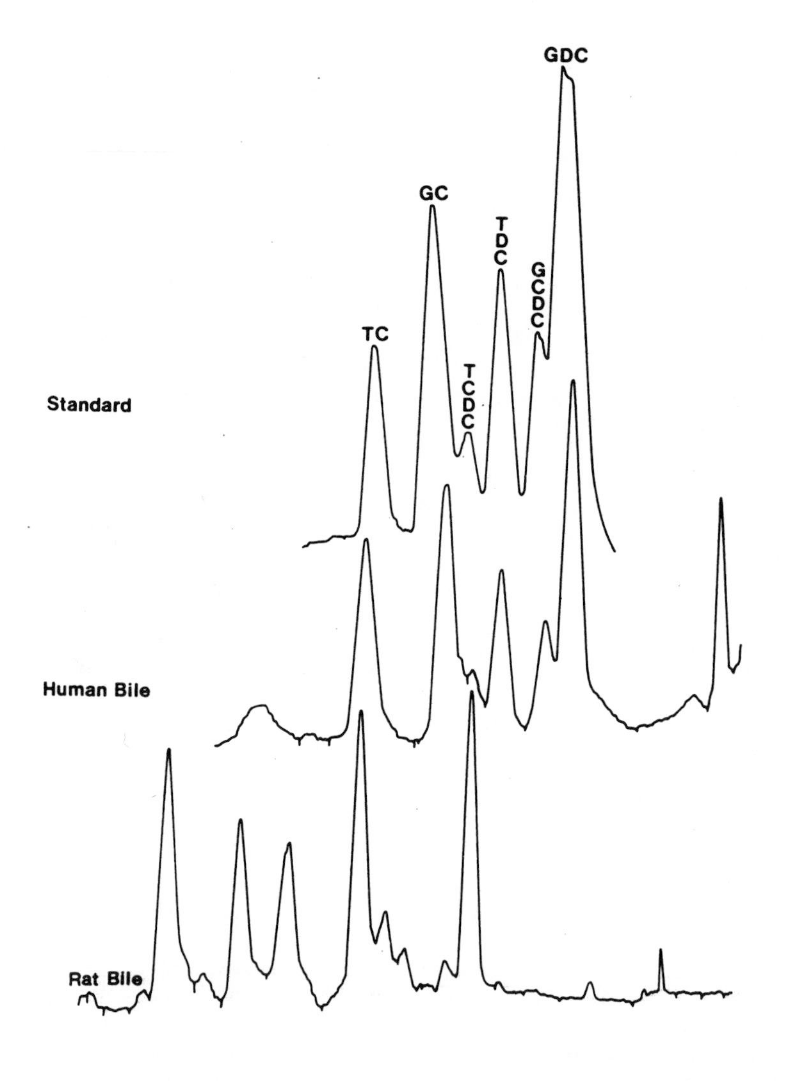

Figure 1. Densitometer scan: Separation of conjugated bile acids using Whatman $KC_{18}F$ reverse-phase thin layer chromatograms. Mobile phase: EtOH:$CaCl_2$ (0.3M ph 3.0):DMSO, 50:50:4.
Top scan: Authentic standards, see text.
Middle scan: Human bile, 5 µl; Dil 1:100.
Lower scan: Rat bile; 5 µl; Dil 1:50.

major peaks more polar than those seen in the human bile, one of which has been tentatively identified as α-tauromuricholic acid (TMC).

The calibration curves plotted for each of the bile acid standards (Figure 2) are linear to 800 ng. An individual curve must be prepared for each compound to be quantitated since each has a different absorption coefficient. A common error in quantitation in analytical methods is to use a single calibration curve for all components in the mix.

The changes in profile, that is, the relation of the individual bile acids to each other during the first 72 hr of a long-term bile fistula can be seen in Figure 3. The changes are similar to those reported by Danielsson et al. (12). Since in the bile fistula most, if not all, of the bile is directed away from the enterohepatic circulation, there are changes in the proportions of the individual conjugated bile acids. The initial recording in Figure 3 is that obtained from the bile immediately after cannulating the bile duct in the rat. TC is the principal bile acid representing almost 50% of the total area under the identified peaks. During the first 24-hr period there is a decrease in all of the bile acids; however, the TC remains at greater than 50% of the total area. The TMC decreases from 23% in the initial sample to less than 10% by the time of the 48- to 72-hr sample. GC, which is 17% initially, is down to about 5% on the third day.

Figure 4 and 5 illustrate the changes in proportion of the four major bile acids when followed over a 20-day period. These figures indicate the percentage of the total area under the peaks versus the number of days after the fistula was placed. Figure 4 is a continuation of the study of the animal illustrated in Figure 3. TC remains the principal bile acid throughout the study. GC, which was decreasing, disappears after the sixth day. TMC remains below the 8% level with little variation. TCDC, which is only 8% of the area on the second day, increases and is in the 15 to 25% range for the period illustrated.

Although the TC is only about 23% of the total peak area on the first day as seen in Figure 5, it increases and remains at the 50% level for the balance of the 20-day study. GC starts at 32% but then decreases and disappears by day eight. TMC starts at 25%, then falls to and remains at 8% or below. TCDC is only 9% of the total area initially, but it increases, then behaves in an erratic manner.

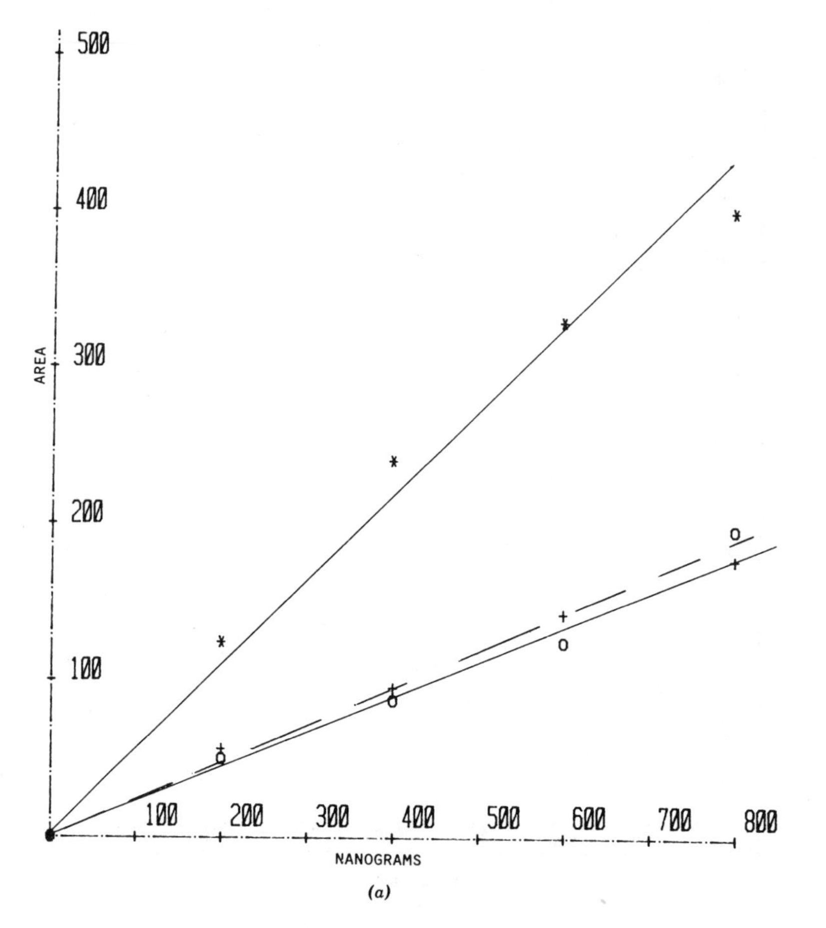

Figure 2. (a) Calibration curve from densitometer scans of authentic taurine conjugated bile acid standards; TLC scanned at 360 nm. (b) On following page, calibration curve of glycine conjugated bile acid standards.

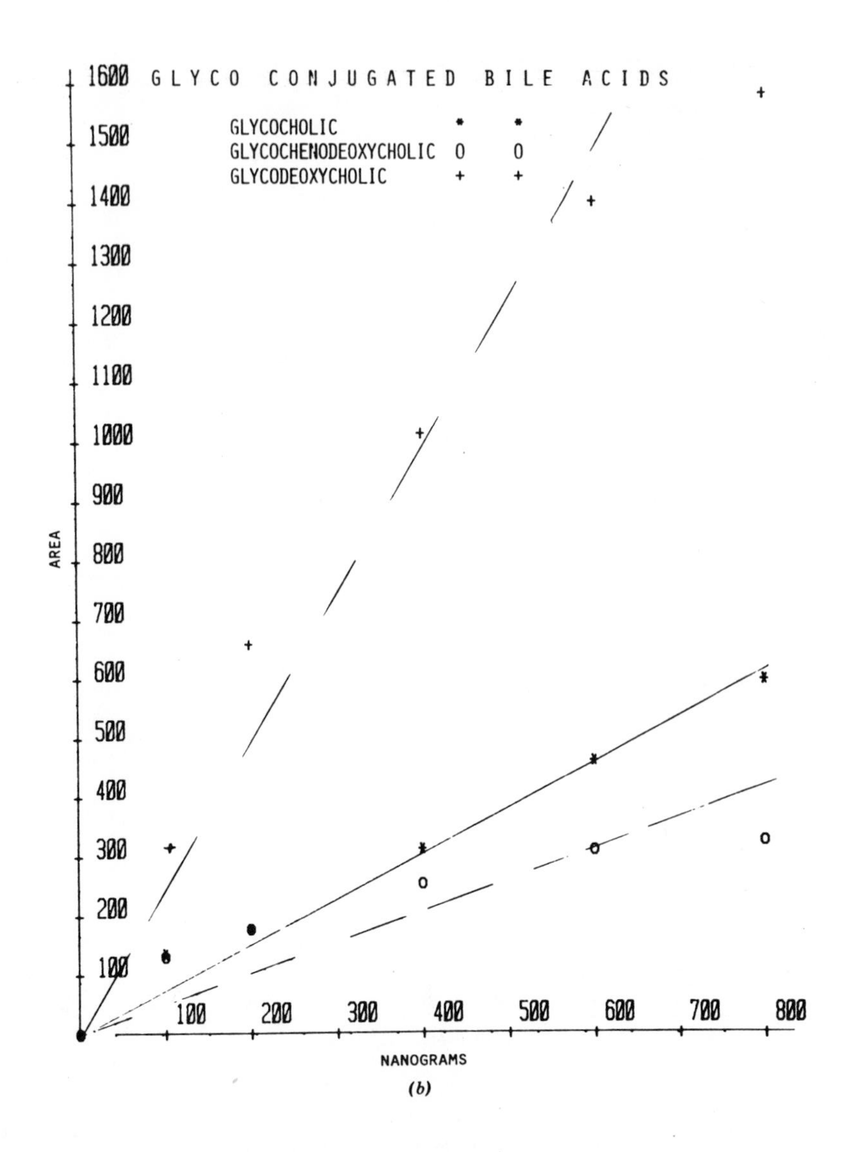

Figure 2 (continued).

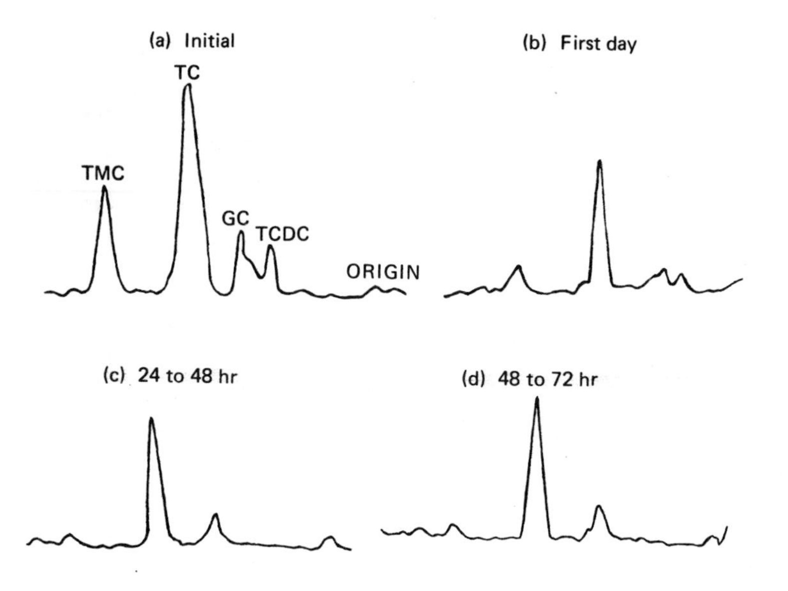

Figure 3. Densitometer scan at 360 nm of rat bile visualized with 10% H_2SO_4 in methanol. (a) Initial sample immediately after bile duct cannulation. (b) First day pool of bile. (c) 24 to 48 hr pool of bile. (d) 48 to 72 hr pool of bile.

DISCUSSION

The use of the rat bile fistula prepared by a modification of the Fisher and Vars method produces a laboratory model that has a relatively long survival period. It enables a study of not only the metabolites of a drug directly but also the effect of the drug on the liver function through changes in bile acid profile. The number of samples available by the use of this method would tax the HPLC instrumentation in the average research laboratory. However, this can be handled with a minimal difficulty by the use of the RPTLC methodology. Quantitation on the chromatogram also aids in the relatively rapid rate of analysis of the bile.

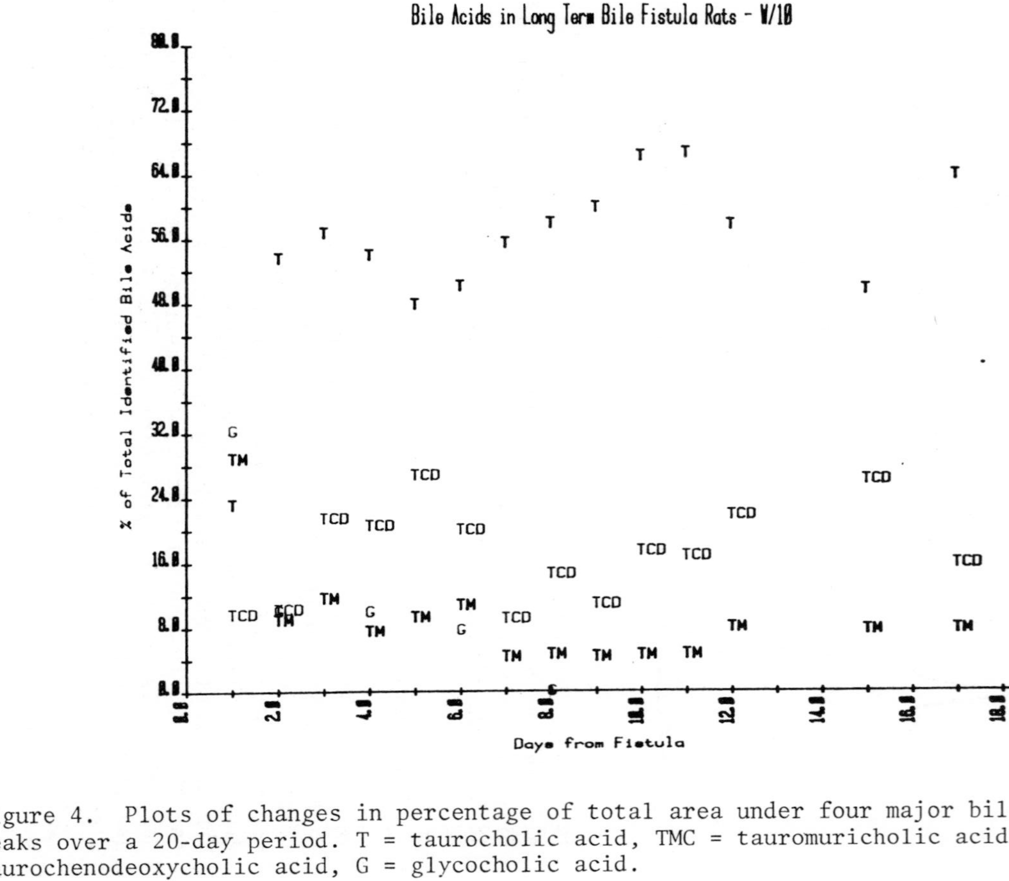

Figure 4. Plots of changes in percentage of total area under four major bile acid peaks over a 20-day period. T = taurocholic acid, TMC = tauromuricholic acid, TCD = taurochenodeoxycholic acid, G = glycocholic acid.

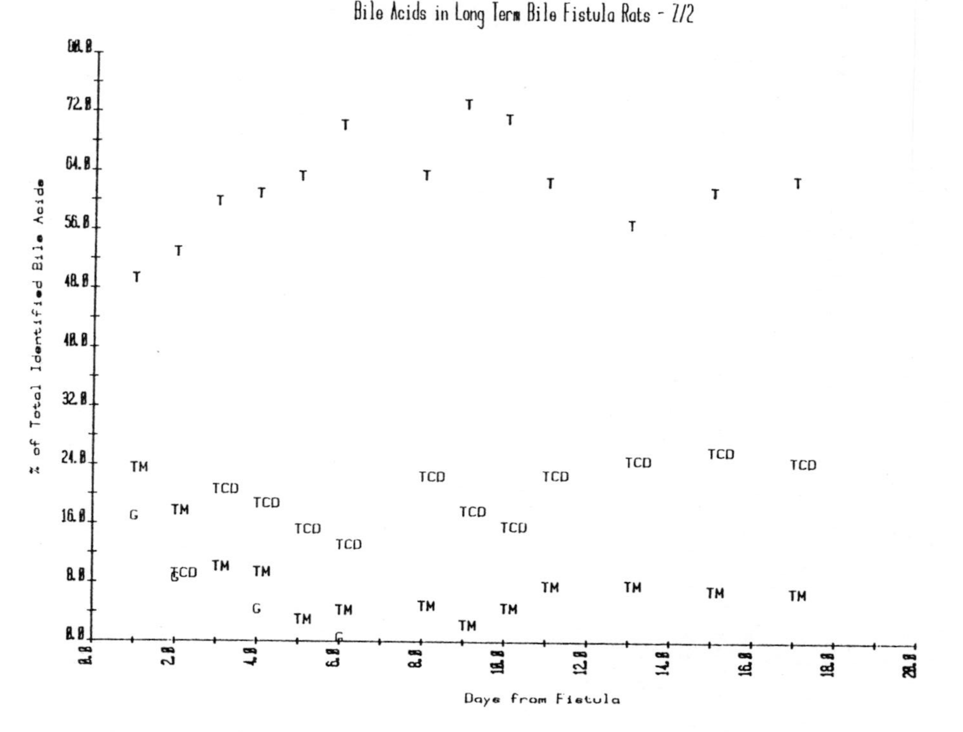

Figure 5. Plots of changes in percentage of total area under four major bile acid peaks over a 20-day period. T = taurocholic acid, TMC = tauromuricholic acid, TCD = taurochenodeoxycholic acid, G = glycocholic acid.

REFERENCES

1. H. Danielsson and J. Sjorall, Ann. Rev. Biochem., 233 (1975).
2. K. Uchida, I. Okuno, H. Takase, Y. Nomura, and M. Kadowaki, Lipids, 13, 42 (1978).
3. I. M. Yousef, G. Kakis, and M. M. Fisher, Can J. Biochem., 50, 402 (1972).
4. H. Greim, D. Trülzsch, P. Czygan, F. Hutterer, F. Schaffner, H. Popper, D. Y. Cooper, and O. Rosenthal, Ann. N.Y. Acad. Sci., 212, 139 (1973).
5. R. Shaw and W. H. Elliott, J. of Lipid Resc., 19, 783 (1978).
6. R. Shaw and W. H. Elliott, Lipids, 13, 971 (1978).
7. C. A. Bloch and J. B. Watkins, J. Lipid Resc., 19, 510 (1978).
8. M. N. Chavez and C. I. Krone, J. Lipid Resc., 17, 545 (1976).
9. J. C. Touchstone, R. E. Levitt, S. S. Levin, and R. D. Soloway Lipids, 15, 386 (1980).
10. B. Fisher, J. W. Zerbean, and H. M. Vars, Surgical Form, 407 (1953).
11. S. S. Levin, J. C. Touchstone, D. Y. Cooper, and H. M. Vars, Fed. Proc., 39, 1720 (1980).
12. H. Danielsson, K. Einarsson, and G. Johansson, Europ. J. Biochem., 2, 41 (1967).

CHAPTER 18

Comparison of TLC, HPLC and GC in Analysis of Bile Acids

Robert E. Levitt, Joseph C. Touchstone, and Roger D. Soloway

INTRODUCTION

Separation and quantitation of bile acids have been useful for study of the physiology and kinetics of the gallbladder and enterohepatic circulation in man. Rapid analysis of the conjugated bile acids as they are found in bile has been difficult until recently. Classical techniques of bile acid analysis employed hydrolysis and derivitization for analysis by gas chromatography, which may lead to loss of material and production of artifacts (1-3). Other methods for separation such as paper, ion exchange, and column chromatography as well as electrophoresis give inadequate separations. Radioimmunoassay has been used and appears to be quite sensitive, but there is considerable overlap in specificity of antisera for each bile acid (4). Finally, a fluorometric and enzymatic method for estimation of total serum or biliary bile acids is available and quite accurate, but this does not give any further information concerning the proportions of individual bile acid conjugates (5-7). The three presently available chromatographic methods that have been used most successfully are gas-liquid chromatography (GLC), high-performance liquid chromatography (HPLC), and, most recently, thin layer chromatography (TLC).

GLC has had wide application in studies of bile acids and gives accurate information, but the individual bile acid conjugates cannot be determined if this information is required. Unlike GLC, both TLC and HPLC are able to separate both the taurine and glycine conjugates of the three major bile acids found in bile: cholic (C), deoxycholic (DC), and chenodeoxycholic (CDC) acids. Furthermore, GLC requires chemical alteration of the bile acids before determinations can be accomplished, while TLC and HPLC do not. It is important that studies of bile acids and their metabolism involve analysis without hydrolysis because this may obscure the proportion of each individual conjugated bile acid.

HPLC has been successful in separating the conjugated bile acids (8, 9). Previous to the methods of Block and Watkins, methods not only involved multiple steps in extraction, purification, and chemical alteration, but also were unable to give complete resolution of the two dehydroxy bile acid conjugates, deoxycholic and chenodeoxycholic acids (10). As noted above, quantification had involved a complex, time-consuming approach. By using modern HPLC equipment and a mobile phase of water, methanol, and glacial acetic acid at a pH of 4.7, Block was able to give complete resolution of all six glycine and taurine conjugates of the major bile acids. The sensitivity they noted was less than that afforded by GLC or fluorometric techniques, but was sufficient for quantification of bile acids in duodenal bile.

Most recently, the use of TLC has been applied to separation of bile acid conjugates. A number of publications have appeared that describe methods that do not give complete resolution of all the individual bile acids (11-14). Attempts to use the mobile phase reported for HPLC with reverse-phase columns were unsuccessful when applied to TLC. Although separations occurred in some cases, resolution was such that scanning by densitometry was not possible. The first reported separation of the conjugated bile acids that made quantitation possible was developed by Touchstone and Levitt (15). This method employed two separate chromatographic systems, one that separated the three glycine bile acid conjugates (GC, GDC, and GCDC) and another that separated the three taurine conjugates (TC, TDC, and TCDC). For separation of the glycine conjugates continuous development was carried out on silica gel G layers using a mobile phase of chloroform, 2-butanone, and ethanol. Separation of taurine conjugates employed KC_{18} reverse-phase layers (12% carbon load) and a potassium dehydrogen phosphate-ethanol mobile phase with conventional development. Chromatograms were sprayed lightly with 10% sulfuric acid in ethanol and heated at 170° for 5 minutes in order to develop fluorescence. The spraying and short heating time result in the formation of stable fluorescence derivatives of bile acids. Scanning could then be performed using a Schoeffel model 3000 spectrodensitometer (Schoeffel Instrument Company, Westwood, New Jersey) equipped with a computer to convert the photomultiplier response to optical density units. The response was channeled into a density computer recorder to provide a continuous recording of the photomultiplier response. A modification of these systems employing two-dimensional development on a single plate in one dimension with silica gel and the other with reverse-phase (C_{18}) sorbents was then used to give separation of all six bile acids on a single chromatographic plate (16).

Finally, modification of the mobile phase to calcium chloride-ethanol made possible separation of all six conjugated bile acids on a single lane (17) with adequate resolution so that quantification by scanning as above was possible (Figure 1). Separation and quantification can be done within 3 hr of beginning analysis and is simple and inexpensive.

In order to compare the various chromatographic techniques for bile acid quantification, the analysis of human gallbladder bile samples was performed by TLC, GLC, and HPLC. Human gallbladder bile samples were obtained at surgery from patients undergoing cholecystectomy for cholelithiasis. Samples were diluted 1:100 in methanol and stored at -5°C until used. For unknown bile samples, total bile acid concentrations were determined by a fluorometric modification of Talalay's procedure (18). GLC was performed according to the method generally used, and HPLC was done according to the method of Block and Watkins (9). The results for 12 samples analyzed by all three methods are shown in Table I. For the purpose of comparison with GLC, which does not distinguish between glycine and taurine conjugates, the HPLC and TLC values for the two conjugates of each major bile acid were added together. Table II shows the correlation between TLC and HPLC for the conjugates.

Table III shows the correlation between methods for each bile acid. The correlations between GLC and TLC and between GLC and HPLC were highly significant for each bile acid, except for the bile acid DC, where there was not good correlation between the results obtained with TLC and HPLC. For cholic acid the methods best correlated were GLC and HPLC, and for CDC the GLC and TLC correlated well. Overall, the correlations between each method and the others were not different.

With its cost and convenience advantages over HPLC and GLC and its ability to easily separate all six conjugated bile acids, reverse-phase TLC is of considerable value. According to the present data, TLC correlates as well with either HPLC or GLC as

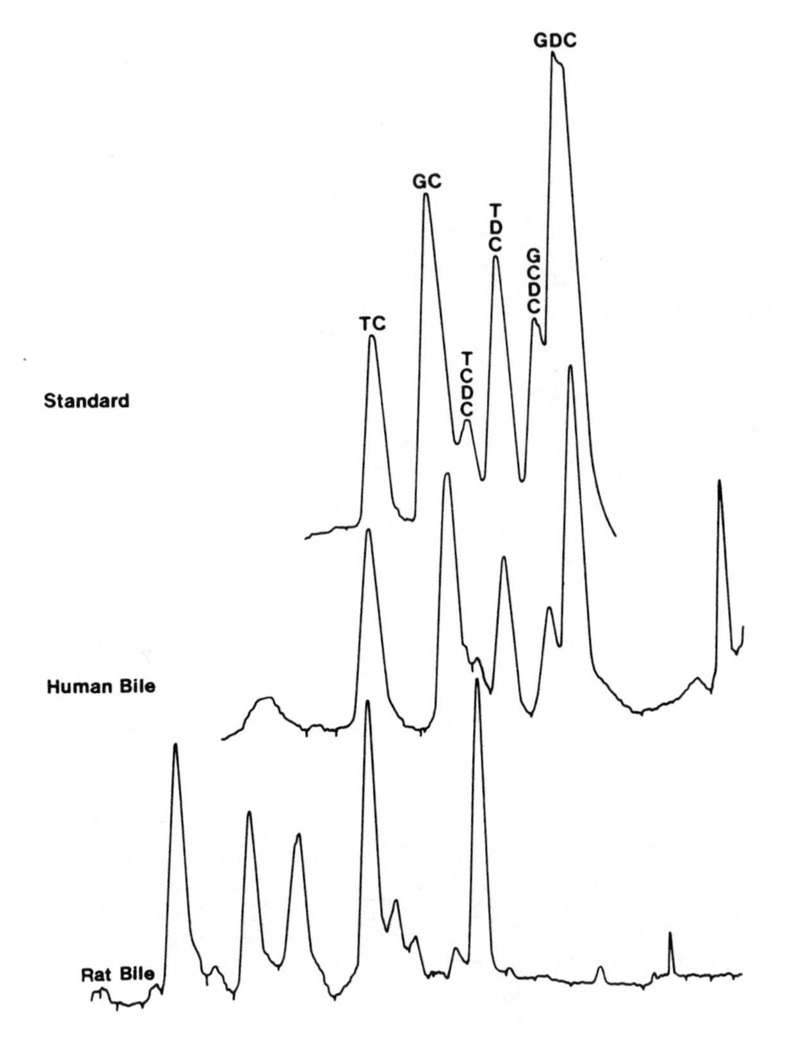

Figure 1. Separation of bile acid by reverse-phase TLC.

TABLE I. COMPARISON OF CHROMATOGRAPHIC METHODS FOR QUANTITATION OF BILE ACIDS

Sample Number	Cholic			Deoxycholic			Chenodeoxycholic		
	TLC	GLC	HPLC	TLC	GLC	HPLC	TLC	GLC	HPLC
1	41.5	40.4	47.3	1.1	1.5	0	66.1	64.8	50
2	23.9	19.3	29.5	7.7	17	13.5	53.7	38.2	30.7
3	54.6	26.8	31.5	24.7	49.5	46	51	52	32.8
4	76.1	56.7	74.6	0	7.2	0	72.3	81	53.8
5	118.2	64.3	71.9	26.1	62.1	71.9	54	70.5	22.7
6	57.3	47.5	58.6	27.1	22.3	19.9	23.8	36.8	29
7	51.8	19.5	40.8	27	50	38.1	24.9	30.1	20.3
8	34.0	13.2	15.7	22.1	43.1	41.4	28.9	19.8	15.9
9	51.7	43.2	42.8	17.2	7.3	0	46	64.1	47.6
10	67.7	24.9	36.3	25.5	37	12.1	37	50.6	46.3
11	49.9	26.4	10.2	16.6	14.3	3.6	22.7	7.3	5.6
12	67.2	67.8	62.7	0	8.7	33.5	49.5	39.7	37.6
	57.9	37.5	43.5	16.3	26.7	23.3	44.2	46.2	33.2
	6.9	5.4	5.9	3.2	5.9	6.6	4.8	6.3	4.5

TABLE II. CORRELATION BETWEEN TLC AND HPLC TESTS FOR INDIVIDUAL BILE ACIDS

Variable	Correlation
GC	0.723[a]
GDC	0.493
GCDC	0.629[b]
TC	0.596[b]
TDC	0.358
TCDC	0.086

[a] $p \leq 0.01$.
[b] $p \leq 0.05$.

either method does with the other. The correlation between the three methods is such that it is not clear that any one method is superior to the other for determining the true quantity of bile acid. Also, because of a lack of extremely high correlations between the methods, data obtained from studies using different methods are not directly comparable. Because of its low cost and simplicity, TLC may be favored over HPLC for measuring concentrations of individual bile acid conjugates.

REFERENCES

1. P. Eneroth and J. Sjovall, in The Bile Acids, Volume 1, P. O. Nair and D. Kritchevsky (Eds.), Plenum Press, New York, 1971, p. 21.
2. R. Shioda, P. D. S. Wood, and L. W. Kensell, J. Lipid Res., 19, 505 (1978).
3. G. Lepage, A. Fontaine, and C. C. Roy, J. Lipid Res., 19, 505 (1978).
4. L. M. Denes and G. Hepner, Clin. Chem., 22, 602 (1976).
5. G. M. Murphy, B. H. Billing, and D. N. Baron, J. Clin. Path., 23, 594 (1970).
6. B. A. Skalhegg, Scand. J. Gastroent., 9, 555 (1974).

TABLE III. CORRELATIONS BETWEEN DIFFERENT METHODS BY BILE ACID AND GROUP

	GLC versus TLC	GLC versus HPLC	TLC versus HPLC
Cholic	0.571^a	0.795^b	0.802^b
Deoxycholic	0.590^a	0.823^b	0.510
Chenodeoxycholic	0.625^b	0.740^a	0.640^c

$^a p \leq 0.01$.
$^b p \leq 0.001$.
$^c p \leq 0.05$.

7. O. Fausa and B. A. Skalhegg, Scand. J. Gastroent., 9, 249 (1974).
8. R. Shaw and W. H. Elliott, Lipids, 13, 971 (1979).
9. C. A. Block and J. B. Watkins, J. Lipid Res., 19, 510 (1978).
10. P. Eneroth and J. Sjovall, The Bile Acids, Volume 1, op cit., pp. 121-171.
11. M. N. Chavez and C. L. Krone, J. Lipid Res., 17, 545 (1976).
12. R. W. Shepard, P. S. Bunting, M. Kahn, and J. G. Hill, et al., Clin. Biochem., 11, 106 (1978).
13. A. K. Batta, G. Salem, and S. Shefar, Chromatogr. 168, 557 (1979).
14. A. F. Hofmann, J. Lipid Res., 3, 127 (1962).
15. J. C. Touchstone, R. E. Levitt, R. D. Soloway, and S. S. Levin, J. Chromatogr., 178, 566 (1979).
16. R. E. Levitt and J. C. Touchstone, J. High Res. Chromatogr., 2, 587 (1979).
17. J. C. Touchstone, R. E. Levitt, S. S. Levin, and R. D. Soloway, Lipids, 15, 386 (1980).
18. P. Talalay, Methods Biochem. Anal., 8, 119 (1960).

CHAPTER 19

Reproducibility and Precision of Quantitation of Skin Surface Lipids by TLC

Michael R. Ruggieri, Kenneth J. McGinley, James J. Leyden, and Joseph C. Touchstone

INTRODUCTION

Quantitative analysis of sebaceous lipids has played an important part in defining the role of sebum in many cutaneous diseases and in determining the effects of various therapeutic modalities on sebaceous gland function. Most of this analysis was done by gravimetric measurement of the ether-soluble material absorbed from the forehead onto a stack of cigarette papers over a 3-hr period (1). Although this technique is somewhat cumbersome for patients and subjects, it does provide fairly reproducible data (coefficient of variation 33%) on the amount of lipid removed from a defined area of a given subject over a 3-hr period.

Downing has described a thin layer chromatographic system (TLC) that separates all seven classes of skin surface lipids (2). Data as percentage of total were obtained by sulfuric acid spraying, slow charring, and adjusting photodensitometric peak areas for efficiency of charring according to percent carbon in each lipid class. Subsequently, this laboratory reported the use of methyl nervonate as an internal standard in the collection fluid, which then permitted quantitation by comparing the area of the internal standard with those of skin surface lipids (3). These workers used this procedure primarily to quantitate the amount of lipid from various body areas, and in subsequent publication (4, 5) have continued to employ the cigarette-paper technique whenever quantitative measurements were made.

It is relatively easy to collect skin surface lipids by extraction with a liquid solvent. It is advantageous to be able to simultaneously obtain both qualitative and quantitative measurements.

In this chapter we describe modifications of the Downing and the Green et al. (3) procedures which provide a simple, reproducible, precise technique for qualitative and quantitative determination of skin surface lipids.

Skin Sampling Technique

Sites were washed 3 hr prior to sampling using the following procedure. The skin site is washed with a gauze pad saturated with Triton X-100 for 30 seconds, followed with a 30-second tapwater rinse and then a 30-second washing with gauze saturated with hexane. The nonionic detergent was chosen since preliminary studies showed that the use of soaps causes deposition of fatty acids, which contaminate the skin lipids. To protect the site from external contamination during the replacement time, a perforated plastic weighing boat was glued to the skin as shown in Figure 1. The perforations prevented excessive sweating and translocation of the sebum secreted to the skin surface.

After 3 hr the plastic boat was removed and an open glass cylinder of 3.8 cm^2 diameter was held firmly against the site. Two ml of hexane containing methyl nervonate as an internal standard (50 µg for lipid-rich area such as the face, 25 µg for the trunk, and 10 µg for the extremities) was placed in the cup, using a volumetric pipette, and the site was scrubbed for 30 seconds using a blunted Teflon* policeman (Figure 1). The sample was then passed through a 0.45-µm millipore filter to remove skin debris and bacteria and placed in Teflon capped glass screw-top vials. The uncapped vials were dried in a vacuum oven at 40°C overnight, then capped and stored at -40°C until ready for chromatography.

Preparation of the Plates

Commercially prepared 20 × 20 cm glass TLC plates coated with a

---

*Dupont trademark.

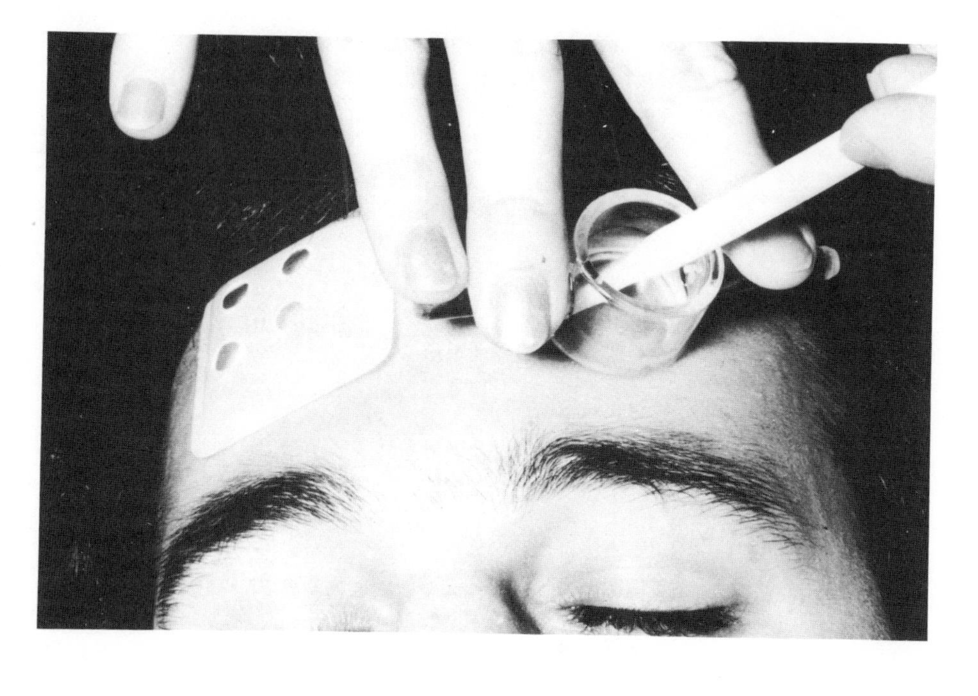

Figure 1. Skin surface extraction technique. A 3.8 cm^2 area of skin is extracted by scrubbing with hexane containing methyl nervonate as an internal standard. The other site is being protected by the inverted plastic weighing boat during the replacement time.

250-µm-thick layer of silica gel G (Analtech, Inc., Newark, Delaware) were used. The plates were scored in the same direction in which the plate was spread with a Schoeffel scoring device so that one 10-mm lane separated every four lanes of 5 mm each (Figure 2). The 10-mm lane was not spotted and served as a reference lane for the two lanes on either side of it during subsequent dual-beam scanning. Plates were cleaned by overnight development in chloroform/methanol 2:1 then activated at 130°C for 60 minutes.

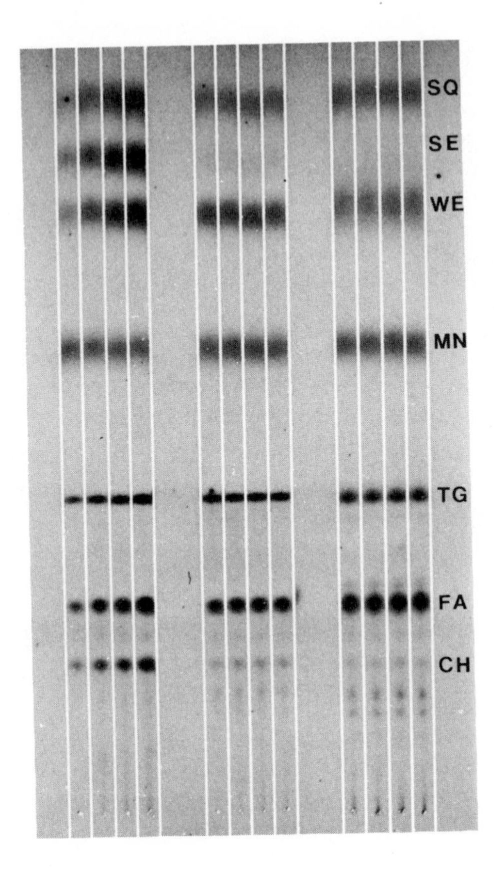

Figure 2. Thin layer chromatogram of natural and synthetic sebum. The photograph was deliberately overexposed to show faint spots. The wide unspotted lane served as the reference lane for the 2 lanes on either side of it during dual-beam densitometric scanning. Samples are, from left to right, 4 levels of external standards, 4 replicates of synthetic sebum, and 4 replicates of skin surface extracts. SQ = squalene, SE = steryl esters, WE = wax esters, MN = methyl nervonate, TG = triglycerides, FA = fatty acids, CH = cholesterol. Spots below cholesterol represent diglycerides and constitute less than 1% of the total.

External Standards

Five µl each of four calibration solutions was spotted on each TLC plate using Drummond microcapillary pipettes. The four external standard solutions were prepared such that 5 µl contained 0.5, 1.0, 1.5, and 2.0 µg of oleic acid, triolein, cetyl oleate, cholesterol oleate, and squalene; the amount of cholesterol in these being one half the others (Figure 3). All these calibration solutions were prepared in a stock solution of methyl nervonate in hexane so that 5 µl of each of the four calibration solutions contained the same amount of methyl nervonate (1 µg) serving as an internal standard. These concentrations of external standards were chosen since standard curves were found to be linear up to 4 µg for all components except cholesterol, which became nonlinear at 2 µg.

Preparation and Application of the Sample

Sample vials were removed from the freezer and redried by placing them in a vacuum oven at 40°C for 1 hr. A suitable volume of hexane was placed in the vial to make the concentration of total lipid close to 1 µg/ml. For lipid-rich areas like the face, the volume of hexane used was 0.2 ml. This was agitated to thoroughly dissolve any lipid adhering to the sides of the vial, and then 5 µl was spotted in the center of a 5-mm lane using Drummond microcapillary pipettes. A stream of air was passed across the surface of the plate to permit rapid evaporation of the carrier solvent. At no time was the zone created by the carrier solvent allowed to encroach upon the scored areas defining the sides of the lanes. This was done by touching the tip of the microcapillary pipette to the layer for short periods two or three times until all of the sample had been spotted within the edges of the lane.

Preparation of the Development Tanks

Plates were developed successively in hexane to 20 cm, in benzene to 20 cm, then in a mixture of ether : hexane : acetic acid (30:70:1) to 8 cm with a 10-minute air-drying time between each development. All tanks were lined with saturation pads on all four sides and equilibrated overnight before chromatography. This process promoted complete saturation with solvent vapors and maintained constant conditions. When the relative humidity rose above 40%, activated molecular sieve (Fisher) was placed in the development tanks to remove all traces of water from the atmosphere.

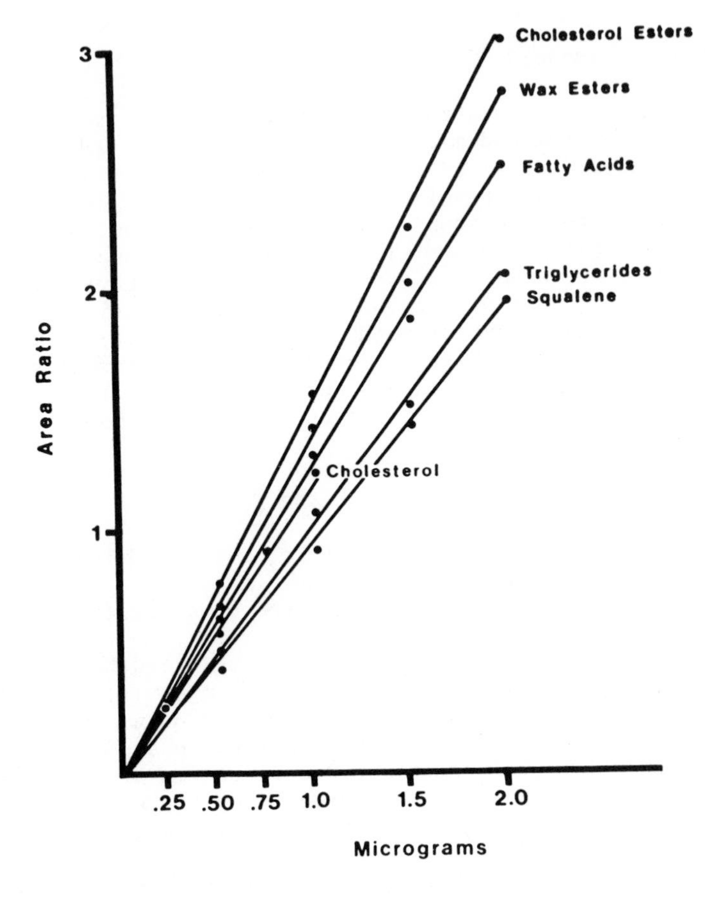

Figure 3. Standard curves for six components of skin surface lipids. Points are means of 10 determinations on separate TLC plates. Area ratio is the ratio of the area of the component to the area of the internal standard.

Charring the Plates

After the plate was developed in all three solvents and dried thoroughly for at least 10 minutes it was sprayed with 50% sulfuric acid. The spray, an even mist using a Supelco glass sprayer attached to a nitrogen tank, was applied uniformily to a point just short of visible wetting of the silica. This step was critical since too light spraying resulted in incomplete charring and too heavy a spray washed components into each other.

Immediately after spraying, the plate was placed on a 20 × 20 cm aluminum slab on top of a Thermolyne hot plate with an aluminum surface. The temperature of the hot plate was set at 600°F and timed for 40 minutes. During this time the plate heated slowly and produced successive decarboxylations, leaving something approaching elemental carbon at the end of the charring period. The temperature of the hot plate should not be too high since this would produce excessive warping and shattering of the glass TLC plates.

Densitometry

After charring, the plates were cooled to room temperature, and the back of the glass was carefully cleaned of any material picked up from the hot plate. The chromatograms were then scanned in the dual-beam mode using transmitted light on a Schoeffel SD 3000 spectrodensitometer interfaced with a Spectra Physics SP4100 computing integrator. The spectrodensitometer monocromator was set at 650 nm with the slit width completely open to allow the broadest bandpass possible. The dimensions of the light beams were 4 mm long and 0.5 mm wide.

Analysis of Data

A "synthetic sebum" mixture was prepared using known quantities of standards of a composition closely resembling that of typical skin surface extract and analyzed using three different methods of calculation.

Method 1--Carbon Density Factors

Peak areas for each component were adjusted by a factor derived by dividing the molecular weight of the respective molecule by the sum of the atomic weight of the carbon atoms in the molecule according to Downing (2). The peak area for cholesterol was reduced by the factor 0.66 as described by Downing. The amount of each component was calculated using the following formula:

$$\text{Amount of component } i = \frac{A_i \times CDF_i}{A_{mn} \times CDF_{mn}} \times \text{amount}_{mn}$$

A_i = area of component i, CDF_i = carbon density factor of component i, A_{mn} = area of methyl nervonate internal standard, CDF_{mn} = carbon density factor of methyl nervonate, and amount_{mn} = micrograms of methyl nervonate.

Method 2--Area Calibration

Standard curves were made by plotting the four concentrations of each of the external standards against the integrated peak area produced. The amount of each component was read directly from the standard curves without regard to the methyl nervonate internal standard.

Method 3--Area Ratio Calibration against External Standard

In this method standard curves were prepared by plotting the four concentrations of each of the external standards against the integrated peak area of the component divided by the methyl nervonate peak area. Since the amount of methyl nervonate in each of the calibration solutions was the same, this area ratio standard curve corrects for any errors in the amount of external standard applied to the plate during the spotting procedure. Figure 3 shows standard curves using this method for each of the six major components of the neutral lipids of the human skin. Each point represents the mean of 10 samples on separate TLC plates. As can be seen, all points lie along a straight line that was not found using the area calibration.

RESULTS

Table I displays representative data on analysis of synthetic sebum using the three different calculation methods. As can be seen, the area ratio method provided the best precision and reproducibility, with the carbon density method nearly as reproducible but not quite as precise. The area calibration method was the least reproducible and least precise of all

TABLE I. QUANTITATION OF SYNTHETIC SEBUM AND PRECISION BY THREE METHODS OF ANALYSIS

Constituents	R_f	Known	Observed[a]		
			Area Ratio	Carbon Density	Area
Cholesterol	0.09	0.20	0.25 (0.04)	0.22 (0.06)	0.25 (0.07)
Oleic Acid	0.19	2.00	1.83 (0.27)	1.76 (0.36)	1.38 (0.30)
Triolein	0.36	2.00	1.93 (0.14)	1.48 (0.35)	1.65 (0.30)
Cetyl Oleate	0.63	2.00	2.11 (0.17)	2.36 (0.27)	1.67 (0.40)
Cholesteryl oleate	0.72	0.20	0.21 (0.03)	0.26 (0.07)	0.20 (0.07)
Squalene	0.82	1.50	1.70 (0.30)	1.27 (0.16)	1.37 (0.26)
Mean		7.90	7.82 (0.63)	7.35 (1.10)	6.52 (1.30)

[a] Mean and standard deviation of eight replicates.

calculation methods. Table I represents data using only one amount of synthetic sebum. Results using three other amount of synthetic sebum provided comparable data to those displayed in Table I.

To test the applications of this method to the analysis of skin surface extracts, samples were obtained from eight young, white, male subjects on duplicate forehead and upper inner arm sites on six consecutive days. The reproducibility of the area ratio method of data calculation is shown in Table II. As can be seen, the variation of a given individual on a day-to-day basis was nearly twice that found for analysis of synthetic sebum in Table I.

DISCUSSION

The major modification of the method of Green et al. (3) involves the use of external standards on each TLC plate to calibrate for efficiency of charring rather than the use of carbon density factors (2). In addition the use of the internal standard corrects for errors in the amount of calibration standards spotted. These modifications minimize effects of plate-to-plate variations in the charring reaction and, as the results demonstrate, provide more reproducible and precise data. In addition, the above described method allows far less sample to be applied to the TLC plate. This simple principle, the application of small sample amounts (less than 2 μg of any one component) is generally applicable to densitometric analysis in TLC because it yields linear standard curves (4). As a result, extensive and complex mathematical manipulation of area measurements (5) to correct for deviations from the Lambert-Beer law (produced by spotting larger amounts) are unnecessary. The use of internal and external standards provides precision and reproducibility in the analysis of human skin lipids by TLC.

REFERENCES

1. J. S. Strauss and P. E. Pochi, J. Invest. Derm., 36, 293 (1961).
2. D. T. Downing, J. Chromatogr., 38, 91 (1968).
3. R. S. Green, D. T. Downing, P. E. Pochi, and J. S. Strauss, J. Invest. Derm., 54, 240 (1970).

TABLE II. REPRODUCIBILITY IN THE ANALYSIS OF SKIN SURFACE EXTRACTS

	Forehead[a]	Upper Inner Arm[a]
Cholesterol	1.79 ± 0.44	0.62 ± 0.19
Fatty Acids	10.30 ± 2.30	1.22 ± 0.69
Triglycerides	19.28 ± 4.08	3.63 ± 1.08
Wax Esters	24.68 ± 4.78	1.80 ± 0.65
Cholesterol Esters	3.88 ± 1.62	0.64 ± 0.33
Squalene	13.60 ± 3.16	1.22 ± 0.52
Total	75.70 ± 11.09	9.13 ± 1.75

[a]Mean ± standard deviation for 96 separate samples; eight subjects, duplicate sites sampled on six consecutive days. Micrograms lipid per square centimeter skin per 3 hr collection.

4. J. C. Touchstone, S. Levin, and T. Muzawec, Anal. Chem., 43, 858 (1971).
5. D. T. Downing and A. M. Stranieri, J. Chromatogr., 192, 208 (1980).

CHAPTER 20

Clinical Applications of Ion-Exchange TLC

Tibor Dévényi

INTRODUCTION

The determination of the amino acid content of biological fluids or tissue homogenates is a frequent task in clinical chemistry. Ion-exchange column chromatography (analyzer-technique) is most widely used because of its unique resolving power. This typical column chromatographic technique, however, does not permit the handling of several samples simultaneously. On the other hand, TLC seems to be the ideal method that--because of its elegant technical simplicity--permits the simultaneous analyses of large-scale series. Conventional TLC represents a much less effective resolving power, in the case of amino acid analyses of biological fluids, or similarly complicated mixtures and has to be used mostly as a two-dimensional chromatographic technique.

The combination of the two methods, that is, the preparation of a resin-coated chromatosheet (or plate), promised a new tool that may combine the advantages of both methods (1):

1. High resolving power.
2. Possibility of simultaneous handling of a large number of samples.

The commercially available, strongly acidic resin-coated chromato-sheet Fixion 50X8* contains a spherical, sulphonated polystyrol-

*Fixion ion-exchange chromatosheets are distributed by Chromatronix, Palo Alto, California.

divynilbenzene copolymer resin (in Na^+ form), similar to a Dowex 50 × 8 resin, with additional silica gel as an adjuvant, and an organic binder. The strongly basic anion exchanger Fixion 2X8 contains a polystyrol-divynilbenzene copolymer having quaternery amine side chains (in acetate form).

The Fixion ion-exchange chromatosheets can be employed with buffer solutions usually applied in ion-exchange column chromatography, but using the conventional technical aids and tools of TLC. Several one-dimensional separation methods were described for the amino acids (2-5), nucleic acid constituents (6-10), antibiotics (11), organic acids (12, 13), drugs (14), and so on, using Fixion chromatosheets. The method was also found to be suitable for large-scale screening purposes in clinical (15-19) and agricultural programs (20-22).

As far as clinical applications are concerned, a simple, one-dimensional separation method may have an important role in

1. Early detection of amino acid disorders.
2. Control of therapy (e. g., protein restriction).
3. Drug monitoring.
4. Activity assays of enzymes involved in amino acid synthesis or degradation.

ONE-DIMENSIONAL SEPARATION OF AROMATIC AND BASIC AMINO ACIDS

A sharp separation of the aromatic and basic amino acids can be obtained on the strongly acidic cation exchanger Fixion 50X8 (in Na^+ form) with a sodium citrate buffer, pH = 5.1 (2, 20). For buffer composition see Table I, and a typical chromatogram is shown in Figure 1.

ONE-DIMENSIONAL SEPARATION OF 16 AMINO ACIDS

Most of the protein-constituting amino acids can be separated with a single development on the strongly acidic resin-coated chromatosheet Fixion 50X8 (in Na^+ form) using a sodium citrate buffer pH = 3.3 (3). A typical chromatogram obtained with a slightly modified buffer (see Table I) described previously (3) is shown in Figure 2.

The improved buffer contains a slightly higher concentration of sodium, 0.1 mol/1 NaCl, and it also contains 5% ethylene glycol for lowering the buffer tension. Figure 3 shows a chromatogram

TABLE I. BUFFERS USED FOR THE ONE-DIMENSIONAL SEPARATION OF AMINO ACIDS AND RELATED COMPOUNDS

pH	Ingredients per 1000 ml		Application	Ref.
2.4	LiCl	12.7 g	$CySO_3H, MeSO_2$	24
	Li-citrate	28.2 g		
	HCl	15 ml[a]		
3.09	Citric acid	84 g	Diaminopimelic acid	31
	NaOH	16 g		
	HCl	12 ml[a]		
3.75	Citric acid	42.0 g	Homocystine	Unpublished
	NaOH	8.0 g		
	NaCl	35.1 g		
	HCl	5 ml[a]		
3.3	Citric acid	84.0 g	One-dimensional separation of amino acids	3
	NaOH	16.0 g		
	HCl	4 ml[a]		
	Improved buffer: ingredients as above, plus			
	NaCl	5.8 g		
	ethylenglycol	50 ml		
4.2	Citric acid	14.1 g	Ornithine from lysine	5
	NaOH	8.0 g		
	NaCl	11.7 g		
	HCl	7.7 ml		
4.4	Citric acid	14.1 g	D-Penicillamine	26
	NaOH	8.0 g		
	NaCl	5.85 g		
	HCl	5 ml[a]		
5.1	Citric acid	14.1 g	Aromatic and basic amino acids	20 (2)
	NaOH	8.0 g		
	NaCl	14.0 g		
	HCl	4.4 ml		
6.0	Citric acid	7.0 g	Tryptophan	4
	NaOH	4.0 g		
	NaCl	87.75 g		

Table I (continued)

pH	Ingredients per 1000 ml		Application	Ref.
6.0	Citric acid	7.0 g	Biogenic amines	32
	NaOH	4.0 g		
	NaCl	134.4 g		

[a]The exact amount of HCl (37%, sp.g. 1.19) necessary for the adjustment of the pH should be checked by a pH meter.

obtained by identical experimental conditions to the one in Figure 2, demonstrating that a heavy overloading of the Fixion chromatosheet (20 samples) does not seriously affect the resolving power of the ion exchanger.

The variation of the buffer composition (pH, molarity, additives, etc.) permit the accomplishment of several special tasks, for instance, the determination of tryptophan (4), asparagine, and glutamine (23), and the separation and determination of the S-containing amino acids in performic acid-treated samples (24).

Detection of Phenylketonuria (PKU) by the Fixion Technique

The one-dimensional separation of the aromatic and basic amino acids (see Figure 1) is a simple procedure. The commercially available chromatosheets (Fixion 50X8) are used directly, without any pretreatment (e. g., equilibration). Samples containing salt or acid can be spotted directly onto the chromatosheets, and chromatography can be performed at room temperature. By these simple technical conditions almost any of the most common amino acid disorders such as PKU, tyrosinemia, histidinemia, lysinemia, and different forms of the maple syrup urine disease, can be detected from a single development.

This procedure is used routinely at the Section of Pediatrics, Péterfy Hospital, Budapest, Hungary and by the Apáthy Children's Hospital, Budapest, Hungary. Figure 4 illustrates the detection of three PKU cases on 10 × 10 cm Fixion 50X8 chromatosheets, from blood samples dried on filter paper. The two children showing the highly elevated Phe concentrations in Figure 4b are first cousins.

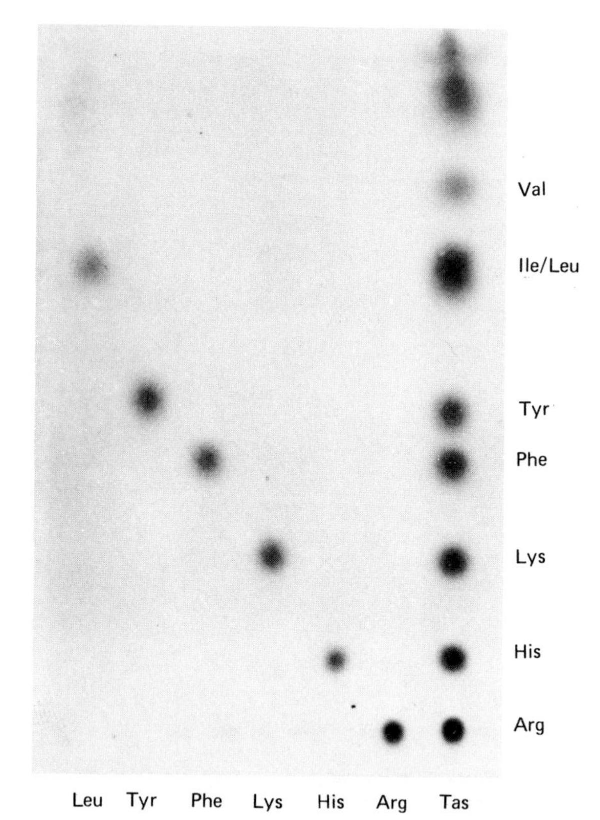

Figure 1. One-dimensional separation of aromatic and basic amino acids on strongly acidic resin-coated chromatosheet. Chromatosheet: Fixion 50X8 (Na^+ form). Eluting buffer: sodium citrate pH = 5.1, Na^+ = 0.44 mol/l, citrate = 0.07 mol/l. Chromatography at room temperature. Detection: ninhydrin-cadmium spray reagent.

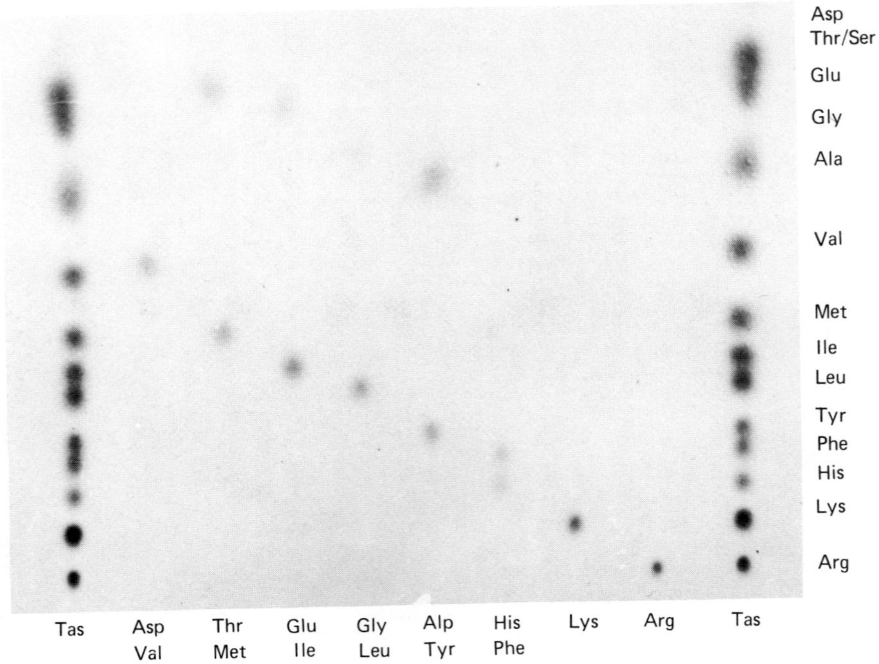

Figure 2. One-dimensional separation of amino acids on strongly acidic resin-coated chromatosheet. Chromatosheet: Fixion 50X8 (Na^+ form). Eluting buffer: sodium citrate pH = 3.3, Na^+ = 0.5 mol/1, citrate = 0.4 mol/1, 5% ethylene glycol Chromatography at 45°C. Detection: ninhydrin-cadmium spray reagent containing 1% pyridine. Before use the chromatosheet has to be prewashed by continuous development with 30-fold diluted buffer.

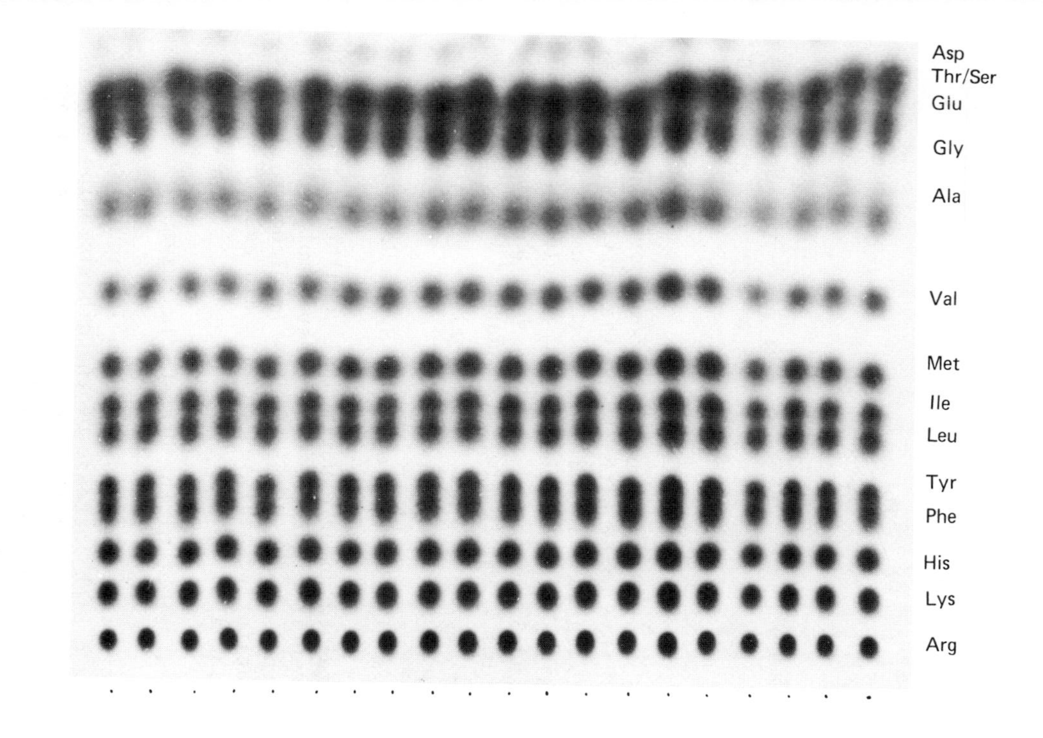

Figure 3. Overloading of a Fixion 50X8 chromatosheet with the application of twenty samples. Experimental conditions as in Figure 2. Manual spotting was performed with 1-μl samples.

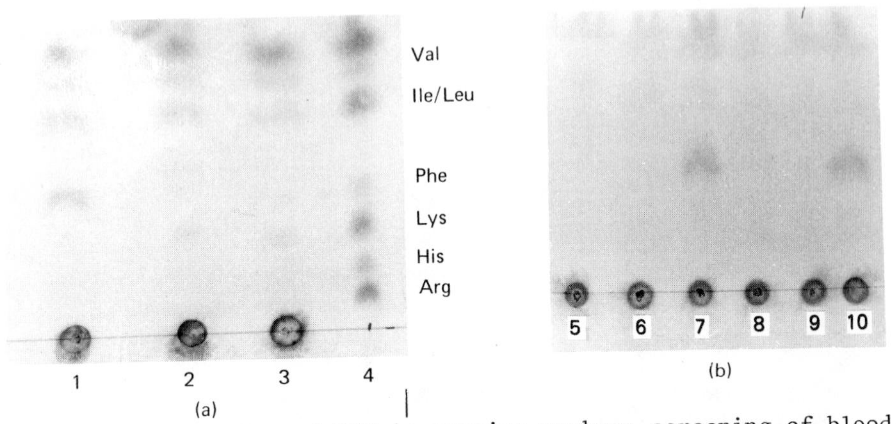

Figure 4. Detection of PKU in routine newborn screening of blood samples dried on filter paper with TLC on Fixion chromatosheets. Chromatosheet: Fixion 50X8 (Na^+ form), 10 × 10 cm in size. Experimental conditions as in Figure 1. (a) Sample 1 is a newborn baby with PKU; 2 and 3 are normal samples, 4 is an amino acid mixture. (b) Samples 7 and 10 are first cousins with PKU. Other samples are normal.

Detection of a Mild Variant of Maple Syrup Urine Disease

A chromatogram obtained under experimental conditions similar to those above is shown in Figure 5. The first sample shows an increase of the branched-chain amino acids (isoleucine, leucine, and valine). Quantitative analyses confirmed a case of a mild variant of MSUD (18, 19).

Detection of a Defect of the Urea-Cycle Enzymes

The chromatogram of urine samples is shown in Figure 6. One of the samples showed a high increase of the dibasic amino acids, which indicated the defect of the urea cycle. A liver biopsy, followed by enzymatic analyses, confirmed the disorder of the carbamoyl-phosphate synthetase enzyme (25). As is apparent in Figure 6, the excretion of the dibasic amino acids has been normalized after protein restriction.

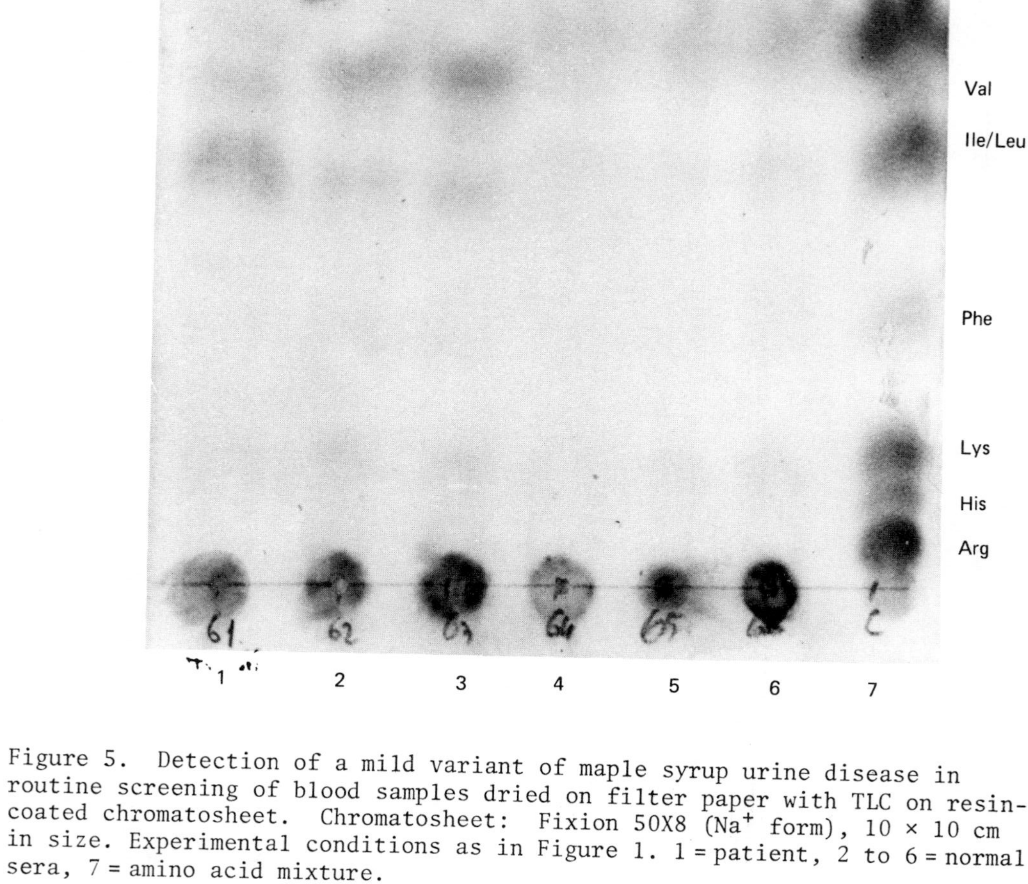

Figure 5. Detection of a mild variant of maple syrup urine disease in routine screening of blood samples dried on filter paper with TLC on resin-coated chromatosheet. Chromatosheet: Fixion 50X8 (Na^+ form), 10 × 10 cm in size. Experimental conditions as in Figure 1. 1 = patient, 2 to 6 = normal sera, 7 = amino acid mixture.

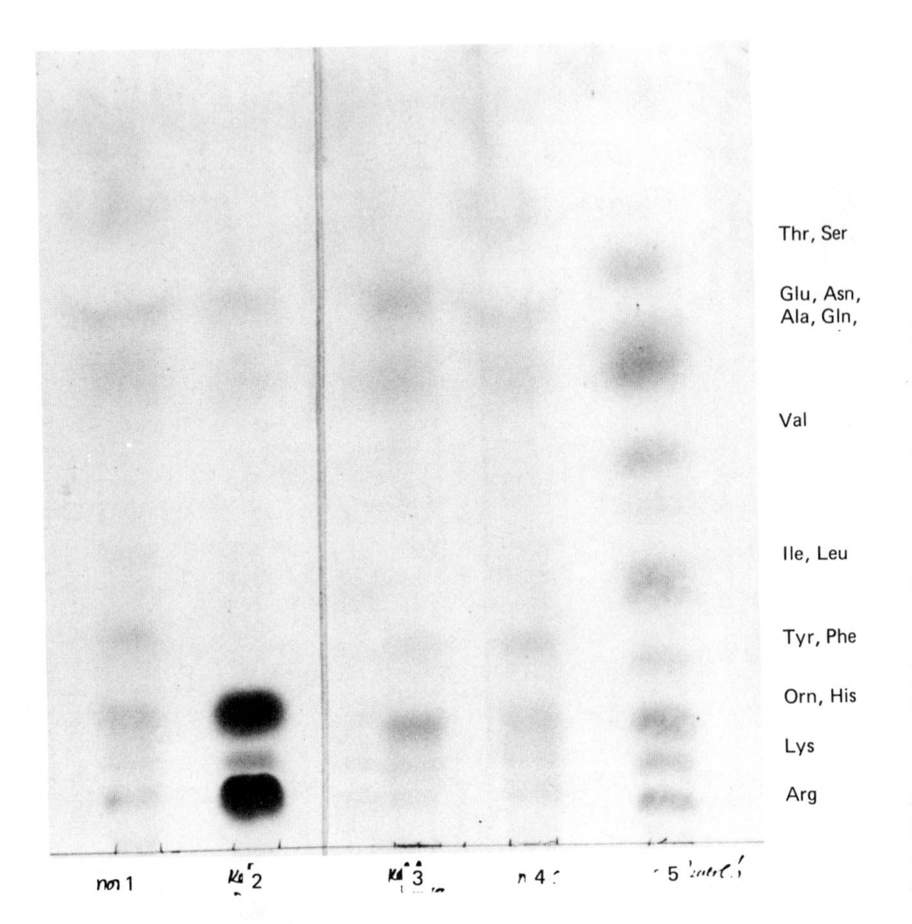

Figure 6. Chromatography of urine samples of a patient before and after protein restriction, suffering carbamoyl-phosphate-synthetase deficiency. Chromatosheet: Fixion 50X8 (Na^+ form). Eluting buffer: sodium citrate pH 3.3 (see Figure 2). Chromatography was performed without prewash, at room temperature. (This is causing incomplete separation.) Samples 1 and 4 are normal urine samples. Sample 2 is patient's urine before protein restriction. Sample 3 is patient's urine after protein restriction Sample 5 is an amino acid mixture.

Drug Monitoring by the Fixion Technique

A good example of drug monitoring is given by the determination of D-penicillamine (DPa) in blood samples dried on filter paper (26). DPa, a metabolite of penicillin was found to be effective in the treatment of various chronic and other diseases. During therapy drug monitoring (determination of DPa concentration in sera) can be performed by the Fixion technique using cation-exchanger chromatosheets with a sodium citrate buffer pH = 4.4, eluting at +4°C. Typical chromatograms are shown in Figure 7. Figure 7a shows a chromatogram of blood samples of a patient treated with a drop infusion, while Figure 7b represents a chromatogram of blood samples from a patient treated with a single i.v. injection. The elimination of the drug can be quantitatively evaluated (26).

Enzyme Assays Using Quantitative Fixion Technique

Quantitative ion-exchange TLC can be used with reasonable accuracy for the measurement of activity of enzymes catalyzing the synthesis or degradation of amino acids (27). The main advantage of this method is the fact that it can be applied in the case of crude tissue homogenates or extracts prepared from biopsy or necropsy material, when conventional (spectrophotometric) methods may fail because of turbidity and so on. Homogenates or extracts can be analyzed in reaction mixtures of 50 to 100 μl volume, analyzing 5- to 10-μl aliquots at different time intervals with ion-exchange TLC. It should be pointed out that from such a crude reaction mixture aliquots can be spotted directly onto the chromatosheet without affecting the separation. An appropriate method is employed for quantification. Several enzyme assays were achieved recently. These include the measurement of arginase, enzymes of the urea cycle, enzymes involved in polyamine, and pyrimidine synthesis, dehydrogenases, transferases, ammonia lyase (27-30). The activity measurements of some urea cycle enzymes are shown in Table II.

SUMMARY

A simple method is presented for the one-dimensional separation of amino acids by ion-exchange TLC. Using a commercially available, strongly acidic, cation-exchanger resin-coated chromatosheet (Fixion 50X8, Na^+ form), the amino acid composition and content of biological fluids (blood, sera, urine) and tissue

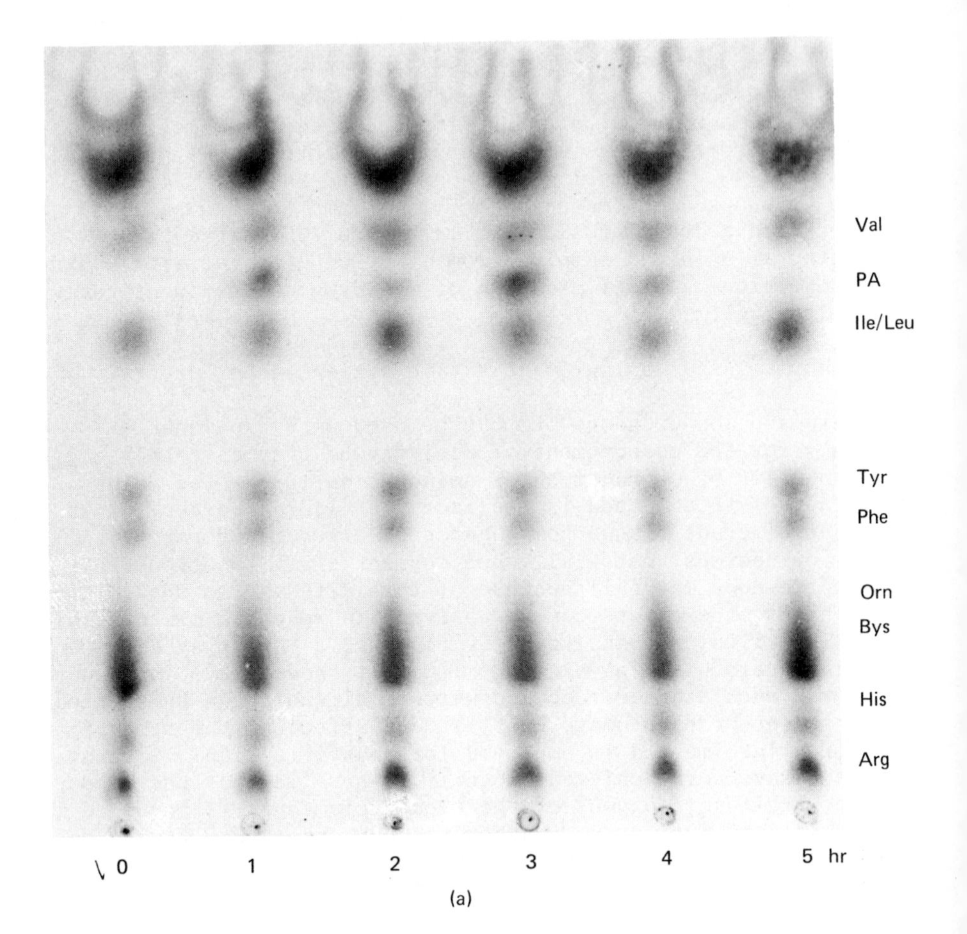

Figure 7. Drug-monitoring by ion-exchange TLC on resin-coated chromatosheets: Determination of penicillamine in blood samples. Chromatosheet: Fixion 50X8 (Na^+ form). Eluting buffer: sodium citrate pH = 4.4, Na^+ = 0.3 mol/l, citrate = 0.07 mol/l. Chromatography at +4°C. Before use the chromatosheets should be prewashed by continuous development with 30-fold diluted buffer.

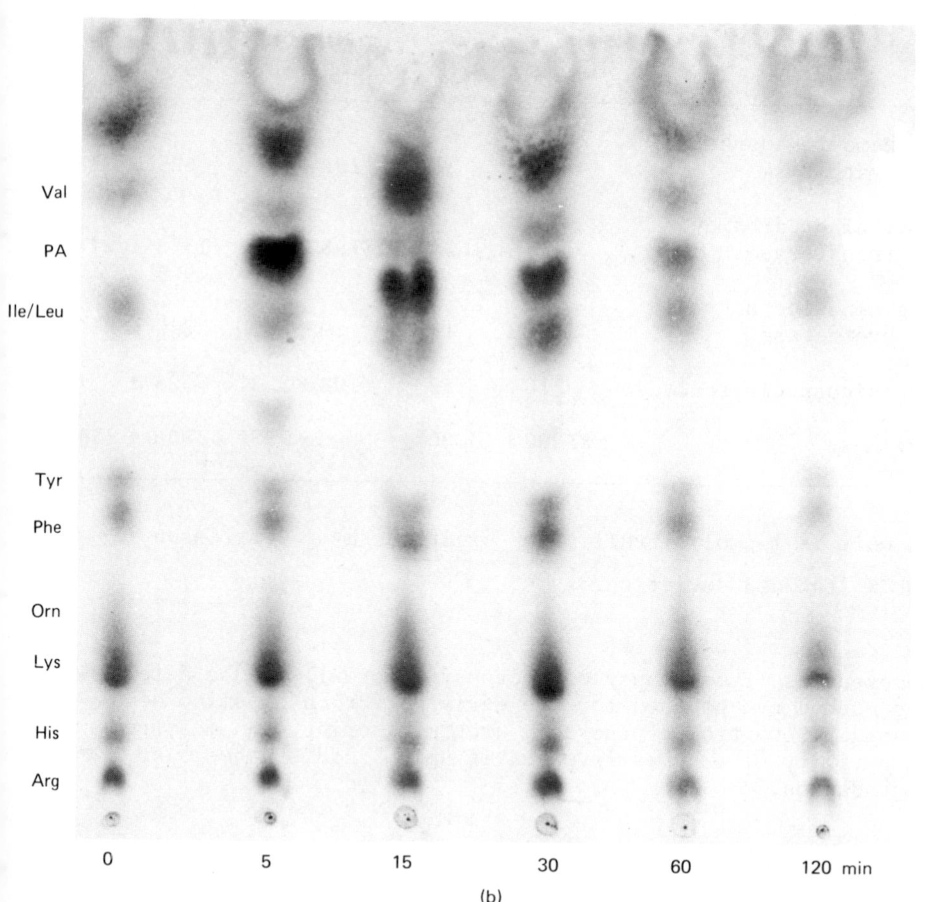

Figure 7 (continued). (a) Penicillamine administration by drop infusion (100 mg/kg body weight). (b) Penicillamine administration by a single i.v. injection (100 mg/kg body weight).

TABLE II. ACTIVITY OF UREA-CYCLE ENZYMES DETERMINED BY THE CV TECHNIQUE

Enzymes	Source of Enzymes: Human Liver, From Biopsy (units ± SD[b])	Human Liver, From Necropsy (units ± SD[b])	Rat Liver (units ± SD[b])
Carbamoyl-phosphate synthetase	264 ± 54	192 ± 82	625 ± 134
Ornithine carbamoyl-transferase	6178 ± 1634	3758 ± 1076	1373 ± 2483
Argininosuccinate synthetase	87 ± 18	39 ± 10	225 ± 28
Argininosuccinate lyase	216 ± 32	98 ± 21	323 ± 63
Arginase	83000 ± 13000	35000 ± 7850	127000 ± 23000

[a] 1 unit is 1 μmol citrulline or ornithine/hr g wet tissue (29).

[b] SD = standard deviation.

homogenates (from biopsy or necropsy material) can be determined. The method can be used for the early detection of amino acid disorders, control of therapy, drug monitoring, and measurement of activity of enzymes involved in amino acid synthesis or degradation.

ACKNOWLEDGEMENTS

The author is indebted to Dr. Judit Kovács (Section of Pediatrics, Péterfy Hospital, Budapest,) and Dr. P. Kiss (Apáthy Children's Hospital, Budapest, Hungary) for the kind permission to present chromatograms from their routine screening program. The author wishes to thank Professor P. Elődi and Dr. T. Karsai (Department of Biochemistry, University Medical School, Debrecen, Hungary) for their help and valuable discussions.

REFERENCES

1. J. Báti, I. Bihari, T. Dévényi, T. Palágyi, and S. Zoltán, Hungarian Patent 3937; Chem. Absti., 56155 (1972).
2. T. Dévényi, Acta Biochim. Biophys. Acad. Sci. Hung., 5, 435 (1970).
3. T. Dévényi, I. Hazai, S. Ferenczi, and J. Báti, Acta Biochim. Biophys. Acad. Sci. Hung., 6, 385 (1971).
4. T. Dévényi, J. Báti, and F. Fábián, Acta Biochim. Biophys. Acad. Sci. Hung., 6, 133 (1971).
5. A. Hrabák and S. Ferenczi, Acta Biochim. Biophys. Acad. Sci. Hung., 6, 383 (1971).
6. J. Tomasz, J. Chromatogr., 84, 208 (1973).
7. J. Tomasz and T. Farkas, J. Chromatogr., 107, 396 (1975).
8. J. Tomasz, Anal. Biochem., 68, 226 (1975).
9. J. Tomasz, Chromatogr., 13, 345 (1980).
10. J. Tomasz, J. Chromatogr., 169, 466 (1979).
11. J. Kádár Pauncz, J. Antibiot. (Japan), 25, 677 (1972).
12. S. Ferenczi and T. Dévényi, Acta Biochim. Biophys. Acad. Sci. Hung., 6, 389 (1971).
13. A. Himoe and R. W. Rinne, Anal. Biochem., 88, 634 (1978).
14. A. Dávid and T. E. Takacsy, J. Pharm. Pharmacol., 27, 774 (1975).
15. T. Dévényi, J. Báti, J. Kovács, and P. Kiss, Acta Biochim. Biophys. Acad. Sci. Hung., 7, 237 (1972).
16. S. Pongor, J. Kovács, P. Kiss, and T. Dévényi, Acta Biochim. Biophys. Acad. Sci. Hung., 13, 117 (1978).
17. J. Kovács, Acta Paediatr. Acad. Sci. Hung., 14, 165 (1973).
18. J. Kovács and P. Kiss, Acta Paediatr. Acad. Sci. Hung., 19, 137 (1978).
19. J. Kovács, Acta Biochim. Biophys. Acad. Sci. Hung., 14, 119 (1979).
20. T. Dévényi, Acta Biochim. Biophys. Acad. Sci. Hung., 11, 1 (1976).
21. B. Georgi, E. G. Hiemann, R. D. Brock, and H. Axmann, Int. Atomic Energy Agency, Vienna, IAES-SM 230/81, 1, 311 (1979).
22. B. Sigurbjörnsson, R. D. Brock, and T. Hermelin, A Joint FAO/IAEA GSF Program on Grain Protein Improvement, Int. Atomic Energy Agency, Vienna, IAES-SM 230/86, 1, 387 (1979).

23. A. Váradi, J. Chromatogr., 110, 166 (1975).
24. A. Váradi and S. Pongor, J. Chromatogr., 173, 419 (1979).
25. J. Kovács and P. Kiss, in preparation.
26. S. Pongor, J. Kovács, P. Kiss, and T. Dévényi, Acta Biochim. Biophys. Acad. Sci., 13, 123 (1978).
27. P. Elődi and T. Karsai, J. Liq. Chromatogr., 3, 809 (1980).
28. T. Karsai and P. Elődi, Acta Biochim. Biophys. Acad. Sci. Hung., 14, 123 (1979).
29. T. Karsai, A. Ménes, J. Molnár, and P. Elődi, Acta Biochim. Biophys. Acad. Sci. Hung., 14, 133 (1979).
30. T. Karsai, A. Ménes, J. Molnár, and P. Elődi, Acta Biochim. Biophys. Acad. Sci. Hung., 13, 181 (1979).
31. S. Pongor and K. Baltiner, Acta Biochim. Biophys. Acad. Sci. Hung., 15, 1 (1980).
32. S. Pongor, J. Kramer, and E. Ungár, J. High Resol. Chromatogr. & Chromatogr. Comm., 3, 93 (1980).

CHAPTER 21

Determination of Urinary Free Cortisol and Cortisone by Sequential TLC and HPLC

Marvin L. Lewbart and Robert A. Elverson

INTRODUCTION

It is generally agreed that the diagnosis of adrenocortical hyperfunction is best accomplished by measurement of urinary free cortisol (1, 2). Most methods presently available, based either on competitive protein-binding (CPB) or radioimmunoassay (RIA) (3), were derived from procedures for blood cortisol. Although not without their limitations (4), the radioisotope-based methods are generally satisfactory for estimation of plasma or serum cortisol levels. However, based on experience detailed in the present paper and from unpublished data (L. Haeffner and M. Lewbart), it was concluded that this technique lacks the necessary specificity for measurement of this steroid in urine, presumably because of both steroidal and nonsteroidal constituents that cross-react with the antiserum for cortisol. This problem has not been addressed previously in the literature, because critical comparisons between radioisotope and HPLC-based procedures have not been made.

The wide, published range of normal values from 0 (5) to 500 μg/(24 hr) (6) frequently makes difficult the diagnosis of adrenal hyperfunction and makes virtually impossible a diagnosis of hypofunction. Indeed, a recent monograph on the adrenal cortex states that "It is not advisable to attempt to measure adrenal hypofunction directly with any of the urinary baseline measurements of the corticosteroids since the difference between a low normal and a definitely decreased value is not readily discernible" (7). Although a number of published procedures for blood cortisol are available that utilize high-performance liquid chromatography (HPLC) (8-10), there are no comparable methods for urinary free cortisol. Older methods involving double-isotope labeling and sequential paper chromatography are both complex and laborious.

Urinary free cortisone excretion has been studied to a very limited extent (11) despite its free but variable interconversion with cortisol in vivo.

A new method for urinary free cortisol and cortisone that involves preliminary sorption on and desorption from an XAD-2 ion-exchange resin cartridge, purification of the eluate by thin layer chromatography (TLC), and quantitation by HPLC in a reverse-phase system is described here.

MATERIALS AND METHODS

A DuPont Prep I Automated Sample Processor supplied with Type W Cartridges (Du Pont Instrument Products Division, Wilmington, Delaware) on Program Number Four was used. The operation was carried out with filtered, dehumidified, compressed air. The wash solvent was methanol/water (60:40, v/v); the eluting solvent was acetone.

For TLC, E. Merck Silica Gel 60 F254 sheets (E. Merck, Darmstadt, West Germany) were used. The sheets were scribed to 1.0-cm channels and washed with acetone for 24 hr. The solvent system was acetone/toluene (50:50 v/v), and the development by ascending development took 20 minutes.

A DuPont Model 850 Liquid Chromatography System equipped with an 860 Absorbance Detector set at 254 nm was employed. The sensitivity routinely employed was 0.005 absorbance units full scale. The analytical column (DuPont Zorbax ODS) measuring 4.6 mm × 25 cm, was connected to a Whatman Co: Pell ODS Guard Column. The mobile phase of methanol/water (50:50 v/v) was pumped at 2.0 ml/min at 50°C.

SOLVENTS AND STANDARDS

All organic solvents were of "glass-distilled" grade (Burdick and Jackson Laboratories, Inc., Muskegon, Michigan).

Reference steroids prednisone (Δ'E), cortisol (F), cortisone (E), and 21-acetoxy-11β,17,20α-trihydroxypregn-4-en-3-one ("triol acetate") were obtained from the Upjohn Company (Kalamazoo, Michigan). The reference steroid, 17-hydroxypregn-4-ene-3,11,20-trione (21-deoxycortisone) was obtained from Sigma Chemical Company (St. Louis, Missouri). Stock solutions of 200 ng/μl in methanol included cortisol, cortisone, prednisone (internal standard), and 21-deoxycortisone (external standard). Addition of 0.5% acetic acid to these stock solutions afforded stability for at least six months at 4°C (10). The TLC standard, triol acetate, was prepared in methanol (2 μg/μl). Working standards, prepared by diluting 50 μl of stock solution to 10 ml with water, were prepared daily. Urine specimens were collected from normal subjects for 24 hr without preservative. Aliquots were stored at -20°C until assay. Creatinine levels were determined by the Jaffe alkaline picrate method to ensure completeness of collection.

PROCEDURE

Prep I Extraction

All samples were run in duplicate. To each Type W cartridge containing a filtration bed of glass microbeads was added 1 ml of urine, 1 ml of wash solvent, and 100 ng of internal standard in 100 μl of water. Each run of 12 samples included one normal control urine to which internal standard was added. An additional 1 ml of wash solvent was dispensed during the extraction. After drying, the acetone eluate residue in the aluminum recovery cup was stored at -20°C.

TLC Step

The eluate residue was applied manually with a micropipet in 50 μl of chloroform/methanol (50:50, v/v) followed by a 25-μl rinse.* Flanking TLC standard (1 to 2 μg) was also applied. Following development and drying in air, the standard spots (R_f 0.32) were located and marked while exposed briefly to a 254-nm source. After a zone 1 cm above and below their centers was delineated, the silica gel was removed immediately with an

---

*Preliminary experience with an 18-channel TLC Multi-Spotter (Analytical Instrument Specialties, Libertyville, Illinois) has shown this instrument to afford comparable recoveries while speeding up considerably this otherwise time-consuming step.

Analtech Adsorbent Scraper (Analtech, Inc., Newark, Delaware). Desorption was effected by vortexing the powder in a small, tapered test tube with 1 ml of methylene chloride/methanol (90: 10, v/v). The suspension was filtered by suction through a 2-ml- fine, scintered glass funnel. An additional 1 ml of eluting solvent was used as a rinse. The eluate was evaporated in a nitrogen stream in a water bath at 40°C. The residues were stored at -20°C.

HPLC

To the TLC eluate residue was added 200 µl of mobile phase containing 200 ng of external standard, and an aliquot was injected through a 50-µl loop.

RESULTS

Retention times for triol acetate, internal, and external standards, cortisol and cortisone are given in Table I. A number of possible interfering steroids were carried through the procedure, namely, progesterone, 17-hydroxyprogesterone, 11-deoxycorticosterone, corticosterone, 11-dehydrocorticosterone, 11-deoxycortisol, aldosterone, testosterone, androstenedione, estrone, estradiol, estriol, equilenin, prednisolone, methylprednisolone, dexamethasone, and triamcinolone. With the exception of prednisolone and aldosterone, none of these compounds interfered with the analysis because they either exhibited different retention times by HPLC or were eliminated in the TLC step.

Standard curves for cortisol and cortisone showed linearity from 0.5 to 3000 ng (Figure 1). Peak heights of cortisol and cortisone from urine samples served as the means of quantitation.

Correction for loss of the analytes was made by comparing the peak height of recovered internal standard with that obtained by direct injection. Correction for evaporation and loop-injection variation was made with the external standard in the same manner. Values were expressed as micrograms per 24 hr per gram of creatinine. The overall percentage recovery of 100-ng amounts of steroids added to 11 replicate urine samples was 77 ± 6.06 (mean ± SD) for internal standard, 78 ± 6.69 for cortisol, and 75 ± 8.14 for cortisone.

Precision was determined on a normal urine. Intraday variability (n = 12) was 5.8% (15.3 ± 0.9) for cortisol and 5.2% (52.3 ± 2.7) for cortisone. Interday variability (n = 35) was

TABLE I. RETENTION TIME FOR STEROID STANDARDS AND ANALYTES

Steroid	Retention Time (Min)
Prednisone	5.1
Cortisone	5.5
Cortisol	6.2
Triol acetate	6.6
21-Deoxycortisone	10.0

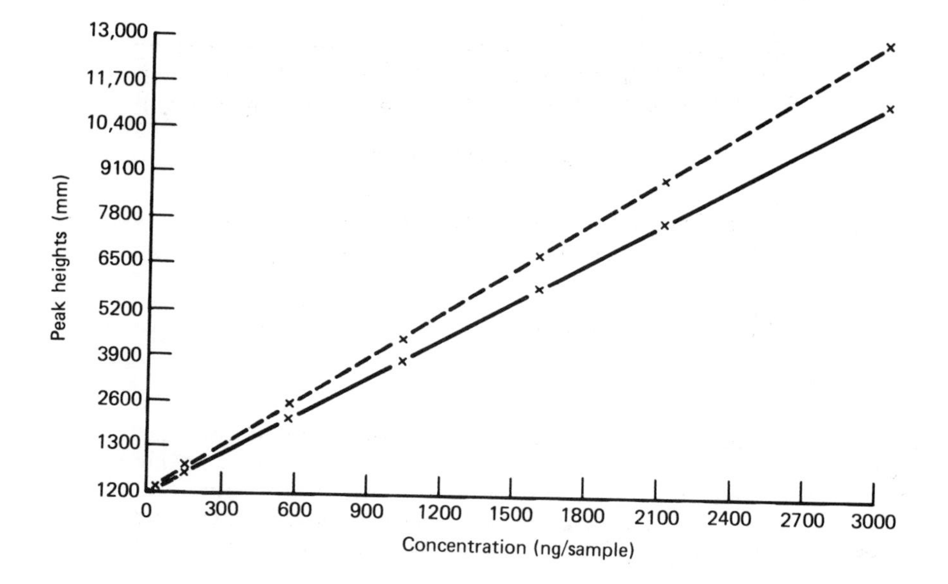

Figure 1. Standard curves of chromatographic peak heights versus cortisol concentrations [$y = 4.182x - 17.575$; $r = 1.00$] and cortisone (--x--) concentrations [$y = 3.693x - 11.358$; $r = 1.00$].

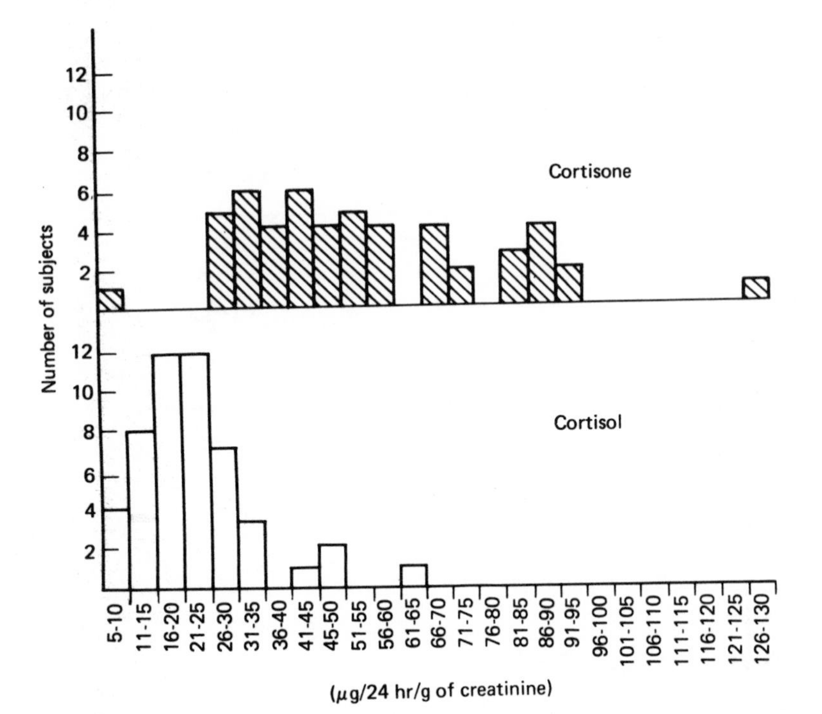

Figure 2. Histogram of urinary free cortisone and cortisol (μg/24 hr g of creatinine) in 51 normal subjects.

17% (19.2 ± 3.3) for cortisol and 13% (51.4 ± 6.8) for cortisone.

HPLC

In 51 normal adults (30 females) the urinary free cortisol excretion was 22 ± 9 (mean ± SD) μg/24 hr g of creatinine and 54 ± 25 for urinary free cortisone. As shown in Figure 2 the frequency distribution of cortisol values approximates a typical Gaussian curve, while the cortisone values are more randomly dispersed.

Radioimmunoassay

Aliquots of the 51 normal urines were subjected to RIA for urinary

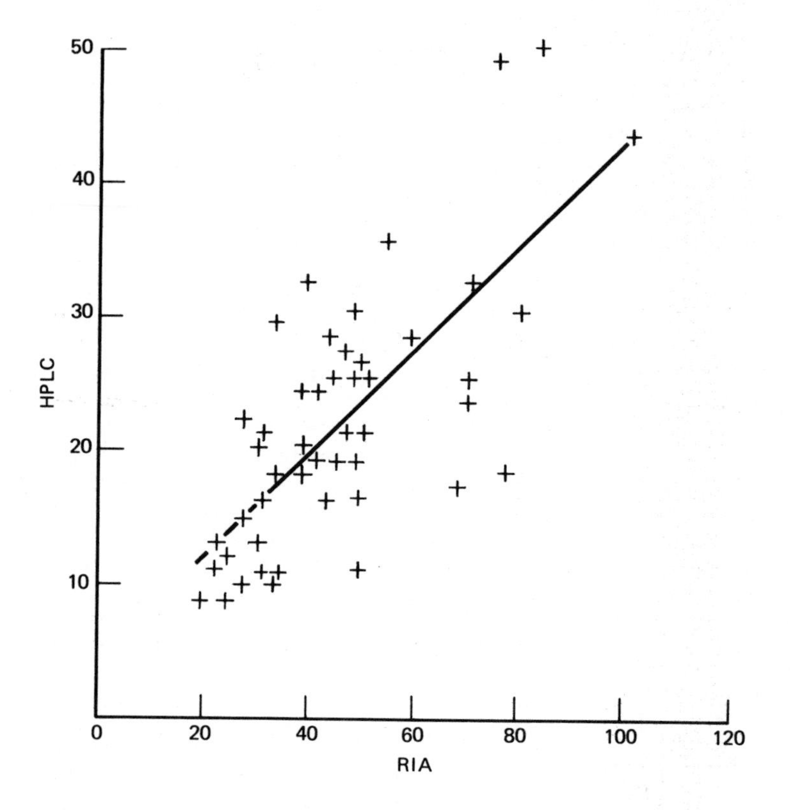

Figure 3. Regression curve for urinary free cortisol [μg/24 hr g of creatinine]. HPLC versus RIA [$y = 0.37x + 4.85$: $r = 0.72$].

free cortisol.* A mean of 45 ± 18 μg/24 hr g of creatinine was obtained. A linear regression analysis of the values generated by the two procedures is shown in Figure 3. The slope of 0.37 indicates that the RIA values average nearly three times greater than those obtained by HPLC. Furthermore, inspection of the

---

*The procedure is a modification of the Cortisol [^3H] Radioimmunoassay Pak[R] (December 1977) supplied by New England Nuclear (Boston, Massachusetts). Preliminary purification on a Sephadex LH-20 column and use of a relatively specific rabbit antiserum prepared against cortisol-21-succinyl-bovine serum albumin are the main features of this method. The normal range is 20 to 90 μg/24 hr. We are indebted to Dr. Andrew R. Matz for performing these analyses.

figure clearly reveals greater variability of the RIA values.

Patient Values

Table II shows comparative levels of urinary free cortisol by HPLC versus RIA ordered routinely in a number of unselected patients with various disorders.

METHOD DEVELOPMENT

As shown in Figure 4, concentrations of up to 50% aqueous methanol as a wash solvent in the Prep I extraction resulted in minimal loss of steroids. Aqueous methanol (40%) was chosen because it washed through the resin more interfering urinary chromagens than did water. In addition, the aqueous methanol wash, by providing a smaller eluate residue, expedited spotting for TLC. A number of possible eluting solvents were investigated. Water-immiscible solvents such as ethyl acetate and methylene chloride/methanol (95:5, v/v) afforded good recoveries of steroids when run through the cartridges by suction, but gave low and inconsistent results in the sample processor. These disparate findings were attributed to a channeling effect in the cartridge resin bed, which apparently is promoted by centrifugal forces generated in the Prep I.

Use of the water-miscible solvents acetone and methanol provided excellent recoveries, but the former removed fewer interfering chromagens.

A manual methylene chloride extraction of urine could be carried out as a substitute for automated sample processing. The Prep I afforded both more extractions per unit time and more consistent recoveries, however.

The use of chloroform/methanol (50:50, v/v) for sample application of Prep I eluates to TLC gave higher, more consistent recoveries than more volatile solvents such as methylene chloride and methylene chloride/methanol (90:10, v/v).

Triol acetate, which does not coelute during HPLC with the standards and analytes, was selected as the TLC standard so that inadvertent cross contamination with analytical channels during scraping would not interfere. The toluene-acetone system was selected because of the nearly identical R_f values of internal standard and analytes, and because it gave satisfactory cleanup of the eluate residue. A variety of eluting solvents for TLC were investigated. These included methanol, water, methanol/water (50:50, v/v), acetone, and methylene chloride/methanol (90:10, v/v).

TABLE II. COMPARISON OF HPLC AND RIA URINARY FREE CORTISOL VALUES (μg/24 hr g of creatinine) FROM AN UNSELECTED PATIENT POPULATION

Patient	Diagnosis	HPLC	RIA
D.G.	Adult-onset adrenogenital syndrome, baseline	35	77
	2 mg dexamethasone suppression, day 1	11	63
	2 mg dexamethasone suppression, day 2	3	10
	First postsuppression day	5	<5
	ACTH stimulation	216	426
F.H.	Diabetes mellitus, baseline	31	113
	ACTH stimulation, day 1	2134	>811
	ACTH stimulation, day 2	3817	>944
C.B.	Panhypopituitarism, diabetes mellitus, baseline	12	13
	ACTH stimulation, day 1	226	604
	ACTH stimulation, day 2	1024	>977
E.M.	Unknown	332	659
D.C.	Diabetes mellitus, hypothyroidism	58	128
V.M.	Obesity, hirsutism, oligomenorrhea	32	43
L.W.	Obesity	7	14
M.S.	Obesity, hypothyroidism	39	236
J.Q.	Eosinophilia, viral infection	17	131

Table II (continued)

Patient	Diagnosis	HPLC	RIA
C.D.	Unknown	23	94
C.M.	Partial 11β-hydroxylase deficiency	87	212
R.J.	Panhypopituitarism, renal disease	12	43
D.H.	Obesity, essential hypertension	6	17
R.P.	Septic shock, ACTH stimulation	4279	>1360
B.P.	Unknown	4	9
C.T.	Diabetes mellitus, hypertension	74	303
D.S.	Hypokalemia	27	95
T.E.	Unknown	11	30

The latter mixture was selected because it removed minimal background UV absorption from the silica gel while affording good recoveries of chromatographed steroids.

Figure 5 shows the cumulative benefit of pre-HPLC cleanup on the urine of a burn patient. There is complete obfuscation of the Δ'E, E, and F peaks when the TLC is omitted (chromatograms 5a and 5b). It is also obvious that the best chromatogram (D) is achieved by combining TLC with the aqueous methanol wash.

DISCUSSION

Development of the present method arose from the need for a rapid, sensitive, and specific means of assessing adrenocortical function in burn patients. It was anticipated that because of multiple drug therapy and tissue destruction, their urine would be consider-

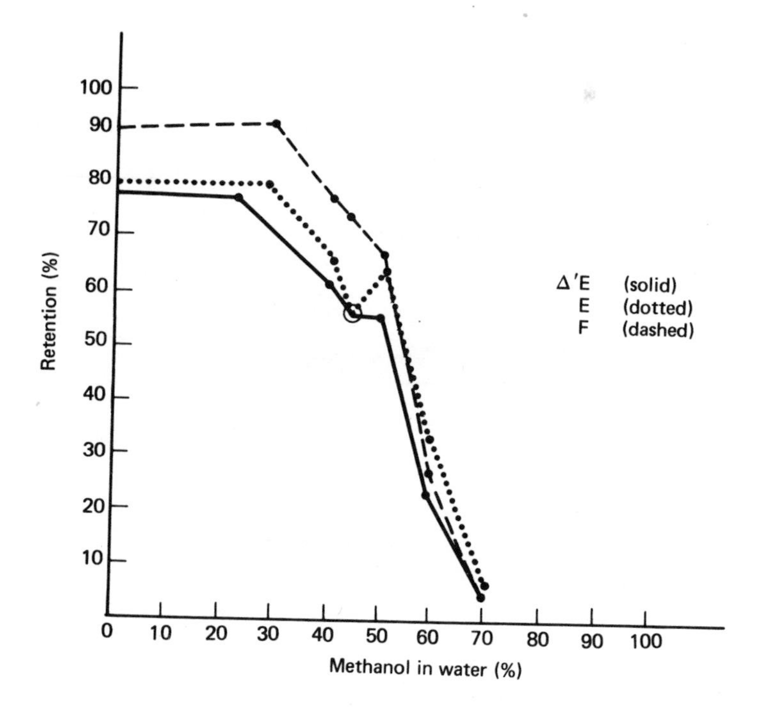

Figure 4. Retention of three steroids by type W cartridges as a function of changing concentration of methanol in wash solvent. Δ'E = solid line, E = dotted line, and F = dashed line.

ably more complex than that of normal individuals or patients with nontraumatic endocrinopathies. Rigorous purification before HPLC was therefore considered essential. The use of 40% aqueous methanol as the Prep I wash solvent served to remove a number of polar, nonspecific impurities that otherwise frequently overlapped with and obscured the internal standard and analyte peaks during HPLC. The need for TLC was established by the demonstration that not only did it remove additional nonspecific UV chromagens, but also eliminated unknown compounds coeluting with the steroids of interest. These substances occur both in normal subjects and in burn patients.

The greater complexity of urine generally as compared with blood is evidenced by the ability to measure cortisol in the

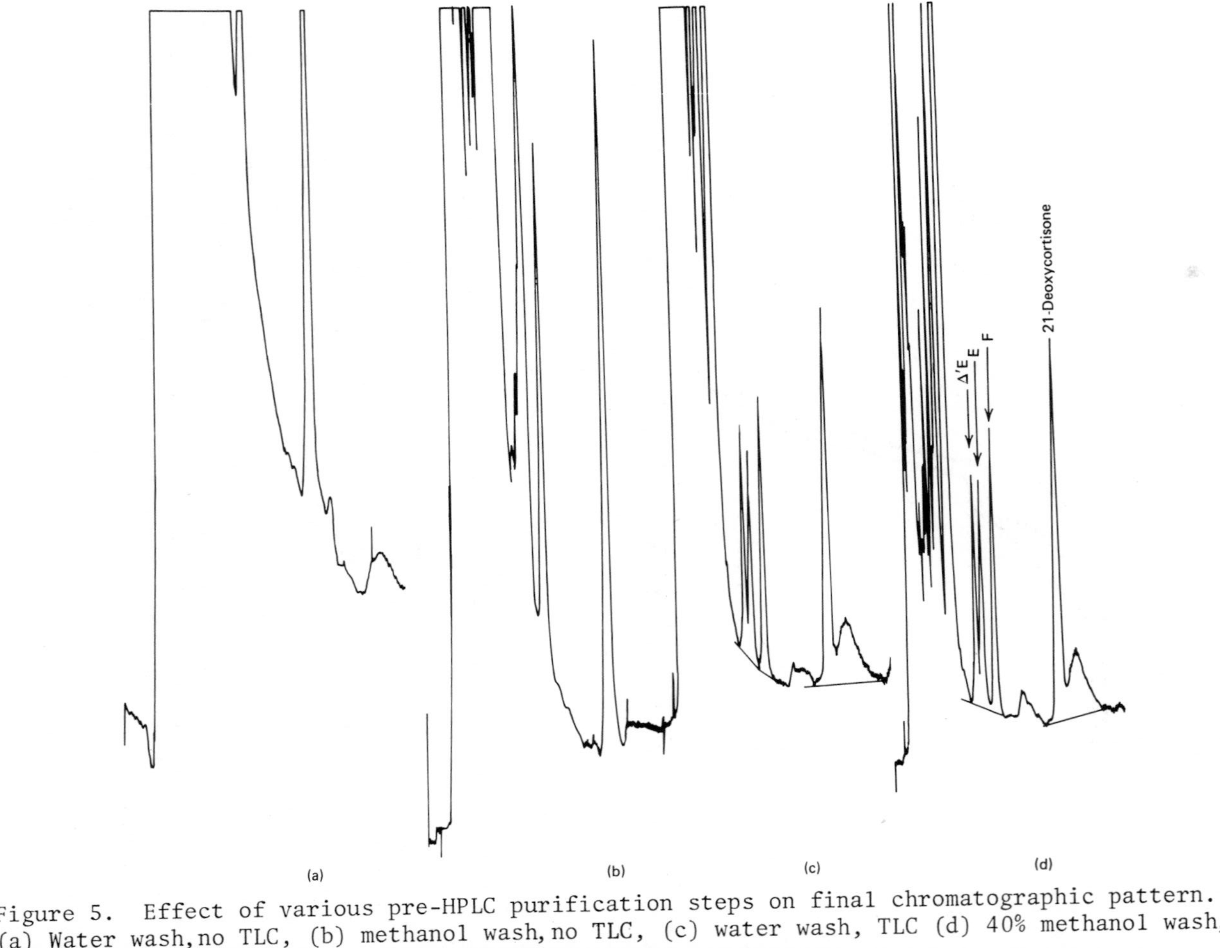

Figure 5. Effect of various pre-HPLC purification steps on final chromatographic pattern. (a) Water wash, no TLC, (b) methanol wash, no TLC, (c) water wash, TLC (d) 40% methanol wash, TLC.

latter fluid without the need for TLC (Lewbart and Elverson, unpublished observations). The greater specificity of the new procedure as compared with a current RIA method for urinary free cortisol both in normal subjects (Figure 3) and in patients (Table II) is clearly evident. The ability to measure precisely both cortisol and cortisone at very low levels permits the diagnosis of adrenal hypofunction either of primary type or secondary to administration of suppressive steroids other than prednisone or prednisolone. The consistent overall recovery of internal standard in the roughly 70 to 90% range compares well with blood-based methods not utilizing TLC (8-10).

The parallel recoveries of both analytes and internal standard justify the use of a factor to correct for manipulative loss. It was observed that recovery of the three steroids added to urine is consistently greater than recovery of pure steroids added to water, presumably because of an ill-defined protective effect afforded by urinary components. The ratio of cortisone to cortisol in normal urine varied from 1.5:1 to 7.0:1. The more random distribution of cortisone versus cortisol values (Figure 2) reflects the fact that the former steroid is not a primary secretory product. The extent of cortisone excretion, therefore, is dependent upon activity of the 11β-dehydrogenase that interconverts it with cortisol. For example, it is well known that the relative amounts of the two steroids can be influenced by thyroid function (12) and pregnancy (13). It has been noted in burn patients (Lewbart and Elverson, unpublished observations) that considerable variation of the absolute levels and ratios occurs, which appears to have both prognostic and therapeutic implications. Also studied was a patient with a cortisone-secreting metastatic adrenocortical carcinoma (made available by Dr. Woodrow B. Kessler). Thus, not only does the measurement of urinary free cortisone afford a more complete index of adrenocortical function, but it also has potential as a new and useful parameter reflecting the physiology of various clinical states.

ACKNOWLEDGMENTS

We wish to thank Dr. John Fenton for performing the creatinine analyses and reviewing the manuscript. We are grateful to Jaki Katz Adler for the art work. We are indebted to Dr. Philip Walsh and Dr. Ming-Chung Wong of Laboratory Procedures, Inc., Division of Upjohn Company, for providing normal urine samples.

Special thanks are also due to Dr. S. Randolph May for making several valuable suggestions concerning preparation of the manuscript.

REFERENCES

1. R. L. Eddy, A. L. Jones, P. F. Guilliland, et al., Am. J. Med., 55, 621 (1973).
2. L. Crapo, Metab., 28, 955 (1979).
3. H. N. Antoniades, Hormones in Human Blood, Harvard University Press, Cambridge, Mass., 1976.
4. J. H. Keffer, A. H. Caldarella, G. E. Reardon, and E. E. Canalis, Critique: Advanced Clinical Chemistry Check Sample No. ACC-31 (1979).
5. C. G. Beardwell, C. W. Burke, and C. L. Cope, J. Endocr., 42, 79 (1968).
6. L. Wise, H. W. Margraf, and W. F. Ballinger, Arch. Surg., 105, 213 (1972).
7. D. H. Nelson, The Adrenal Cortex: Physiological Function and Disease, W. B. Saunders, Philadelphia, 1980, p. 81.
8. G. E. Reardon, A. M. Caldarella, and E. Canalis, Clin. Chem., 25, 122 (1979).
9. P. M. Kabra, L. Tsai, and L. J. Marton, Clin. Chem., 25, 1293 (1979).
10. F. J. Frey, B. M. Frey, and L. Z. Benet, Clin. Chem., 25, 1944 (1979).
11. J. C. Brouillet and V. R. Mattox, J. Clin. Endocr. Metab., 26, 453 (1966).
12. L. Hellman, H. L. Bradlow, B. Zumoff, et al., J. Clin. Endocr. Metab., 21, 1231 (1961).
13. B. E. P. Murphy, J. Clin. Endocr. Metab., 44, 1214 (1977).

CHAPTER 22

TLC in the Study of the Metabolism of Tiotidine Sulfoxide in the Rat

Deborah P. DiZio, Anthony F. Heald, and Joseph O. Malbica

INTRODUCTION

Pharmacological studies have shown that tiotidine, an H_2-receptor antagonist, is a potent inhibitor of gastric acid secretion in rat, dog, and man (1). Studies in rat and dog using ^{14}C-tiotidine have shown that elimination is mainly via excretion of tiotidine and tiotidine sulfoxide in urine and feces (2).

Sulfoxides are known to undergo reduction to sulfides in vivo. Cimetidine sulfoxide reduces in the gut to cimetidine (3). Likewise, sulindac, a sulfoxide, is reduced to a sulfide, the active pharmacophore (4). Thus a study was undertaken to determine whether tiotidine could be detected in excreta after administration of radio-labeled tiotidine sulfoxide to rats.

METHODS

Tiotidine (I) and tiotidine sulfoxide (II) were synthesized by ICI as were the ^{14}C-labeled compounds. The position of the radioactive carbon is indicated by an asterisk in Figure 1. Two groups of four male HLA/Wistar rats were dosed with 2.8 mg/kg of ^{14}C tiotidine sulfoxide (10 μCi/rat). Group I was dosed orally by gavage. Group 2 was dosed i.v. via the tail vein.

Urine was collected continuously into containers over dry

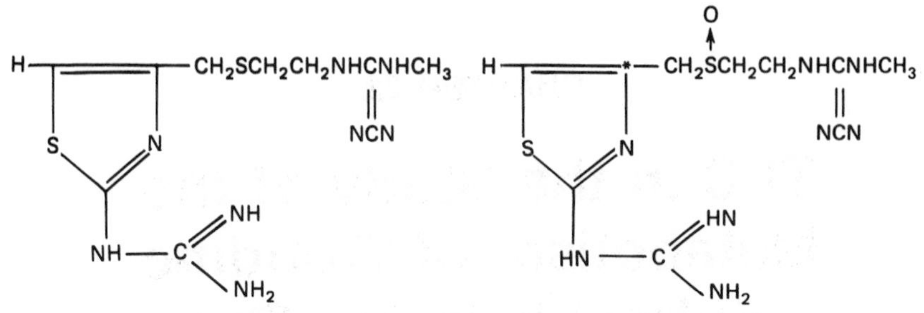

Figure 1. The structure of Tiotidine (I) and Tiotidine Sulfoxide (II). The position of the radioactive carbon atom is indicated by *.

ice/methanol. Feces were collected at room temperature. Containers were changed at 12, 24, 48, and 72 hr postdose. Samples were stored at -10°C until assayed.

Aliquots of urine were taken for liquid scintillation counting (LSC) in Liquiscint* to determine total radioactivity excreted. Urine was spotted directly onto TLC plates as described below and developed in solvent systems 1 and 2.

Feces were homogenized (Figure 2). Aliquots of fecal homogenates were weighed and combusted in a Harvey oxidizer using Oxosol** as the CO_2-trapping LSC cocktail. Also, 1 g of homogenate was mixed with 3 ml of water and made basic with NH_4OH. Slurry was extracted twice with ethyl acetate, then twice with ethyl acetate/isopropanol (9:1). The extracts were pooled and concentrated to a suitable volume for TLC in solvent systems 1 and 2. Subsequently, three 1-g aliquots of fecal homogenate spiked with approximately 6 μg of ^{14}C tiotidine and three 1-g aliquots of fecal homogenate spiked with approximately 20 μg of ^{14}C tiotidine sulfoxide were taken and treated as experimental samples. These control homogenates were used to obtain correction factors for the recovery of tiotidine and tiotidine sulfoxide from feces. Results indicate that 51% of tiotidine and 73% of tiotidine sulfoxide were recovered by this procedure.

---

* Trademark, National Diagnostics, Somerville, New Jersey.
** Trademark, National Diagnostics, Somerville, New Jersey

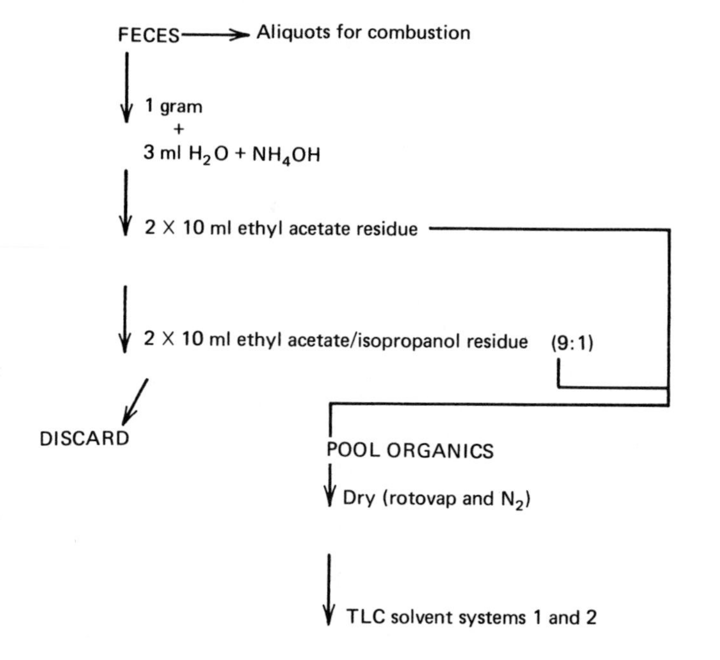

Figure 2. Scheme for method.

Urine and fecal extracts were streaked onto 0.25-mm silica gel plates over previously applied 100 µg tiotidine and 60 µg tiotidine sulfoxide. Since tiotidine is susceptible to oxidation on silica gel TLC plates, oxidation of radioactive tiotidine in experimental samples was overcome by prestreaking with tiotidine. Additionally, prestreaking allowed easy detection of tiotidine and tiotidine sulfoxide under UV illumination and gave more reproducible results and better resolution of the two compounds. Table I lists the solvent systems and R_f values for tiotidine and tiotidine sulfoxide. Developed plates were then scraped in 0.5-cm bands into scintillation vials using a zonal scraper. Liquiscint was added, and vials were counted in a scintillation counter.

The counter was directly interfaced to a computer for report analysis. Quench curves were generated for the counter. These curves expressed efficiency of counting as a function of either samples channel ratio (SCR) or external source channels ratio

TABLE I. THIN LAYER CHROMATOGRAPHY SOLVENT SYSTEMS

Solvent System Number[a]	Plate Type	Composition	R_f Tiotidine	R_f Tiotidine Sulfoxide
1	Silica (0.25 mm Anasil H)	Diethyl ether : ethanol : conc. NH_4OH 12:7:1	0.60	0.29
2	Silica (0.25 mm Anasil H)	Ethyl acetate : methanol : conc. NH_4OH 6:1:1	0.41	0.28

[a]Plates developed in a 23 × 27 × 7 cm glass TLC chamber. Chambers equilibrated with solvents through use of saturation pads.

(XSCR). The coefficients of these curves were stored in computer memory.

After all the vials of one TLC plate had been counted, the computer searched all the data and determined the five lowest counting samples, which were taken as blanks. The average of these values was calculated and subtracted from each sample. A minimum detectable count (MDC) was calculated using the following formula:

$$\text{MDC} = \text{mean of blanks} + \left[\begin{pmatrix}\text{standard error of} \\ \text{difference for blanks}\end{pmatrix}\right] \times (t)$$

and

$$\begin{matrix}\text{Standard error of} \\ \text{difference}\end{matrix} = \begin{pmatrix}\text{standard deviation} \\ \text{of blanks}\end{pmatrix}\left(\sqrt{\frac{1}{N} + \frac{1}{n}}\right)$$

where t = student's t value for p = .05 for N - 1 degrees of freedom (2.132 is value for TLC)--single sided since only values greater than blanks are of interest, N = number of blanks (5 for TLC), and n = number of replications per sample (1 for TLC). Thus

$$\text{MDC} = \text{mean of blanks} + (t)\begin{pmatrix}\text{standard deviation}\\ \text{of blanks}\end{pmatrix}\left(\sqrt{\frac{1}{N}+\frac{1}{n}}\right)$$

$$= \text{mean of blanks} + (2.132)\begin{pmatrix}\text{standard deviation}\\ \text{of blanks}\end{pmatrix}(1.10)$$

$$= \text{mean of blanks} + (2.345)(\text{standard deviation of blanks})$$

The rationale of MDC is that there is some variation about background; the MDC is a statistical attempt to deal with this variation (5). Any samples that did not exceed the MDC were considered below detectable limits.

The distintegrations per minute (dpm) values for each vial was calculated from observed counts per minute (cpm) using both quench correction curves. The dpm values generated from both curves were averaged, except in the case where gross cpm was 4 or less, above background cpm. In this case, the dpm value calculated using the SCR quench curve was discarded, since this method of quench correction is unreliable at low levels of radioactivity.

Figure 3 is an example of a thin layer chromatogram report, pages 1 and 2. In this instance, the report was generated by an Intertechnique LEM II computer interfaced to a SL4000 liquid scintillation counter. Raw data were printed out as page 1. Page 2 is a record of the cocktail, isotope, quench curve coefficients, and background counts, as well as the minimal detectable count.

Page 3 (Figure 4) records sample number (A), gross counts per minute (B), SCR disintegrations per minute (C), XSCR disintegrations per minutes (D), and average disintegrations per minute (E). Additionally, the average disintegrations per minute were totaled (F), and this total was divided into each sample dpm to obtain a dpm percent (G). Figure 5 shows the dpm percent in histogram form printed as the final page of the report.

Thus, by matching the R_f values of UV detectable unlabeled carriers with R_f values of radioactive peaks, detection and quantitation of tiotidine and tiotidine sulfoxide were possible.

RESULTS

Tiotidine was found in urine and feces from rats dosed intravenously and orally (Table II). More tiotidine sulfoxide than tiotidine was detected in the urine; in feces, however, the situation was reversed. The greatest amount of tiotidine was

```
EXP #    6274        ANAL DT: 1/16/80        CNT DT: 1/18/80                PAGE 1

1103 10.00 99.00000 25.00000 166.0000 1.173200
1104 10.00 96.00000 26.00000 190.0000 1.195900
1105 10.00 210.0000 67.00000 315.0000 1.203900
1106 10.00 1332.000 360.0000 1688.000 1.194500
           696.0000 197.0000 953.0000 1.194900
                    68.00000 387.0000 1.186300
                             272.0000 1.197100
                                      1.197100
```

LEM II Printout of Page 2
of TLC Report

```
EXP #    6274        ANAL DT: 1/16/80        CNT DT: 1/18/80                PAGE 2

CKTL= 1 ISTP= 1 IFORM= 2
DATE: 3/23/79
SCR COEFFS  A=0.9008  B=2183.2
            C=0.6081  D=0.0000
            E=0.0000  K1=10.391
            K2=2.2503 K3=0.0000
ESCR COEFFS A=0.9048  B=3.9170
            C=-0.034  D=0.0000
            E=0.0000  K1=4.4231
            K2=0.9814 K3=0.0000
MULT FACTOR=1.000000

 BKGD CNT CHN 1=          9.200001
 BKGD CNT CHN 2=          2.460000
 BKGD CNT CHN 3=          16.40000

 MDC IN CHN# 3,5.00 BLANKS USED
                  1    17.8662 ** USED FOR MD DETECTION **
                  2    17.5198
                  3    17.3774
                  4    17.2978
                  5    17.2465
```

Figure 3. LEM II printout of pages 1 and 2 of TLC report.

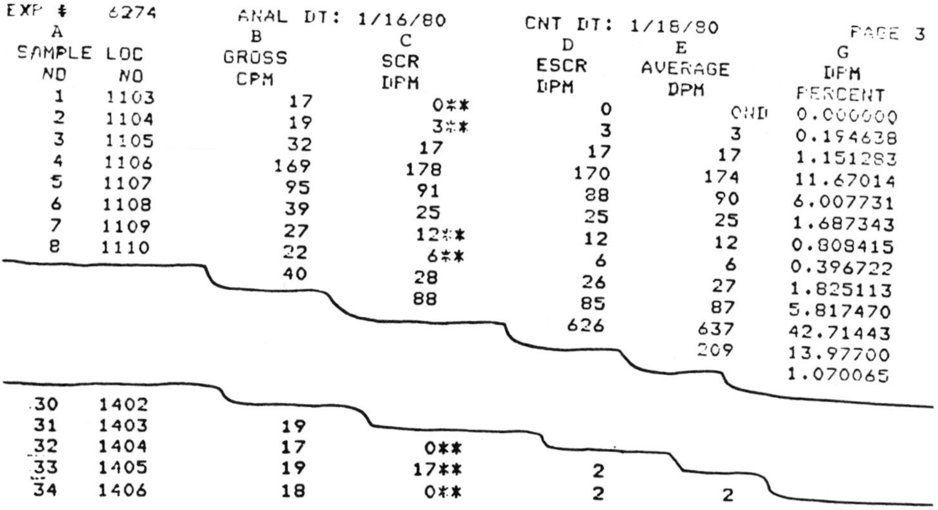

EXP # 6274 ANAL DT: 1/16/80 CNT DT: 1/18/80 PAGE 3

A SAMPLE NO	LOC NO	B GROSS CPM	C SCR DPM	D ESCR DPM	E AVERAGE DPM	G DPM PERCENT
1	1103	17	0**	0	OND	0.000000
2	1104	19	3**	3	3	0.194638
3	1105	32	17	17	17	1.151283
4	1106	169	178	170	174	11.67014
5	1107	95	91	88	90	6.007731
6	1108	39	25	25	25	1.687343
7	1109	27	12**	12	12	0.808415
8	1110	22	6**	6	6	0.396722
		40	28	26	27	1.825113
			88	85	87	5.817470
				626	637	42.71443
					209	13.97700
						1.070065
30	1402					
31	1403	19				
32	1404	17	0**			
33	1405	19	17**	2		
34	1406	18	0**	2	2	

TOTAL=1491.773 F

Figure 4. LEM II printout of page 3 of TLC report.

detected in feces of the orally dosed rats. The least amount of tiotidine was found in urine of intravenously dosed rats.

There was a low recovery of total radioactivity after oral administration. As 23.7% of the dose was found in the last fecal collection and 3.4% in the last urine collection, it is probable that additional collections would increase total radioactivity recovery. However, the low recovery does not affect the conclusions drawn from the study.

DISCUSSION

The presence of tiotidine in urine and feces from rats after oral and intravenous administration of tiotidine sulfoxide indicates that this metabolite is converted, in part, to tiotidine in vivo. Cimetidine sulfoxide, a compound similar to tiotidine sulfoxide, has been reported to be reduced to cimetidine by incubation in

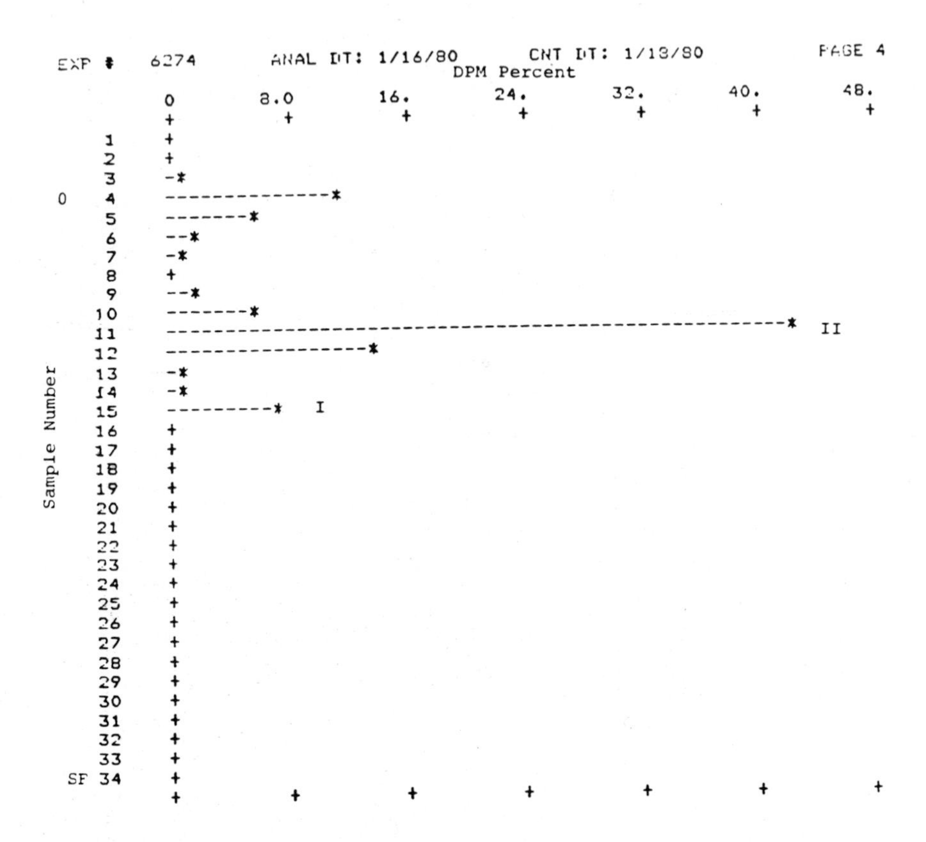

Figure 5. LEM II printout of TLC report. Histogram of DPM percentage of TLC rat urine developed in mobile phase 2. O = origin, SF = solvent front, I = radioactive peak corresponding to tiotidine, II = radioactive peak corresponding to tiotidine sulfoxide.

fecal homogenates of rat (3) and man (6). Since after dosing with the sulfoxide the largest amount of tiotidine appears in rat feces, it appears likely that tiotidine sulfoxide was reduced

TABLE II. 72 HOUR EXCRETION OF TIOTIDINE, TIOTIDINE SULFOXIDE AND TOTAL RADIOACTIVITY AFTER ORAL OR INTRAVENOUS ADMINISTRATION OF TIOTIDINE SULFOXIDE (2.8 mg/kg) IN MALE RATS

	Percent of Administered Dose (Mean ± S.E.M.)	
	Oral (n = 3)[a]	Intravenous (n = 4)
Urine		
Tiotidine	2.3 ± 0.7	0.5 ± 0.3
Tiotidine sulfoxide	6.2 ± 1.7	56.4 ± 2.9
Total ^{14}C	15.1 ± 3.5	63.6 ± 3.3
Feces		
Tiotidine	26.7 ± 6.8	9.2 ± 1.2
Tiotidine sulfoxide	14.2 ± 2.2	4.1 ± 0.5
Total ^{14}C	46.2 ± 9.4	16.4 ± 1.2
Urine + Feces		
Tiotidine	29.0 ± 6.6	9.7 ± 1.4
Tiotidine sulfoxide	20.4 ± 1.7	60.4 ± 2.9
Total ^{14}C	61.3 ± 8.2	80.0 ± 3.0

[a]Results from one animal omitted because of fecal contamination of urine.

to tiotidine by fecal microorganisms. Absorption, distribution, and excretion, as well as biliary and metabolism studies, indicate significant biliary excretion of tiotidine and tiotidine sulfoxide, thereby extending residence time of both compounds in rat gut (2). Tiotidine found in the intestine could thus be absorbed.

Alternatively, it is known that liver microsomes can catalyze sulfoxide reduction (7). It is therefore possible some tiotidine could also be formed in the liver.

The bioavailability of tiotidine resulting from tiotidine sulfoxide reduction is difficult to quantitate. Quantitation is obscured by many processes taking place at the same time. These are enzymatic and nonenzymatic oxidation of tiotidine to sulfoxide, secretion of tiotidine and tiotidine sulfoxide in bile, reduction of sulfoxide by intestinal microflora and liver microsomes, and poor absorption of tiotidine and sulfoxide from the gut.

This study illustrates that TLC is an accurate and reproducible analytical tool in drug metabolism. It was used to quantitate tiotidine and tiotidine sulfoxide in urine and feces. Thus, new information about the metabolism and excretion of these compounds was obtained.

ACKNOWLEDGMENTS

We thank expert technical assistance of Dora Santarelli, Teresa Larkin, and S. Adair Hearn. We are grateful to Mike Fleming and Ron Lego for the computer programming.

REFERENCES

1. T. O. Yellin, S. H. Buck, D. J. Gillman, D. F. Jones, and J. M. Wardleworth, Life Sci., 25, 2001 (1979).
2. F. R. Zuleski, M. D. Melgar, D. P. DiZio, and D. E. Reed, Fed. Proc. (1980).
3. D. C. Taylor, P. R. Cresswell, and D. C. Bartlett, Drug Metab. Dispos., 6, 21 (1978).
4. D. E. Duggan, K. F. Hooke, E. A. Risley, T. Y. Shen, and C. G. VanArman, J. Pharm. Exptl. Ther., 201, 8 (1977).
5. D. E. Case, S. H. Ellis, and P. R. Reeves, ICI Commun. (1975).
6. A. Grahnen, C. vonBahr, B. Lindstrom, and A. Rosen, J. Eur. J. Clin. Pharmacol., 16, 335 (1970).
7. P. G. C. Douch and L. L. Buchanan, Xenobiot. 9, 675 (1979).

CHAPTER 23

TLC Detection and Fluorometric Determination of Quaternium-15 in Cosmetics

Harris H. Wisneski

INTRODUCTION

Cosmetics must be effectively preserved during their manufacture, storage, and use to control microbial contamination. The quaternary amine quaternium-15 is one of a number of water-soluble compounds used as preservatives in cosmetics that have aqueous systems or oil-water emulsions. Under certain conditions, these preservatives have the common property of undergoing hydrolytic cleavage to form formaldehyde. This property forms the basis for methods of detection and determination of these compounds.

Previous analytical work with quaternium-15 included a qualitative TLC method developed by Liem (1) in 1977 and a polarographic method listed in a bulletin issued by the manufacturer (2).

In the present study, quaternium-15 was identified and differentiated from other potentially interfering formaldehyde-releasing preservatives in representative products such as bath oils, lotions, and shampoos by thin layer chromatography. An aluminum oxide-coated plate on which the standards had been applied was developed twice in a solution of hexane-tert-butyl amine (15 + 2). The structures of these preservatives, including quaternium-15, and their separation by TLC are presented in Figures 1 through 4 and Table I.

Two different TLC indicating reagent sprays were used for spot visualization. The first is based on the use of Hantzsch

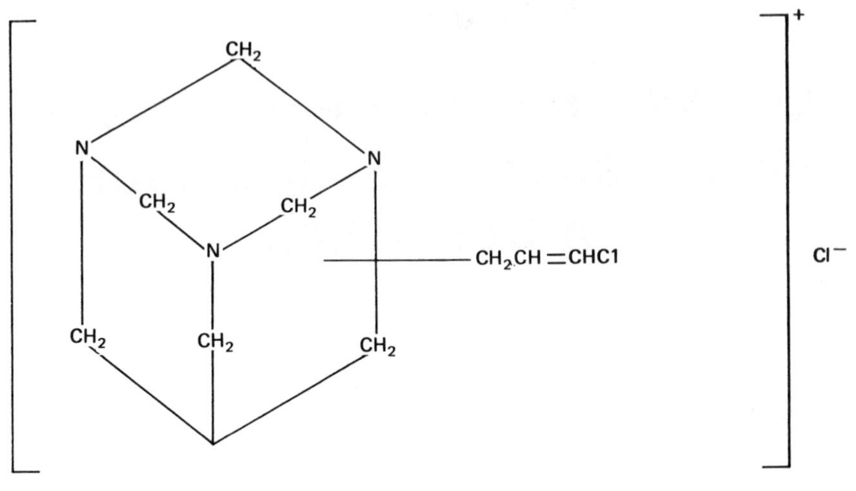
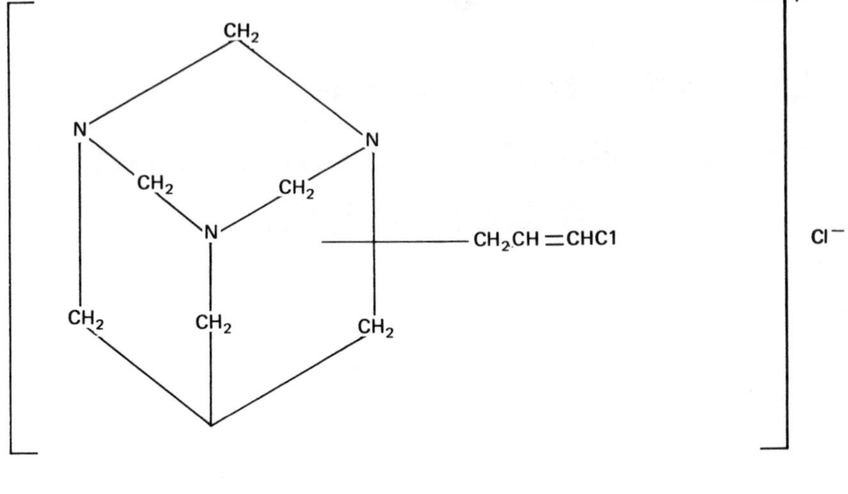

Figure 1. Structure of quaternium-15.

$$HOCH_2-C(Br)(NO_2)-CH_2OH$$

Figure 2. Structure of 2-bromo-2-nitropropane-1,3-diol.

reaction for the preparation of substituted pyridines from β-diketones, primary amines, and aldehydes.

When 2,4-pentanedione, formaldehyde, and ammonia (from ammonium acetate) react in a solution buffered at pH 6 (ammonium acetate/acetic acid), 3,5-diacetyl-1,4-dihydrolutidine, a substituted pyridine, is produced after the plate is heated at 110°C for 10 to 15 minutes. This product appears as a strongly

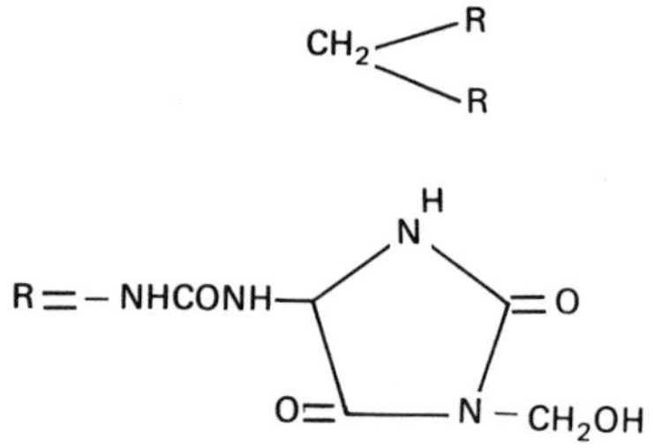

Figure 3. Structure of imidazolidinyl urea.

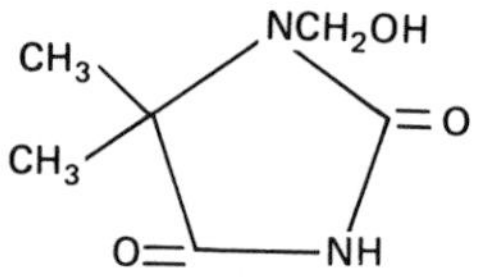

Figure 4. Structure of 1-hydroxymethyl-5,5-dimethylhydantoin.

fluorescent yellow spot when viewed under short-wave ultraviolet light. The formaldehyde is formed by hydrolytic cleavage of each preservative. The selection of pH 6 ensures specificity for formaldehyde as opposed to other aldehydes. A minimum of 0.1 μg of quaternium-15 can be seen on the plate using this reagent. Figure 5 depicts this reaction.

An alternative spray reagent based on the Dragendorf reaction of quaternary salts to form colored products serves as a confirmatory indicator. The reagent is modified with a larger proportion of hydrochloric acid to compensate for the basicity introduced on the plate by residual traces of tert-butyl amine used in the developing solution. The stock solution consists of bismuth nitrate, 25% nitric acid, potassium iodide, and 6N hydrochloric acid. Water, hydrochloric acid, stock solution, and sodium hydroxide are added in that order to prepare the spray reagent. After the plate is developed, dried for 30 minutes, and sprayed,

TABLE I. TLC DATA[a]

Preservative	R_f
Quaternium-15	0.5
2-Bromo-2-nitropropane-1,3-diol	0.1
Imidazolidinyl urea	0.0
1-Hydroxymethyl-5,5-dimethylhydantoin	0.0

[a]2,4-Pentanedione indicator.

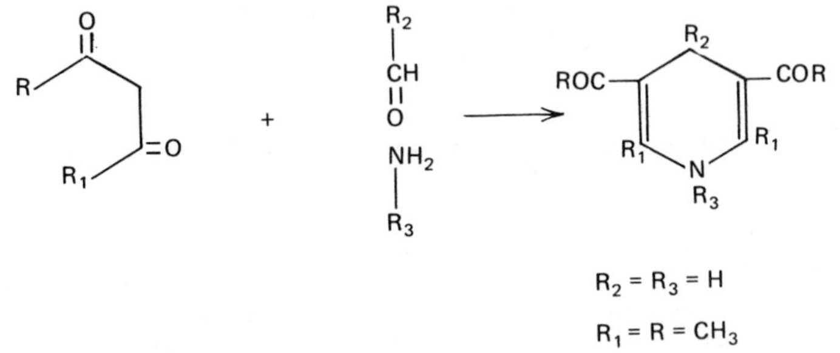

Figure 5. The Hantzsch reaction.

it is placed in a chamber saturated with hydrochloric acid fumes for 5 minutes to enhance sensitivity. A minimum of 4 mg of quaternium-15 can be detected as an orange spot against yellow background. The other preservatives do not react, but the sensitivity for detection of quaternium-15 is sacrificed. In order to evaluate possible interferences from other amines and quaternary salts with this reagent and developing system, a plate was spotted with a number of such standards. Results may be seen

TABLE II. TLC DATA[a] QUATERNIUM-15, AMINES, AND QUATERNARY AMMONIUM SALTS

Amine or Quaternary Amine	R_f	Color[b]
Stearyl dimethyl benzyl ammonium chloride	0.0	OY
p-Bromoaniline	None	None
Tri-n-hexylamine	0.9	O
m-Nitroaniline	0.3	Y[c]
Dimethylamine hydrochloride	0.0	Y
Aniline hydrochloride	0.0	OY
4-Amino-diphenylamine[d] hydrochloride	0.0	G
	0.3	B
	0.7	B
Quaternium-15	0.5	O

[a]Dragendorf indicator.

[b]OY--Orange-yellow, O--orange, Y--yellow, G--grey, B--black.

[c]Color present before application of reagent.

[d]Three different spots noted.

in Table II.

The quantitative determination of quaternium-15 in cosmetics was based on the use of the same Hantzsch reagent used as an indicator spray. From the structural formula, 6 mol of formaldehyde is theoretically available from 1 mol of quaternium-15. This value was verified experimentally by heating known concentrations of quaternium-15 and a series of formaldehyde standards with the same concentration of Hantzsch reagent. The intensities from the fluorometric emissions of the lutidine product in all solutions (410 excitation, 510 emission wavelength) were determined after heating. A graph was plotted of emission intensity versus concentration of standard formaldehyde. The intercept of the quaternium-15 emissions with the curve indicated the concentrations of formaldehyde liberated by known concentrations of quaternium-15. Results appear in Table III.

Based on the release of 6 mol of formaldehyde per mol of quaternium-15, this preservative was measured in cosmetics against

TABLE III. MOLES FORMALDEHYDE RELEASED PER MOLE OF QUATERNIUM-15

Quaternium-15 (μg/ml)	HCHO, Theoretical[a] (μg/ml)	HCHO, Found (μg/ml)	Moles HCHO/Mole Quaternium-15
0.500	0.358	0.345	5.78
0.822	0.589	0.575	5.86
0.998	0.716	0.758	6.34
1.000	0.717	0.695	5.81
1.020	0.731	0.721	5.92

[a]Based on 6 mol theoretical HCHO (1 mol quaternium-15); micrograms quaternium-15 × 0.717 = micrograms HCHO (theoretical).

formaldehyde standards. One gram of cosmetic was initially diluted in absolute methanol. Insolubles were allowed to settle. An aliquot of supernatant solution was removed and diluted to volume in 10% methanol. An aliquot of the final sample dilution was incubated with an aliquot of the Hantzsch reagent. Aliquots from at least three standard formaldehyde solutions ranging from 0.1 to 1.0 μg/ml were also incubated with the reagent. The range of standard formaldehydes selected produced a linear curve. If the sample intensity was too high it was further diluted, and the incubation with reagent was repeated, so that sample emission was adjusted to lie within the linear range of the standards. A curve of standards was plotted versus their fluorometric emission intensities.

The intercept of sample with standard curve was determined, and the corresponding concentration of formaldehyde was plotted; then the percent quaternium-15 in the sample was plotted:

$$D = \frac{10^4\ (F)(V)}{0.717\ S},$$

where D is the percentage of quaternium-15 in the sample, F is micrograms of formaldehyde found per milliliter, V is the dilution volume in milliliters, and S is the sample weight in grams.

The efficiency of this method was verified by determining

TABLE IV. RECOVERIES OF QUATERNIUM-15 FROM BATH OIL[a]

Sample (1.0 g)	Added %	Added μg/ml[b]	Found μg/ml[b,c]	Rec. (%)
Bath oil	0.2	1.00	0.99	99
			0.94	94
			0.89	89
				Average 94
Bath oil	0.1	1.02	0.97	97
			0.92	90
			0.92	90
				Average 92
Bath oil	0.05	0.82	0.75	91
			0.75	91
			0.75	91
				Average 91

[a]Triplicate determinations.

[b]Concentration of final dilutions.

[c]Micrograms HCHO found per milliliter divided by 0.717 = microgram quaternium-15 found per milliliter.

the percent recoveries of known added amounts of quaternium-15 from cosmetic sample. As shown in Table IV and V, average recoveries over the range of ordinary use, namely, 0.05 to 0.2%, ranged from 91 to 104%.

The concentration of quaternium-15 cannot be determined accurately by the present method in cosmetic samples also containing other preservatives such as free formaldehyde or other formaldehyde releasers.

REFERENCES

1. D. H. Liem, Cosmet. Toiletries, 92, 59 (1977).

TABLE V. RECOVERIES OF QUATERNIUM-15 FROM LOTION[a]

	Added		Found	
Sample (1.0 g)	%	μg/ml[b]	μg/ml[b,c]	Rec. (%)
Lotion	0.2	1.00	0.92	92
			0.94	94
			0.94	94
				Average 93
Lotion	0.1	0.50	0.47	94
			0.43	86
			0.50	100
				Average 93
Lotion	0.05	0.50	0.52	104
			0.52	104
			0.52	104
				Average 104

[a]Triplicate determinations.

[b]Concentration of final dilutions.

[c]Micrograms HCHO found per milliliter divided by 0.717 = micrograms quaternium-15 found per milliliter.

2. Method MLS-AM-67-13 (1967) Dow Chemical Co., Midland, Mich.

CHAPTER 24

Determination of Urine Residues by Cellulose TLC

Joel J. Thrasher

INTRODUCTION

During the analysis of regulatory samples from food storage and processing establishments, a great number of samples may give a positive test for urea when the xanthydrol test (1) is used for detection of mammalian urine. However, a positive reaction with this test cannot be interpreted as conclusive proof of urine contamination, because urea is a component of many industrial products, such as fertilizer, and is often a component of paper and paperboard products used as food packaging materials (2). Because of this, a more definitive test for urine using thin layer chromatography (TLC) was developed. The new test is based on the fact that mammalian urine contains allantoin (5-ureido hydantoin) and urinary indican (potassium salt of indoxyl sulfate) as well as urea (3). Allantoin, however, may also be found in cereal grains (4) and some botanicals. The level of allantoin in human urine is very low (5) and has not been detected in dried urine residues by this test; thus the natural occurrence of both allantoin and urea must be considered when the TLC test results are interpreted. Urinary indican, on the other hand, is found only in mammalian urine. This paper describes the method for determination of this substance in residues using thin layer chromatography.

METHODS AND MATERIALS

The method consists basically of a warm acetone extraction of the sample, concentration of the extract, transfer of the concentrate to a cellulose TLC plate, chromatographic separation with the mobile phase *n*-BuOH-MeOH-H_2O (2:2:1), and spraying with the chromogenic agents.

The following reagent grade chemicals are required: acetone, methanol, *n*-butanol, 1,4-dimethylaminobenzaldehyde, sodium acetate, and hydrochloric acid. Thirty-milliliter beakers and 15-ml conical tubes are used in the extraction and concentration steps, respectively. Thin layer plates coated with either fibrous cellulose (MN 300) or microcrystalline cellulose (Merck) or their equivalents may be used. Merck plates require about 30% more time for 10-cm development. Old plates of either type should be prewashed or developed their full length with methanol and dried in a forced draft oven 60°C before use.

The chromatogram, dried at 80°C, is sprayed with a 2% solution of 1,4-dimethylaminobenzaldehyde in MeOH-HCl (3:1) and again heated to 80°C. The yellow urea spot produced by the spray changes to a light orange spot upon heating, while the allantoin spot remains a pale yellow. When the plate is sprayed above the urea standards with a saturated aqueous solution of sodium acetate, the indican undergoes the fluorindal reaction (6), which produces a highly fluorescent orange-red compound under long-wave ultraviolet light.

Urinary indican, the primary identifying metabolite for urine residues, is a photosensitive compound. It is degraded to some extent by light in the dry state but more so in solution. Because of this photosensitivity, precautions should be taken when the unknown sample and the reference standards are handled. Urinary indican is also very susceptible to the effects of relative humidity during chromatographic development. At high relative humidity the urea and indican spots may run together. At about 50% relative humidity the indican spot is centered between urea and allantoin. At low relative humidities the indican spot is near the low allantoin spot. The adverse effects of relative humidity are minimized by substituting a sandwich chamber in place of the conventional tank and by making a change in the mobile system. Table I shows the R_f of the three urine metabolites obtained during a collaborative study (7) in which the conventional tank and sandwich systems were used.

This collaboratively studied method is the only known TLC test specific for urine. It requires minimal equipment and is

TABLE I. REPRODUCIBILITY OF METABOLITE SPOT SEPARATIONS (R_f) (100)

Tank	Conventional		Sandwich	
	$\overline{X}^a$	SD	$\overline{X}^a$	SD
Urea	74.0	4.3	59.8	2.9
Indican	64.2	8.6	78.7	3.7
Allantoin	49.5	5.4	39.5	3.5

[a] Values are averages of six collaborators.

very sensitive.

This method has been approved by the Association of Official Analytical Chemists as an official first-action method for the analysis of suspect urine residues (8).

REFERENCES

1. Official Methods of Analysis, 13th ed., AOAC, Arlington, Va., Sec. 44.161, 1980.
2. Code of Federal Regulations, Title 21, Sec. 182.90, 1980.
3. L. Greene (Ed.), Biology of the Laboratory Mouse, McGraw-Hill, New York, 1966, p. 347.
4. R. Fosse, A. Brunel, and E. Thomas, C. R. Acad. Sci., 193, 7 (1931).
5. B. L. Oser (Ed.), Hawk's Physiological Chemistry, 14th ed., Blakiston, New York, 1965, p. 1160.
6. I. Smith, Chromatographic and Electrophoretic Techniques, Wiley-Interscience, New York, 1960, p. 197.
7. J. Thrasher, J. Assoc. Off. Anal. Chem., 63, 189 (1980).
8. Changes in Methods, J. Assoc. Off. Anal. Chem., 63, 410 (1980).

CHAPTER 25

Discrepancy of Results of Radiochemical Assays on Tc_{99m} Medronate

John J. Coupal, Euishen E. Kim, and Frank H. DeLand

INTRODUCTION

Diagnostic bone imaging is a relatively noninvasive method for determining location and extent of bone disease, either metabolic or neoplastic. The method involves intravenous injection of a radiopharmaceutical having an affinity for osseous tissue. The two-dimensional spatial biodistribution of the administered radiopharmaceutical or its metabolized radionuclide in the total body is displayed by an external gamma camera for evaluation of disease by the clinician.

An early radiopharmaceutical for that use was inorganic pyrophosphate labeled or tagged with the gamma-ray-emitting technetium-99m. Pyrophosphate, a condensation product of phosphoric acid, can chelate ^{99m}Tc from pertechnetate anion in the presence of reductant stannous ion. Several organic diphosphonic acids or diphosphonates (the latter being sodium salts of the respective diphosphonic acid; i. e., $-(PO_3HNa)_2$ have proven to be ligands clinically superior to pyrophosphate (1). The first diphosphonate used clinically was 1-hydroxyethylidene-1,1-diphosphonic acid known as HEDP or etidronic acid (Figure 1). A ligand found to be yet superior to HEDP is methylene diphosphonic acid, also known as MDP or medronic acid (Figure 1). All those chelates are somewhat adversely affected by oxygen to yield a clinically unsatisfactory radiopharmaceutical

A clinical dose of radiopharmaceutical typically contains

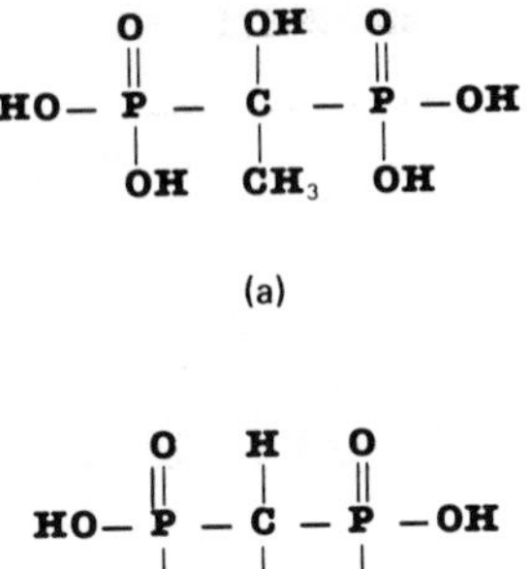

Figure 1. Structures of ligands used for bone imaging. (a) 1-Hydroxyethylidene-1,1-diphosphonic acid, or HEDP. (b) Methylene diphosphonic acid, or MDP.

a few milligrams of ligand, a fraction of a milligram of tin, and only a few nanograms of ^{99m}Tc. Nevertheless, it has been found feasible to perform a radiochemical chromatographic assay on the radiopharmaceutical by quantitation of the characteristic gamma photon emitted by the ^{99m}Tc.

Because of the short radioactive half-life of ^{99m}Tc (6.0 hr), it is important to be able to rapidly assess radiochemical purity of the radiopharmaceutical at any point in its useful "shelf life" which is approximately 8 hr. It was the purpose of this study to compare a standard thin layer chromatographic (TLC) technique with a newer and more rapid TLC procedure for quantitating chelate and radiochemical contaminants in ^{99m}Tc-medronate preparations.

MATERIALS AND METHODS

^{99m}Tc-medronate doses over a nine-week period were prepared in the following manner. Commercial stannous medronate kits from either of two manufacturers (products A and B) were used. Each product A vial contained a nominal 8.0 mg medronic acid and 0.447 mg stannous ion. Each product B vial contained 10 mg medronic acid and 0.626 mg stannous ion. The contents of either vial were reconstituted using ^{99m}Tc-sodium pertechnetate solution (eluted from a fission molybdenum-99 generator, Minitec, E. R. Squibb & Sons, Princeton, New Jersey), which had been diluted to 2.5 ml with 0.9% sodium chloride injection, USP, containing either normal dissolved oxygen (DO) content (6.7 mg DO/1, Invenex Laboratories, Grand Island, New York) or low dissolved oxygen content (3.8 mg DO/1, Ackerman Nuclear, Inc., Glendale, California). No component of the final radiopharmaceutical preparation contained an added preservative or antioxidant stabilizer. We designated product A containing normal dissolved oxygen saline (i. e., A + N) to be the control radiopharmaceutical. Details on patient dosing, patient imaging, and statistical analyses are presented elsewhere (2).

An aliquot of each radiopharmaceutical preparation was assayed for radiochemical constituents by each of two methods. The first method (3) was instant thin layer chromatography--silica gel (ITLC-SG, Gelman Sciences, Inc., Ann Arbor, Michigan) modified to employ certified methyl ethyl ketone (MEK, Fisher Scientific Co., Fair Lawn, New Jersey) (in place of acetone) and 0.9% sodium chloride solution (normal saline, Abbott Laboratories, North Chicago, Illinois) separately for development. In that procedure, the 5-µl sample applied to the strip was first developed ascendingly with applied spot wet to 12 cm by MEK. The intact chromatogram was then permitted to air dry. The same chromatogram was then developed ascendingly to 6 cm by normal saline. The resulting chromatogram was cut into 0.25-in. segments, which were assayed for ^{99m}Tc radioactivity in an automatic gamma counter (Packard Instrument Co., Downers Grove, Illinois). Using that procedure, the radiochemical contaminants of (1) reduced unbound (colloidal) ^{99m}Tc and (2) ^{99m}Tc-pertechnetate were found at the origin and at the MEK solvent front, respectively. The desired ^{99m}Tc-medronate chelate migrated with the saline front. The percentages of each of those three radiocomponents was then computed. That standard method required about 35 minutes to complete.

The second method employed a commercially available miniaturized TLC kit (TECH Quality Control Testing System, Ackerman Nuclear, Inc.). Here, the radiopharmaceutical spot is applied 1.4 cm from the bottom of a 5.6-cm-long by 0.6-cm-wide strip of solid phase and developed ascendingly to the top of the strip situated in a 7-cm-tall and 2.4-cm outside-base-diameter screw-capped bottle to constitute procedure "TECH A". The specific nature of the solid phase and developer is not disclosed by the manufacturer. At completion of development, the strip is dried and cut into two segments that are assayed for ^{99m}Tc radioactivity as described previously. The TECH A procedure quantitates the amount of ^{99m}Tc-pertechnetate present. Simultaneously with TECH A, an aliquot of the same radiopharmaceutical preparation is applied at 1.15 cm from the bottom of a different solid phase strip having identical linear dimensions (TECH B). It is developed ascendingly to the top of the strip in a separate bottle (of same dimensions as the first) containing a different solvent. Here again, the nature of solid phase and developer are not revealed. The dried chromatogram is cut into two segments and assayed as before. The TECH B procedure quantitates reduced unbound (colloidal) ^{99m}Tc present. The sum of TECH A and TECH B percentages is subtracted from 100 to reveal the percentage of the desired ^{99m}Tc-medronate chelate present in the preparation. That miniaturized technique shortens the time required for completion to 15 minutes, largely because of the more rapid developing time and radioassay time.

RESULTS

The results of the radiochemical assays for ^{99m}Tc-medronate by the standard ITLC method appear in Table I. By student t-test, we found no statistically significant difference ($p \geq 0.05$) in mean amount of any of the three radiochemical components when each of the three possible radiopharmaceutical combinations was compared with corresponding ones in the control radiopharmaceutical. By coefficient of variation (standard deviation × 100/mean), it is apparent that the control radiopharmaceutical preparations (A + N) possessed a more consistent day-to-day content of the desired chelate than did any of the other three preparations. In general, product B appeared to be more variable than product A. Surprisingly, both ligand products reconstituted with low dissolved oxygen saline yielded a range of chelate quantitites 50% greater than did the corresponding products reconstituted with normal dissolved oxygen saline. Quantities of the radiochemical contaminants present in all those preparations are presented elsewhere (2).

Results of radiochemical analyses performed by the more rapid method appear in Table II. We found no significant difference in mean quantity of any radiochemical component when compared by student t-test with the corresponding component in the control preparations. The agreement in measured component quantities among the four possible radiopharmaceutical types is striking.

When compared with the standard method, the rapid method

TABLE I. ^{99m}Tc-MDP MEASURED BY STANDARD ITLC METHOD

Radiopharmaceutical Components[a]	^{99m}Tc-MDP Content Mean ± SD(%)	Observed Range of ^{99m}Tc-MDP Content(%)	Coefficient of Variation(%)
A & N (control)	93.9 ± 1.7	91.5 to 96.1 (4.6)	1.8
A & L	93.2 ± 2.0	89.8 to 96.7 (6.9)	2.1
B & N	92.9 ± 2.3	90.4 to 97.1 (6.7)	2.5
B & L	91.9 ± 3.5	86.5 to 97.0 (10.5)	3.8

[a]A = stannous medronate from manufacturer A, B = stannous medronate from manufacturer B, N = physiological saline containing normal dissolved oxygen, and L = physiological saline containing low dissolved oxygen.

consistently overestimated the mean quantity of desired chelate and generally underestimated the quantity of colloidal ^{99m}Tc and ^{99m}Tc-pertechnetate. By subjective assessment, the degree of neither bone-delineation nor soft-tissue (extraosseous) radioactivity on bone image was judged to be different for the three radiopharmaceutical combinations when compared with that of the control by a clinician unaware of the particular ligand and DO combination injected. However, subjectively assessed overall bone image quality was judged to be superior from product A than from product B, regardless of the DO level of saline employed. Unfortunately, neither TLC procedure was able to reveal differences in radiochemical components between products A and B to corroborate readily apparent performance characteristics between the two ligand products clinically.

TABLE II. MINIATURIZED ("RAPID") RADIOCHEMICAL ASSAY

Radiopharmaceutical Ingredients[a]	Radiochemical Components (Mean ± SD)		
	^{99m}Tc-Medronate (%)	Reduced Hydrolyzed Tc (%)	Pertechnetate TcO_4 (%)
A & N (control)	97.8 ± 1.0	2.1 ± 0.9	0 ± 0.1
A & L	97.5 ± 0.5	2.4 ± 0.5	0 ± 0
B & N	97.7 ± 1.5	2.3 ± 1.4	0 ± 0.1
B & L	97.9 ± 0.5	2.1 ± 0.5	0 ± 0.1

[a]A = stannous medronate from manufacturer A, B = stannous medronate from manufacturer B, N = physiological saline containing normal dissolved oxygen, and L = physiological saline containing low dissolved oxygen.

DISCUSSION

The radiochemical assay results by a standard ITLC method appear to be in more agreement with clinical bone image findings than do those by the rapid method employed. The standard method, however, still leaves much to be desired in terms of sensitivity and specificity. Therefore, there remains a need for a rapid TLC method providing information on radiochemical constituents that agrees well with results from the clinical performance of the chelate preparation. With such a technique, it would be easy to identify the occasional defective radiopharmaceutical and eliminate the possibility of its administration to a patient. The discrepancy in results and full-scale and miniaturized radiochemical assay systems has been reported previously (4).

The rapid method employed in this study was performed with one deviation from the manufacturer's directions. Developing solvents were generally reused and supplemented with additional solvent when the level was too low for a chromatographic separation. This was done to simulate the situation in the typical smaller nuclear medicine laboratory, where convenience and economy of reagent costs are often paramount. Such reuse of reagents is not specifically cautioned against by the manufacturer.

Finally, it was of concern to us that the routine chromatographic assays employed could not detect differences revealed clinically between products A and B. That situation appears to be yet another piece of evidence for an interaction between certain radiopharmaceuticals and the patient's pharmacologic-physiologic-pathologic status that is only slowly being understood.

SUMMARY

1. The consistent desirable clinical performance of ^{99m}Tc-medronate preparations can be predicted in vitro. Such performance is associated with a numerically high and precise quantity of ^{99m}Tc-medronate chelate as measured by standard TLC method.
2. The concentration of reduced unbound technetium and ^{99m}Tc-pertechnetate contaminants at the levels found in this study did not adversely affect the clinical bone image quality.
3. A rapid miniaturized TLC method was found to be neither sensitive nor specific enough to serve as an assay system for judging suitability for injection of a ^{99m}Tc-medronate radiopharmaceutical. Any user of a rapid TLC method for such purpose must first assure its adequacy for the task at hand.

REFERENCES

1. G. Subramanian, J. G. McAfee, R. J. Blair, F. A. Kallfelz, and F. D. Thomas, J. Nucl. Med., 16, 744 (1975).
2. J. J. Coupal, E. E. Kim, and F. H. DeLand, J. Nucl. Med., 22, 153 (1981).
3. C. Williams, Section IV in Radionuclide Handling and Radiopharmaceutical Quality Assurance Workshop Manual, U.S. Department of H.E.W., Nuclear Medicine Laboratory, Cincinnati, Oh., Nov. 14, 1975, p. 5.
4. T. Kawada, W. Wolf, and S. F. Botros, Radiochemical purity: Mini-Radiochromatography or Maxi-Radiochromatography? unpublished.

CHAPTER 26

Mycotoxin Analysis by TLC

Peter M. Scott

INTRODUCTION

Since the last review on TLC of mycotoxins other than aflatoxin for the first symposium on Clinical and Environmental Applications of Quantitative TLC (1), considerable progress has been made in the application of liquid chromatography (LC) to mycotoxin analysis. In some cases, for example, for aflatoxins M_1 and M_2 in milk (2), LC is being used on a routine basis. It is also offered as an alternative means of quantifying aflatoxins in the AOAC method for cottonseed products (3). Gas-liquid chromatography (GLC) has become the procedure of choice for measurement of the trichothecenes. However, TLC is still the most widely used technique for analytical chromatography of mycotoxins (3, 4). The aflatoxins and many of the other known mycotoxins are fluorescent under ultraviolet (UV) light, while others such as patulin require spray reagents for their detection (3). This paper updates the previous one (1), with the addition of the aflatoxins, and presents on a selective basis the best available TLC detection procedures for those individual mycotoxins for which at least some attempt at development of a method for their analysis in foods or grains has been made, without going into details of extraction and cleanup.

As an indication of the sensitivity of conventional TLC for detection of the more important mycotoxins, limits of visual detection on silica gel layers by either fluorescence or after

TABLE I. TLC DETECTION LIMITS (VISUAL) FOR FLUORESCENT MYCOTOXINS ON SILICA GEL LAYERS[a]

Mycotoxin	Fluorescence Color under UV Light (L = Long wave, S = Shortwave)		Detection Limit (ng)
Aflatoxins B_1, G_1	Blue, green	(L)	0.1
Aflatoxins B_2, G_2	Blue, green	(L)	0.04
Aflatoxin M_1	Blue	(L)	0.2
Ochratoxin A	Blue-green	(L + S)	2
Sterigmatocystin	Brick-red	(L)	40
Zearalenone	Greenish-blue	(S)	20
Alternariol, alternariol methyl ether	Blue	(L + S)	20
Citrinin	Yellow	(L)	5
Citreoviridin	Yellow	(L)	
PR toxin	Green	(L after S)	25
Rubratoxin B	Blue	(L, after heating)	500

[a]Adapted from Scott (1).

use of spray reagents are given in Tables I and II. These limits are approximate only and will vary with the UV lamp, chromatographic conditions, and the observer. By using high-performance thin layer chromatography (HPTLC), much lower sensitivities can be obtained (in the low picrogram range for naturally fluorescent mycotoxins by instrumental scanning, Table III) (5). For the purposes of this paper, however, discussion will be restricted to the conventional TLC in use in most mycotoxin laboratories. Another recent innovation is the use of thin layers of corn starch, on which markedly lower limits of detection by fluorescence were observed for aflatoxins B_1, B_2, G_1, and G_2, aflatoxins M_1 and M_2, zearalenone, zearalenol, and ochratoxins A,B,B-ester and C (Table IV) (6). Whether this technique will prove of practical

TABLE II. DETECTION OF MYCOTOXINS USING SPRAY REAGENTS (SILICA GEL LAYERS)[a]

Mycotoxin	Reagent (S = Shortwave, L = Longwave)	Color or UV Fluorescence of Spot	Visual Detection Limit (ng)
Sterigmatocystin	$AlCl_3$, heat, UV (S)	Yellow	5
Zearalenone	Fast Violet B Salt/pH 9/H^+	Mauve	5
Patulin	MBTH, heat, UV (L)	Yellow-brown	10
	Aniline, heat, UV (L)	Greenish-yellow	4
Penicillic acid	Acidic anisaldehyde, heat, UV (L)	Blue	10
	MBTH, heat, UV (L)	Yellow	50
Tenuazonic acid	Acidic anisaldehyde, heat, UV (L)	Green	500
T-2 toxin	Acid, heat, UV (L)	Blue	54
Diacetoxyscirpenol	Acid, heat, UV (L)	Blue	50
Penitrem A	$FeCl_3$	Greenish-blue	
	p-Dimethylaminobenzaldehyde, HCl	Blue	1000
Cyclopiazonic acid	Oxalic acid (in layer), heat	Purple	100
	Ehrlich reagent	Purple-blue	10
Cyclochlorotine	Cl_2, *o*-tolidine, KI	Yellow	
Cytochalasins A, B	65% H_2SO_4, heat, UV (L)	Blue	
Mycophenolic acid	Diethylamine, UV (L)	Blue	

[a]Adapted from Scott (1).

TABLE III. HIGH PERFORMANCE TLC OF MYCOTOXINS[a]

Mycotoxin	Detection Limit (ng) by Instrumental Scanning: UV-Visible	Fluorescence
Sterigmatocystin	0.2	
Zearalenone	1.0	0.05
Citrinin	1.0	0.01
Ochratoxin A	1.0	0.005
Patulin	0.2	
Penicillic acid	2.0	
Luteoskyrin	0.5	
Aflatoxins B_1, G_1, M_1, M_2	0.5	0.005
Aflatoxins B_2, G_2	0.2	0.002

[a]Reference: Lee et al. (5).

TABLE IV. TLC OF MYCOTOXINS ON CORN STARCH[a]

Mycotoxins	Detection Limit (ng)
Aflatoxins B_1, B_2, G_1, G_2	0.001
Aflatoxins M_1, M_2	0.01
Zearalenone, zearalenol	1
Ochratoxin A,B,B-ester,C	0.1

[a]Reference: Mišković and Perišić-Janjić (6).

use for food analysis remains to be seen; one problem noted was that slight changes in mycotoxin concentration had a considerable effect on R_f values. Solvents used for TLC on starch were generally nonpolar (e. g., toluene) in contrast to the polar solvent systems such as toluene/ethyl acetate/formic acid (5:4:1), chloroform/acetone (9:1), and chloroform/methanol (93:7) employed with silica gel layers. Also of interest is a recent report that layers of "silufol," a Czechoslovakian product, gave 4 to 5 times more sensitivity than silica gel G or Adsorbosil 1 for several fluorescent mycotoxins (7).

AFLATOXINS

That the aflatoxins, metabolites of *Aspergillus flavus* and *A. parasiticus*, include the most carcinogenic of the known mycotoxins is compensated for by their being the most intensely fluorescent and hence the most readily detected by TLC (Table I). It was fortunate that TLC had become available by the time aflatoxins were first being isolated. Initially, various chloroform-methanol mixtures were used for TLC on silica gel layers but improved resolution was only apparent when acetone was used as the more polar component of the mobile phase. The aflatoxins are resolved in the increasing order of polarity B_1, B_2, G_1, G_2, and M_1. A brief summary of the TLC of aflatoxins is given in Table V, but for a more complete account, *The Offical Methods of Analysis of the Association of Official Analytical Chemists* (AOAC) (2) and the review by Nesheim (8) should be consulted. The benzene-ethanol-water system, which is actually the top layer of a two-phase system, is used in the AOAC methods for peanuts and peanut products when interference in the G_1-G_2 area of the chromatogram is suspected (2). Indeed, the importance of adequate confirmation of identity cannot be overemphasized. For example, interferences in products containing fruit or fruit components (9, 10) or feeds containing an ethoxyquin conversion product (11) will require additional tests such as two-dimensional TLC or formation of the hemiacetal of aflatoxins B_1, G_1, or M_1 with trifluoroacetic acid, either on the isolated toxins or, more simply, at the origin of the TLC plate (2). A widely used convenient test for negative confirmation is to spray the chromatogram with acid, which changes the fluorescence color of the aflatoxins from blue or green to yellow, although it must be borne in mind that interferences may also undergo this change.

Many of the technical details worked out for TLC of the aflatoxins are also applicable to the other mycotoxins, although the aflatoxins and, in certain countries, ochratoxin A and patulin are the only mycotoxins that are at present subject to regulatory control. For example, in Canada nuts and nut products

TABLE V. TLC OF AFLATOXINS

Adsorbent	Silica gel
Solvents	$CHCl_3$/acetone (9:1), not equilibrated
	C_6H_6/ethanol/H_2O (40:6:3)
(for M_1)	$CHCl_3$/acetone/isopropanol (85:10:5)
Standards	0.5 μg B_1 or G_1/ml C_6H_6-CH_3CN (98:2)
	0.5 μg M_1/ml C_6H_6-CH_3CN (9:1)
	0.1 μg B_2 or G_2/ml C_6H_6-CH_3CN (98:2)
Detection	Longwave UV
Overall detection limit of AOAC methods	About 1 μg/kg
	About 0.1 μg M_1/l or kg dairy products
Fluorodensitometry	Excitation 365 nm, emission 420 to 460 nm
Natural occurrence	Nuts and nut products, corn, cottonseed, copra, etc. (B_1, B_2, G_1, G_2)
	Dairy products (M_1, M_2)

may not contain more than 15 μg/kg of total aflatoxins. The AOAC prescribes procedures for preparation, calibration, and storage of reference standards, determination of their chromatographic purity, and testing of the silica gel used for TLC. The coefficient of variation within one laboratory for TLC determination by visual comparison of fluorescence intensity of the spots is about 20% (fluorodensitometry is more precise with a coefficient of variation of 5 to 15%) (8). However, when the TLC step alone was tested between laboratories in the Smalley Alfatoxin Check Sample Program, the coefficients of variation averaged 65% for analysis of a working standard according to McKinney (12). This gives some idea of the problems analysts still have with semi-quantitative TLC, although in collaborative studies of methods for foodstuffs subsequently recommended as referee methods, overall coefficients of variation were 15 to 40% (13). No improvement in precision has been discerned from collaborative studies for fluorodensitometric over visual measurements (13), so use of either way of reading aflatoxin spots is permitted in official methods.

OCHRATOXINS

Ochratoxin A is a potent nephrotoxin. Because of its occurrence in Danish and Swedish cereals it has been found as a residue in pig kidneys, a situation that in Denmark has resulted in a regulatory action level of 10 µg/kg in the kidney (14, 15). Evidence of contamination of foodstuffs by ochratoxin A in the Balkans was provided by a recent survey that used TLC for the final analytical step (16); thus ochratoxin A might be a determinant of human nephropathy endemic in the Balkan region. This survey and other work (17) also extend the application of the AOAC method for barley to corn, wheat, wheat bread, and milo.

Much of the recent analytical effort on ochratoxins has been directed toward LC, and the basic TLC parameters remain as summarized in Table VI. Exposure of the TLC plate to ammonia fumes improves the detection limit for ochratoxin A from 0.5 to 0.25 ng as measured by fluorodensitometry according to Trenk and Chu (18). Ochratoxins A and B are carboxylic acids and scarcely migrate in neutral solvent systems unless oxalic acid is incorporated into the silica gel layer as reported by Gorst-Allman and Steyn (19).

STERIGMATOCYSTIN

Sterigmatocystin is not a very fluorescent compound, and it is necessary to use aluminum chloride, with heating, as a spray reagent (Table VII). The fluorescence intensity of the spot slowly goes down with time (20). The official AOAC method is only applicable to barley and wheat (3), but alternative TLC methods for corn, oats, and other agricultural commodities are available (20-23). Two-dimensional TLC is a valuable technique and was used to show the natural occurrence of sterigmatocystin in cheese ripening in warehouses and naturally molded with *Aspergillus versicolor* (24). Procedures for confirmation of sterigmatocystin involve derivatization with trifluoroacetic acid or aqueous hydrochloric acid or formation of the acetate, subsequently followed by TLC (3, 23, 24).

TABLE VI. TLC OF OCHRATOXINS[a]

Adsorbents	(a) Silica gel; (b) oxalic acid impregnated silica gel
Solvents	(a) C_6H_6/AcOH/MeOH (18:1:1), not equilibrated Toluene/EtAc/HCOOH (5:4:1) Hexane/acetone/AcOH (18:2:1) for esters (b) $CHCl_3$/acetone (9:1), $CHCl_3$/MeOH (98:2)
Standards	1 to 5 μg/ml C_6H_6/AcOH (99:1)
Detection	Long- and short-wave UV: blue-green fluorescence (blue with NH_3 or $NaHCO_3$)
Overall detection limit of AOAC methods	12 μg/kg barley; 20 μg/kg green coffee
Fluorodensitometry	Ochratoxin A, excitation 340 nm, emission 475 nm
Natural occurrence (ochratoxin A)	Grains, white beans, green coffee beans, flour, swine tissue.

[a]Adapted from Scott (1).

ZEARALENONE

Assay methods for the estrogenic *Fusarium* mycotoxin zearalenone have been reviewed by Shotwell (25) and by Bennett and Shotwell (26). Zearalenone is detected by fluorescence under short-wave UV light in the AOAC method for zearalenone in corn (3, 27), which has recently been shown to be applicable to grain sorghum by Shotwell et al. (28), but greater sensitivity and specificity are possible with a diazonium salt spray reagent such as Fast Violet B Salt according to Scott et al. (29) (Table VIII). Harrach and Palyusik (30) evaluated 14 spray reagents for visual-

TABLE VII. TLC OF STERIGMATOCYSTIN[a]

Adsorbent	Silica gel
Solvents	C_6H_6/AcOH/MeOH (18:1:1) C_6H_6/AcOH (9:1) $CHCl_3$/acetone (85:15)
Standard	5 μg/ml C_6H_6
Detection	Spray 20% $AlCl_3 \cdot 6H_2O$ in EtOH, heat 80° for 10 minutes, short-wave UV: yellow fluorescence
Overall detection limit of methods	20 to 60 μg/kg grains, soybeans, green coffee, fruits, vegetables
Fluorodensitometry	$AlCl_3$ derivative, excitation 360 nm, emission 500 nm
Natural occurrence	Feed grain, rice, pecans, cheese

[a]Adapted from Scott (1).

ization of zearalenone and preferred 4-methoxybenzene diazonium fluoroborate for specificity in grain and feed analysis. The last three years have seen the publication of some 13 papers on the LC methods for zearalenone and recently the first minicolumn screening method was developed by Holaday (31), who had noted that fluorescence of zearalenone on alumina was several times greater than on silica gel when observed under long-wave UV light. Thin layer chromatograms and fluorescence of zearalenone and an *Alternaria* metabolite, alternariol methyl ether, are reported to be similar, but the compounds can be separated in benzene/ethanol/0.4*N* NaOH (96:4; 5 drops) and differentiated by the latter's failure to form a colored spot with Fast Violet B Salt (32).

TABLE VIII. TLC OF ZEARALENONE[a]

Adsorbent	Silica gel
Solvents	$CHCl_3$/EtOH (95:5) $CHCl_3$/MeOH (93:7) C_6H_6/AcOH (95:5)
Standard	50 µg/ml C_6H_6-CH_3CN (98:2) 5 µg/ml toluene
Detection	Short-wave UV: greenish-blue fluorescence (blue under long-wave UV after spraying 20% ethanolic $AlCl_3$, 130°/5 minutes) Spray 0.7% Fast Violet B Salt, then pH 9 buffer, then 50% H_2SO_4, 120° for 5 minutes: mauve spot
Overall detection limit of methods	100 µg/kg corn (AOAC), 20 µg/kg cornflakes
Fluorodensitometry	Excitation 313 nm, emission 443 nm
Natural occurrence	Grains, especially grain sorghum, corn, and corn products; pecans

[a]Adapted from Scott (1).

PATULIN

The last mycotoxin for which an AOAC method exists is patulin, frequently found in apple juice as a result of *Penicillium expansum* rot of apples (Table IX). While development in the area of LC is slowly progressing, several new papers concerning TLC of patulin have recently appeared. Two-dimensional TLC has found continued favor as a means of final separation of patulin from potential interferences in extracts of fruits and fruit products as reported by Fritz et al. and others (33-37). Fluorodensitometric measurement of spots visualized by various means, including

TABLE IX. TLC OF PATULIN

Adsorbent	Silica gel
Solvents	Toluene/EtAc/HCOOH (5:4:1) Hexane/ether (1:3) $CHCl_3$/MeOH (95:5) $CHCl_3$/acetone (9:1)
2D TLC	(1) C_6H_6/MeOH/AcOH (19:2:1) or $CHCl_3$/acetone (9:1) (2) toluene/EtAc/HcOH (5:4:1) or other combinations of solvent systems
Standard	10 μg/ml $CHCl_3$
Detection	Spray 0.5% 3-methyl-2-benzothiazolinone hydrazone (MBTH) hydrochloride, heat 130° for 15 minutes, long-wave UV: yellow-brown fluorescence Spray 10% aniline (v/v) in dilute acetic acid, heat 80° for 10 minutes, UV: greenish-yellow fluorescence
Overall detection limit of methods	$\geq$20 μg/l apple juice (AOAC), 40 μg/kg corn, 25 μg/l wine
Densitometry	SIL G-25 UV 254, reflectance, 273 nm, 5 ng detection limit
Fluorodensitometry	(MBTH sprayed spot), 2D TLC, excitation 300 to 400 nm, 25 ng detection limit
Natural occurrence	Apple and other fruit juices, bread, silage

fluorescence quenching (33), spraying and heating with 3-methyl-benzothiazolinone hydrazone (MBTH) hydrochloride according to Ough and Corison (36), and spraying and heating with aniline in dilute acetic acid (37) have been successfully carried out. The latter reagent allowed visual detection of as little as 4 ng patulin following 2D TLC while MBTH is the reagent used in the AOAC method (3). MBTH gives rise to a yellow-brown fluorescence, which is less intense after TLC development in neutral solvent systems than in an acidic system. It is thus useful to add 5% of formic acid to the spray reagent.

The search has continued for a simple, sensitive TLC confirmatory test for patulin. The aniline imine has been formed, which was then chromatographed and detected on the TLC plate with fluorescamine after liberation of aniline by hydrogen chloride (38); this technique is sensitive, as little as 5 ng patulin can be detected, and measurements can be made in the 10 to 100 ng range. By comparison, patulin phenylhydrazone and 2,4-dinitrophenylhydrazone formation are much less sensitively detected by TLC (39, 40). Patulin acetate can be formed on the origin of the TLC plate and 20 ng detected after development with chlorine/benzidine as described by Paul (41). These latter reagents, incidentally, offer good sensitivity for patulin itself, but their use is precluded by the carcinogenicity of benzidine.

RECENT DEVELOPMENTS IN THE TLC OF OTHER MYCOTOXINS

TLC conditions for penicillic acid, alternariol and its methyl ether, tenuazonic acid, citrinin, trichothecenes, tremortins, verruculogen, rubratoxin B, cyclochlorotine, citreoviridin, luteoskyrin, cyclopiazonic acid, PR toxin, moniliformin, cytochalasins, and ergot alkaloids were summarized at the First Symposium on Environmental Applications of Quantitative Thin Layer Chromatography (1). New developments on these mycotoxins are discussed below.

No spray reagent giving improved detectibility of penicillic acid has been found and acidic anisaldehyde, which gives rise to a blue fluorescent spot on heating, remains the reagent of choice, although other aldehydes have been tested (42). The effect of hydrogen chloride on the ammonia derivative of penicillic acid has also been investigated (43). The acidic anisaldehyde reagent produced a green fluorescence with the *Alternaria* toxin tenuazonic acid but the amount detectable on silica gel layers, which give rise to a greatly elongated spot, was still limited to 500 ng, the same as observed by fluorescence quenching on oxalic acid-impregnated silica gel as described by Scott and Kanhere (44). These impregnated layers offer a significant improvement in compactness of spots not only of tenuazonic acid but also of another acidic mycotoxin citrinin. A 10-fold improvement in detection limit to as little as 0.8 ng of standard citrinin was observed, with method detection limits for grains lowered to 70 to 95 µg/kg (45, 46); mobile phases may be neutral, for example, chloroform/methanol/hexane (128:2:70). Another solution to eliminate streaking of citrinin is to incorporate ethylenediamine tetraacetic acid in the silica gel layer to complex contaminants (47); streaking of luteoskyrin is also avoided by this technique (48).

The TLC of trichothecenes was included in a review on their analysis by Eppley (49). Recently a new spray reagent was reported, 4-(*p*-nitrobenzyl)pyridine, which has specificity for the 12,13-epoxy group. A blue spot on a colorless background was produced on spraying with or dipping in the reagent, heating at 150°C, and then treating with tetraethylenepentamine solution (50). Detection limits for the 12 trichothecenes examined were 20 to 200 ng per spot. TLC thus remains inferior to GLC for analysis of trichothecenes as far as sensitivity is concerned as reported by Romer (51).

It is of interest to mention here that penitrem A was detected by TLC in a sample of moldy cream cheese involved in the intoxication of two dogs as described by Richard and Arp (52). Cyclopiazonic acid has been detected in 11 out of 20 retail Camembert cheese crusts at concentrations of up to 1.5 µg/g (53). Three newly discovered indolic tremorgenic toxins, janthitrems A, B, and C, are highly fluorescent under long-wave UV light, unlike all previously discovered *Penicillium* tremorgens according to Gallagher et al. (54). Also two nonindolic tremorgens from *Aspergillus terreus*, territrems A and B, have a blue fluorescence resembling that of the aflatoxins; indeed they have similar R_f values to aflatoxins B_1 and B_2 in the commonly used solvent system chloroform/acetone (9:1) (55).

Improved method sensitivity for the analysis of rubratoxin B in corn has been noted in a recent multimycotoxin method in a report by Whidden et al. (56). Ten mg/kg could be detected using a new mobile phase, acetonitrile/acetic acid (100:2), but recoveries were poor at this level. The fluorescence formed when rubratoxin B on a TLC plate was heated to 200°C for 10 minutes (57) changed in color and contrast compared to other nearby fluorescent spots when exposed to ammonia.

MULTIMYCOTOXIN TLC

Rapid screening for the simultaneous analysis of several mycotoxins in the same sample of an agricultural commodity is an important and developing area of analytical research in the mycotoxins. Attempts are being made to apply LC to this problem (58, 59), but although sensitivity is generally better than for TLC, much additional effort is needed. Whereas the eye can readily detect and differentiate, say, a blue and yellow fluorescent spot on a TLC plate, simultaneous instrumental analysis by LC may require more than one detector. Thus TLC generally remains the procedure of choice for multimycotoxin detection, bearing in mind that some tradeoff in method sensitivity is usually necessary. Table X summarizes the available methods for various agricultural commodities. Several systems for one-dimensional TLC of several mycotoxins have been developed as reported by Scott (1). The mobile phase toluene/ethyl acetate/formic acid (5:4:1 or 6:3:1), for example, has proved of wide applicability and requires little maintenance for reproducible results other than occasionally topping up the level in the solvent trough. Typical R_f values illustrating relative polarities of several mycotoxins in this acidic system and in a neutral solvent system are shown in Table XI. A universal spray reagent that gives reasonable sensitivity for nonfluorescent mycotoxins has not been found, but various sulfuric acid sprays, followed by heating and examination under UV light have been used; for the trichothecenes this is the reagent of choice. Recent methods for multimycotoxin analysis, whether of standards alone (46) or extracts of feeds and foodstuffs (74), have if anything become more complex with respect to number of mobile phases recommended or number of spray reagents to be employed. The two-dimensional TLC of Patterson and Roberts (75) is simple and permits detection of low levels of aflatoxin B_1, ochratoxin A, zearalenone, sterigmatocystin, and T-2 toxin in grains and feeds following overnight dialysis. Finally, TLC is invaluable in screening fungal extracts for known mycotoxins. As an example of this from our own laboratory, fungi isolated from tomato were examined for production of 29 toxins, with the limitation that the isolates identified as *Fusarium*, *Penicillium*, *Geotrichum*, and *Alternaria* were tested only for those mycotoxins known to be metabolites of these genera (76). Limits of detection are not normally a serious consideration in this type of work.

REFERENCES

1. P. M. Scott, in Thin Layer Chromatography: Quantitative Clinical and Environmental Applications, J. C. Touchstone and D. Rogers (Eds.), Wiley, New York, 1980, p. 251.
2. J. F. Foos and J. D. Warren, A Rapid Cleanup Procedure for the Determination of Aflatoxin M_1 and M_2 in milk by HPLC. Abstr. 94th Ann. Meeting AOAC, Oct. 20-23, Washington, D. C., 1980, p. 54.
3. L. Stoloff, in Official Methods of Analysis of the AOAC, 13th edi., AOAC, Washington, D. C., 1980, Sec. 26.001-26.132, Chapter 26.
4. L. Stoloff, J. Assoc. Offic. Anal. Chem., 63, 247 (1980).
5. K. Y. Lee, C. F. Poole, and A. Zlatkis, Anal. Chem., 52, 837 (1980).
6. D. Mišković and N. Perišić-Janjić, Chromatogr., 12, 33 (1979).
7. L. S. Lvova, L. V. Kravchenko, and A. P. Shuligna, Priklad. Biokhim. Mikrobiol., 15, 143 (1979).
8. S. Nesheim, Method of Aflatoxin Analysis, in National Bureau of Standards Special Publication 519, Trace Organic Analysis: A New Frontier in Analytical Chemistry, Proc. 9th Materials Res. Symp., April 10-13, 1978, Gaithersburg, Md., 1979, p. 355.
9. P. Lafont and J. Lafont, Cah. Nutr. Diét., 10, 57 (1975).
10. A. V. Jain and R. C. Hatch, J. Assoc. Offic. Anal. Chem., 63, 626 (1980).
11. R. T. Gallagher and H. M. Stahr, J. Agric. Food Chem., 28, 133 (1980).
12. J. D. McKinney, 94th Ann. Meeting AOAC, Oct. 20-23, Washington, D. C., 1980, p. 56.
13. P. L. Schuller, W. Horwitz, and L. Stoloff, J. Assoc. Offic. Anal. Chem., 59, 1315 (1976).
14. K. Hult, E. Hökby, S. Gatenbeck, and L. Rutqvist, Appl. Environ. Microbiol., 39, 828 (1980).
15. E. Josefsson, Vår Föda, 31, 415 (1979).
16. M. Pavlović, R. Plestina, and P. Krogh, Acta Pathol. Microbiol. Scand. B, 87, 243 (1979).
17. Y. Shishido, Shiryo Kenkyu Hokoku (Tokyo Hishiryo Kensasho), 5, 123 (1980); Chem. Abstr., 93, 24557h (1980).
18. H. L. Trenk and F. S. Chu, J. Assoc. Offic. Anal. Chem., 54, 1307 (1971).
19. C. P. Gorst-Allman and P. S. Steyn, J. Chromatogr., 175, 325 (1979).

TABLE X. MULTIMYCOTOXIN METHODS FOR AGRICUL

Method (Ref.)	Commodity	Aflatoxin	Ochra-toxin A	Ochra-toxin C	Zearalenone
Eppley (60)	Various	<32 (B_1)	<55		<500
	Corn	1-3 (B_1 or G_1)	50		200
Vorster(61)	Grains, peanuts	4 (B_1)	20	20	
Stoloff et al. (62)	Grains	20 (B_1 or G_1)	45-90	50-100	200-500
Scott et al. (63)	Grains	+	+		
Scott et al. (63)	Grains	+	+		+
Thomas et al. (64)	Corn	2 (B_1)			100
Hagan and Tietjen (65)	Oils	+	+	+	+
Roberts and Patterson(66)	Feed	3 (B_1)	80		1000
Wilson et al. (67)	Corn	2 (B_1 or G_1)	20		200-300 (Corn, Beans, Peanuts)
	Beans	2 (B_1 or G_1)	20		
	Peanuts	2 (B_1 or G_1)	40		
Takeda et al. (68)	Grains	<10 each	<40		
Balzer et al. (69).	Corn	2 (B_1)	40		200

TURAL COMMODITIES, DETECTION LIMITS (μg/kg)[a]

Sterig-matocystin	Patulin	Penicillic Acid	Citrinin	T-2 Toxin	Diacetoxy-scirpenol	Other Mycotoxins
100						
60	400-1000					
		+	100			
+						
+						
330	600		+	+	4000	Penitrem A +
		300-400	100-200			
		300-500	400-500			
		1000	ND			
<40			<80			

Table X (continued)

Method (Ref.)	Commodity	Aflatoxin	Ochra-toxin A	Ochra-toxin C	Zearalenone
Josefffson and Möller (70)	Cereals	5 (each)	10		35
Yamamoto (71)	Flours	5 (B_1)			
Takeda et al. (72)	Grains, peanuts	10	40-60		300-500
Siriwardana and Lafont (32)	Cheese	1 (B_1 and M_1)			
Gimeno (45)	Grains, feeds, etc.	4-5 (B_1, G_1)	140-145	140-145	410-500
Lvova et al. (7)	Grains	1-5	20		80
Patterson and Roberts (75) (2D-TLC)	Grains, feeds	0.1-0.3 (B_1)	5-10		20-100
Whidden et al. (56)	Corn	+ (B_1)	+		+

[a] + means that toxin is detectable, but limit is not determined.

Sterig-matocystin	Patulin	Penicillic Acid	Citrinin	T-2 Toxin	Diacetoxy-scirpenol	Other Mycotoxins
10	50					
40						
40-60			80-200	500	800	Luteoskyrin 100 Rugulosin 100 Fusarenone X 300-500 Neosolaniol 500
20	20	30				Mycophenolic acid 20
140-145	750-800	3400-3650	70-95	750-950	2400-2600	Penitrem A 14,000-14,300
20						
10-20				200		
+	+	+			+	Rubratoxin B 10,000

TABLE XI. TYPICAL R_f VALUES OF SOME MYCOTOXINS ON SILICA GEL G[a]

	R_f	
	Tol./EtAc/90% HCOOH (6:3:1)	$CHCl_3$/Acetone (9:1)
Citrinin	0-0.48	0-0.20
Luteoskyrin	0.60	0-0.20
Nivalenol	0.05	0.0
Butenolide	0.12	0.08
Kojic acid	0.15	0.02
Aflatoxin G_2	0.08	0.17
Aflatoxin G_1	0.19	0.25
Aflatoxin B_2	0.18	0.33
Aflatoxin B_1	0.32	0.33
Diacetoxyscirpenol	0.31	0.35
T-2 toxin	0.32	0.39
Patulin	0.37	0.25
Penicillic acid	0.43	0.23
Gliotoxin	0.43	0.32
Ochratoxin A	0.60	0-0.23
Zearalenone	0.59	0.62
Sterigmatocystin	0.71	0.74

[a]Reference: Ďuračková et al. (73).

20. R. Schmidt, K. Neunhoeffer, and K. Dose, Fresenius Z. Anal. Chem., 299, 382 (1979).
21. A. K. Athnasios and G. O. Kuhn, J. Assoc. Offic. Anal. Chem., 60, 104 (1977).
22. G. M. Shannon and O. L. Shotwell, J. Assoc. Offic. Anal. Chem., 59, 963 (1976).
23. V. Thurm, P. Paul, and C. E. Koch, Nahrung, 23, 111 (1979).
24. H. P. Van Egmond, W. E. Paulsch, E. Deijll, and P. L. Schuller, J. Assoc. Offic. Anal. Chem., 63, 110 (1980).
25. O. L Shotwell, in *Mycotoxins in Human and Animal Health*, J. V. Rodricks, C. W. Hesseltine, and M. A. Mehlman (Eds.), Pathotox Publishers, Park Forest South, Ill., 1977, p. 403.

26. G. A. Bennett and O. L. Shotwell, J. Amer. Oil Chem. Soc., 56, 812 (1979).
27. O. L. Shotwell, M. L. Goulden, and G. A. Bennett, J. Assoc. Offic. Anal. Chem., 59, 666 (1976).
28. O. L. Shotwell, G. L. Bennett, M. L. Goulden, R. D. Plattner, and C. W. Hesseltine, J. Assoc. Offic. Anal. Chem., 63, 922 (1980).
29. P. M. Scott, T. Panalaks, S. Kanhere, and W. F. Miles, J. Assoc. Offic. Anal. Chem., 61, 593 (1978).
30. B. Harrach and M. Palyusik, Acta Vet. Acad. Sci. Hung., 27, 77 (1979).
31. C. E. Holaday, J. Amer. Oil Chem. Soc., 57, 491A (1980).
32. L. M. Seitz, D. B. Sauer, H. E. Mohr, R. Burroughs, and J. V. Paukstelis, J. Agric. Food Chem., 23, 1 (1975).
33. W. Fritz, C. Buthig, and R. Engst, Nahrung, 23, 159 (1979).
34. C. E. Koch, V. Thurm, and P. Paul, Nahrung, 23, 125 (1979).
35. S. Lindroth and A. Niskanen, J. Food Sci., 43, 446 (1978).
36. C. S. Ough and C. A. Corison, J. Food Sci., 45, 476 (1980).
37. M. G. Siriwardana and P. Lafont, J. Chromatogr., 173, 425 (1979).
38. J. C. Young, J. Environ. Sci. Health B, 14, 15 (1979).
39. R. A. Meyer, Nahrung, 22, 75 (1978).
40. E. E. Stinson, C. N. Huhtanen, T. E. Zell, D. P. Schwartz, and S. F. Osman, J. Agric. Food Chem., 25, 1220 (1977).
41. P. Paul, Nahrung, 22, K21 (1978).
42. S. Neelakantan, T. Balasubramanian, R. Balasaraswathi, G. I. Jasmine, and R. Swaminathan, J. Food Sci. Technol., 15, 125 (1978).
43. D. Veselý and D. Veselá, Chem. List., 74, 189 (1980).
44. P. M. Scott and S. R. Kanhere, J. Assoc. Offic. Anal. Chem., 63, 612 (1980).
45. A. Gimeno, J. Assoc. Offic. Anal. Chem., 63, 182 (1980).
46. C. P. Gorst-Allman and P. S. Steyn, J. Chromatogr., 175, 325 (1979).
47. R. D. Stubblefield, J. Assoc. Offic. Anal. Chem., 62, 201 (1979).
48. K. Y. Lee, C. F. Poole, and A. Zlatkis, Anal. Chem., 52, 837 (1980).
49. R. M. Eppley, J. Amer. Oil Chem. Soc., 56, 824 (1979).
50. S. Takitani, Y. Asabe, T. Kato, M. Suzuki, and Y. Ueno, J. Chromatogr., 172, 335 (1979).
51. T. R. Romer, T. M. Boling, and J. L. MacDonald, J. Assoc. Offic. Anal. Chem., 61, 801 (1978).
52. J. L. Richard and L. H. Arp, Mycopath., 67, 107 (1979).

53. J. Le Bars, Appl. Environ. Microbiol., 38, 1052 (1979).
54. R. T. Gallagher, G. C. M. Latch, and R. G. Keogh, Appl. Environ. Microbiol., 39, 272 (1980).
55. K. H. Ling, C.-K. Yang, and H.-C. Huang, Appl. Environ. Microbiol., 37, 358 (1979).
56. M. P. Whidden, N. D. Davis, and U. L. Diener, J. Agric. Food Chem., 28, 784 (1980).
57. A. W. Hayes and H. W. McCain, Food Cosmet. Toxicol., 13, 221 (1975).
58. E. Josefsson and T. Möller, J. Assoc. Offic. Anal. Chem., 62, 1165 (1979).
59. D. C. Hunt, A. T. Bourdon, and N. T. Crosby, J. Sci. Food Agric., 29, 239 (1978).
60. R. M. Eppley, J. Assoc. Offic. Anal. Chem., 51, 74 (1968).
61. L. J. Vorster, Analyst, 94, 136 (1969).
62. L. Stoloff, S. Nesheim, L. Yin, J. V. Rodricks, M. Stack, and A. D. Campbell, J. Assoc. Offic. Anal. Chem., 54, 91 (1971).
63. P. M. Scott, W. van Walbeek, B. Kennedy, and D. Anyeti, J. Agric. Food Chem., 20, 1103 (1972).
64. F. Thomas, R. M. Eppley, and M. W. Trucksess, J. Assoc. Offic. Anal. Chem., 58, 114 (1975).
65. S. N. Hagan and W. H. Tietjen, J. Assoc. Offic. Anal. Chem., 58, 620 (1975).
66. B. A. Roberts and D. S. P. Patterson, J. Assoc. Offic. Anal. Chem., 58, 1178 (1975).
67. D. M. Wilson, W. H. Tabor, and M. W. Trucksess, J. Assoc. Offic. Anal. Chem., 59, 125 (1976).
68. Y. Takeda, E. Isohata, R. Amano, M. Uchiyama, Y. Naoi, and M. Nakao, J. Food Hyg. Soc. Japan, 17, 193 (1976).
69. I. Balzer, C. Bogdanic, and S. Pepeljnjak, J. Assoc. Offic. Anal. Chem., 61, 584 (1978).
70. B. G. E. Josefsson and T. E. Möller, J. Assoc. Offic. Anal. Chem., 60, 1369 (1977).
71. K. Yamamoto, J. Nara Med. Ass., 26, 245 (1975).
72. Y. Takeda, E. Isohata, R. Amano, and M. Uchiyama, J. Assoc. Offic. Anal. Chem., 62, 573 (1979).
73. Z. Ďuračková, V. Betina, and P. Nemec, J. Chromatogr., 116, 141 (1976).
74. A. Gimeno, J. Assoc. Offic. Anal. Chem., 62, 579 (1979).
75. D. S. P. Patterson and B. A. Roberts, J. Assoc. Offic. Anal. Chem., 62, 1265 (1979).
76. J. Harwig, P. M. Scott, D. R. Stolz, and B. J. Blanchfield, Appl. Environ. Microbiol., 38, 267 (1979).

CHAPTER 27

HPTLC Analysis of Mycotoxin Standards Using Multiple-Continuous Development Techniques

Kwan Y. Lee and Albert Zlatkis

INTRODUCTION

Mycotoxins are toxic fungal metabolites found as contaminants in agricultural and food products. The control of these hazardous contaminants in food material requires adequate analytical methods for each mycotoxin. Usually, several mycotoxin-producing fungi are isolated from the same foodstuff. Therefore, there is no real basis for deciding which mycotoxin to look for in any particular sample showing heavy contamination with molds. Since it is time-consuming to analyze all the mycotoxins separately, a rapid screening method for their simultaneous analysis would be preferred.

Several multimycotoxin screening methods have been reported (2-18, 38, 39), including the use of liquid chromatography, HPLC, gas chromatography, and even field ionization mass spectrometry. However, owing to its simplicity, TLC is the most frequently used method for multimycotoxin analysis. But because most of these TLC procedures were qualitative, too time-consuming, and provided poor resolution and sensitivity for multimycotoxin analysis, a search for an improved technique has continued.

In recent years, technical improvements in the manufacture of TLC plates and associated accessories brought about the development of a more precise, sensitive, and quantitative technique, namely, high-performance thin layer chromatography (HPTLC) (19-21). This technique lends itself readily to the analysis of complex multicomponent systems such as mycotoxin mixtures.

HPTLC has resulted from the application of advanced technology to the complete thin layer chromatographic system. Such a complete system includes optimized HPTLC plates, improved spotting techniques, flexible methods of solvent development, and special means for quantitative spectrophotometric detection.

Described here is the application of a complete HPTLC system to the analysis of mycotoxins, highlighting the advantages of a powerful development technique, namely, multiple-continuous development. Results obtained from the simultaneous analysis of 13 mycotoxins using this technique are compared to those obtained by conventional TLC and HPLC. Also, the recent application of circular and anticircular development systems to mycotoxin analysis is discussed.

The HPTLC System

Basic to the improvement of separations obtainable by HPTLC is the commercial availability of HPTLC plates coated with a very uniform small-particle silica gel adsorbent. The new HPTLC plates are prepared by optimizing the silica gel particle size and the gel layer thickness. The reduced thickness and greater uniformity of the gel layer allow for spotting smaller amounts of sample. Thus more samples can be spotted per plate because of the overall reduction of each sample spot size. With the high efficiency of the HPTLC plate, most separations can be performed within a sample migration distance of 3 to 6 cm with a 5- to 10-fold reduction in analysis time compared to conventional TLC.

One of the main hindrances to converting to the new HPTLC plates was the difficulty in applying small sample volumes (nanoliters) on the plates while maintaining optimum separation performance. Mechanical spotting applicators have been developed to meet the need for applying precise, reproducible samples occupying spots of less than 1.5 mm in diameter (20, 21, 37).

The final hurdle to be overcome in the practical application of HPTLC was the development of appropriate quantitative detection instrumentation and techniques. Spectrophotometric scanning and detection systems, such as those used in this work, are now commerciall available and have been designed specifically for HPTLC and also for conventional TLC.

In the circular case, the sample is applied to the center of the plate or in applications along a small concentration ring surrounding the center. The developing solvent is forced up to the surface of the plate at a constant flow rate by a motor-driven syringe. Solvent then flows radially outward from the center. As the sample components travel outward with the solvent, they broaden radially. This means the spot remains compact in the direction of the moving solvent, thus increasing the effective resolution, especially for the lower R_f values. This increase in resolution in the lower R_f range can be as much as four to five times over that obtained with linear chromatography under the same plate and solvent conditions.

With anticircular HPTLC, the samples are spotted on the outside periphery of the plate just inside the scribed circle. The solvent enters the layer at the circular line and flows toward the center. Here, as in the conventional linear system, the solvent speed decreases with the square of the distance. However, since the gel area to be wetted in this development mode also decreases with the square of the antiradial distance, the linear speed of solvent migration is significantly faster than the circular and conventional linear modes.

Circular and anticircular HPTLC have unique advantages useful in solving separation problems that cannot be solved by linear development approaches. In circular HPTLC, analytical conditions such as solvent flow can be controlled much better than in the linear or anticircular case. However, as compounds, especially those with high R_f values, move away from the center, their spots broaden radially as well as longitudinally, and their detection sensitivity is thus decreased. The low R_f spots, though, are compact and show higher resolution power than that obtainable by linear or anticircular development.

The primary advantages of anticircular development are (1) it provides the fastest solvent velocities of the three systems discussed, (2) a larger number of samples per plate can be spotted, and (3) great resolution is obtained for components of high R_f. Lower R_f value components are not as well-resolved as in the circular case, and so the two systems compliment each other.

EXPERIMENTAL

Mycotoxin Standards and Reagents

Sterigmatocystin, zearalenone, citrinin, ochratoxin A, patulin, penicillic acid, luteoskyrin, aflatoxin B_1, B_2, G_1, and G_2 were obtained from Supelco, Inc. (Bellefonte, Pennsylvania). Aflatoxin M_1 and M_2 were obtained as 10 µg/ml solutions in chloroform from the same source. Individual mycotoxin standard solutions were prepared at a concentration of 1.0 mg/ml in chloroform/methanol (1:1). A standard mixture of mycotoxins containing sterigmatocystin (2.0 mg/ml), citrinin (0.5 mg/ml), patulin (1.5 mg/ml), penicillic acid (5.0 mg/ml), luteoskyrin (3.0 mg/ml), aflatoxins B_1 and G_1 (0.1 mg/ml), aflatoxins B_2 and G_2 (0.03 mg/ml), aflatoxin M_1 (0.15 mg/ml), and aflatoxin M_2 (0.1 mg/ml) in chloroform/methanol (1:1) was prepared. All mycotoxin solutions were stored in the dark at 5°C. All dilutions of the standards were made with chloroform/methanol (1:1).

Certified ACS grade toluene, ethyl acetate, formic acid (88%), chloroform, and methanol were obtained from Fisher Scientific Company (Pittsburgh, Pennsylvania). Ethylenediamine tetracetic acid disodium salt (EDTA) and dye mixture were obtained from MCB (Norwood, Ohio).

These instruments can provide ultraviolet/visible absorption spectra and fluorescence excitation and emission spectra for the separated components on a plate. Usually the practical detection range of mycotoxins is in the 10 to 50 ng range for UV/visible measurements, and in the 100 to 500 pg range for fluorescence.

HPTLC Development Techniques

The overall speed of analysis for HPTLC makes possible the use of several powerful development techniques and systems that normally require too much time to be practical for conventional TLC. In general, the development techniques that are available for HPTLC are conventional-linear, horizontal, multiple (22-25, 28, 29, 35, 36), and continuous (25, 26, 34). The development systems to which these techniques may be applied are linear, circular (20, 21, 31), and anticircular (31, 32). The combination of the HPTLC plate and these development techniques and systems has led to greatly improved separation power and an increased number of samples per plate. The interested reader is referred to references cited above for a detailed discussion of the various development modes. However, a brief description of some of the more novel (and potentially powerful) techniques and systems follows.

Most users of TLC are familiar with conventional-linear and horizontal development techniques. It is generally considered that in the linear development mode, the maximum number of components that can be resolved in one development is about 10. For more complex mixtures, the resolution can be improved over that obtained in the single linear development mode if the sample is separated by multiple or continuous development.

In the multiple development mode, the plate is developed several times, and the solvent is evaporated from the plate, at the end of each development stage, before putting the plate back into the developing solvents. Redevelopment of the plate using the same solvent causes spot reconcentration to take place. On the second and subsequent passes, the solvent flows over the trailing edge of the spot before reaching the leading edge, contracting the spot. This method of development is very flexible, as quantitative measurements can be made at the end of any development stage as soon as a component becomes separated from the rest of the sample. Multiple development with different polarity of solvent systems of increasing polarity provides improved separation power for complex mixtures that themselves have a range of polarities. In this case, spot reconcentration effects are less pronounced or nonexistent.

In the continuous development mode, the solvent is permitted to migrate a short, predetermined distance on the plate, at which point it is continuously evaporated. Now the solvent migration velocity is constant, depending on the capillary action of the plate and the volatility of the solvent. A less polar solvent system may be used to improve difficult separations without sacrificing analysis time. Resolution is improved because of the short migration distance of the separated components, which show less diffusion. Unlike the multiple development techniques, no reconcentration forces are brought into play.

A combination of both development techniques--multiple and continuous development--is a very powerful separation tool when used in conjunction with an HPTLC plate. This technique will be discussed as it applies to the analysis of mycotoxins in a later section.

Figure 1 graphically depicts the difference in developing solvent flow among the linear, circular, and anticircular development systems. In the linear case, the samples are spotted on a line near the bottom of the plate. Solvent from the development tank is drawn up the plate by capillary action, and its speed is therefore inversely proportional to the square of the distance traveled. The separated spots begin to spread as they travel because of diffusional broadening associated with chromatographic action.

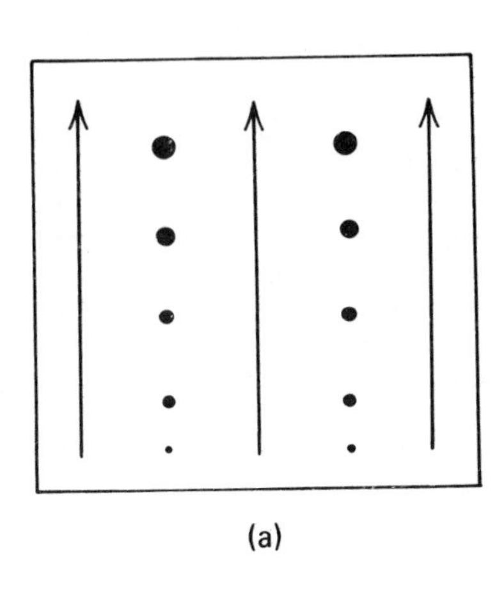

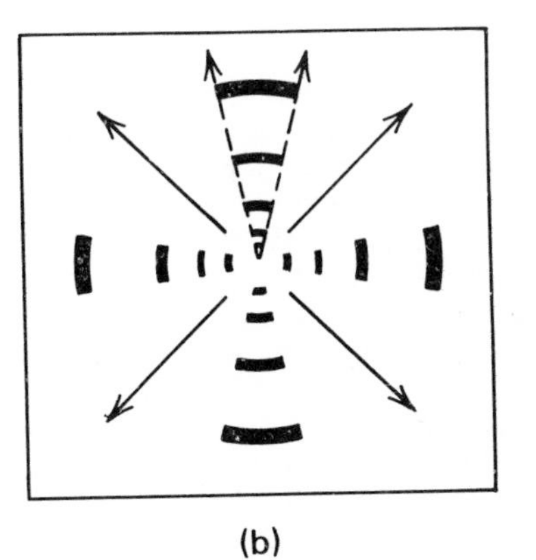

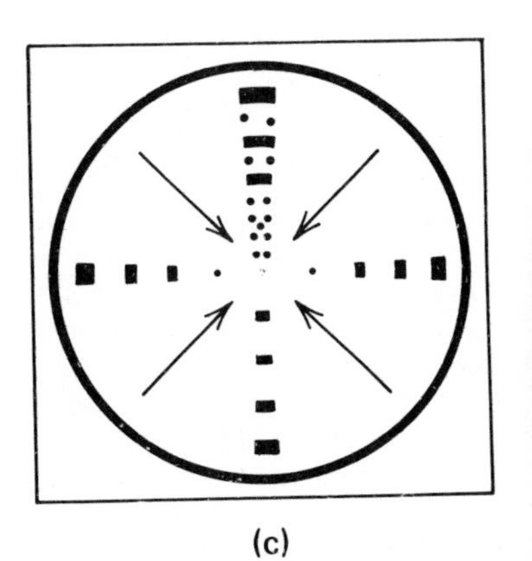

Figure 1. Schematic diagram of (a) linear, (b) circular, and (c) anticircular development mode of HPTLC.

Chromatography

Chromatography was performed on 10 × 10 cm high-performance thin layer plates coated with silica gel 60 (E. Merck, Darmstadt, West Germany). Prior to use, the plates were impregnated with EDTA by horizontal development in a saturated chamber containing a 10% w/v aqueous solution of EDTA. The wet plates were dried at room temperature for 30 minutes, at 85°C for 2 hr and then allowed to cool to room temperature. EDTA-impregnated plates were stored in a desiccator. Samples were applied to the TLC plate using a 200-nl Pt-Ir capillary (Antech, Bad Dürkheim, West Germany) attached to an EVA-Chrom Applicator (W + W Electronic Scientific Instrument Company, Basle, Switzerland), and the CAMAG Nano-Applicator (CAMAG, Switzerland). Sample application and densitometric measurements were made in subdued light (red filter) to avoid decomposition of the mycotoxins. The spot size was less than 1.0 mm in diameter, and the plate capacity was 18 samples spotted 0.5 cm apart.

For the linear HPTLC analysis, the HPTLC plates were developed in the second position of a 10.5 × 24.0 × 4.0 cm Regis SB/CD chamber (Regis Chemical Company, Morton Grove, Illinois) containing 25 ml of toluene/ethyl acetate/formic acid (30:6:0.5). The plate was developed continuously for 5.0 minutes (migration distance 4.0 cm), and then it was removed and dried in cool air. The continuous development procedure was repeated three more times, and then the solvent system was changed to toluene/ethyl acetate/formic acid (30:14:4.5) and the plate developed a further three times (8.0 minutes each). Altogether, a total of seven multiple continuous developments with two solvent systems was used to give a baseline separation of the 13 mycotoxins investigated.

For the circular and anticircular HPTLC analysis, the HPTLC plates were developed using the CAMAG U-Chambes System (CAMAG, Switzerland), and Anticircular Developing System (CAMAG, Switzerland).

Spectrophotometric Detection

In situ spectrophotometric scanning of the developed HPTLC plates was performed in a reflectance mode with a microoptic KM-3 spectrophotometer (Carl Zeiss, New York) and CAMAG TLC/HPTLC Scanner (CAMAG, Switzerland). Slit width and slit length were 0.5 and 3.5 mm, respectively. Six different wavelengths were used to measure all 13 mycotoxins: sterigmatocystin at 324 nm; zearalenone, citrinin, ochratoxin at λ(excitation) = 313 nm, λ(emission) = 400 nm (fluorescence mode); penicillic acid at 240 nm; patulin at 280 nm; luteoskyrin at 440 nm; aflatoxins at λ(excitation) = 365 nm, λ(emission) = 430 nm (fluorescence mode).

Chromatogram and peak area measurements were made with a Minigrator and SP 4100 Computing Integrator (Spectra Physics, Santa Clara, California).

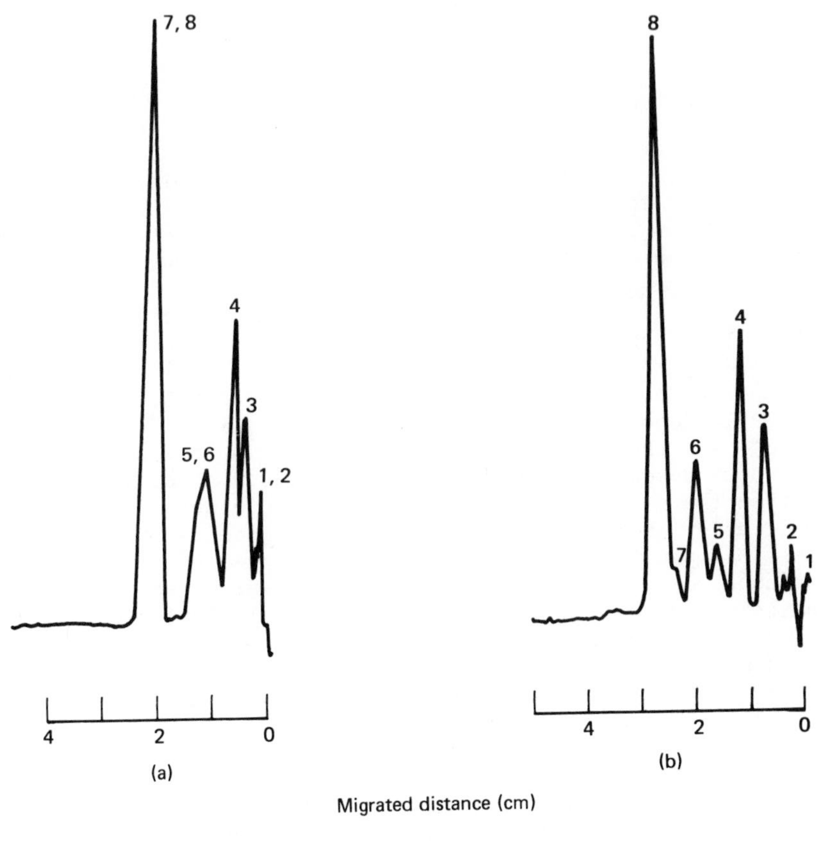

Figure 2. Comparison of development modes: (a) single (saturated chamber) development (1 time, 12 minutes) and (b) triple-continuous development (3 times, 4 minutes each--toluene) of dye mixtures. UV = 436 nm.

RESULTS AND DISCUSSION

Multiple-Continuous Development

To illustrate the power of the combination of multiple and continuous developing techniques, a dye mixture was analyzed. Figure 2 is a comparison of the separation obtained using a single development versus that obtained using triple-continuous development with the same solvent system (toluene) over the same total time period. The single development took 12 minutes, while the second plate was developed continuously for 4 minutes, three times. Scanning the plate at 436 nm, it was found that only five peaks were resolved in the single development case, as compared to the resolution of eight peaks in the triple-continuous development case. The dye mixture separation illustrates the power of multiple-continuous development, but the real potential of this technique is certainly evident in the analysis of mycotoxin mixtures.

Linear Multiple-Continuous HPTLC Analysis of Mycotoxins

Because of the complexity of the mycotoxin mixture, neither single, multiple, nor continuous development alone would provide enough resolution in a reasonable period of time to allow for quantitative analysis of all 13 mycotoxins. Multiple development gave some separation for 7 mycotoxins, but the 6 aflatoxins could not be properly resolved. When multiple and continuous development techniques were combined, however, all 13 mycotoxins were resolved and could be quantitatively analyzed.

Figure 3 shows the result of a single-continuous development (5 minutes) with solvent system I (toluene/ethyl acetate/formic acid at 30:6:0.5). The left-hand chromatogram shows good separation of citrinin, zearalenone, and sterigmatocystin using fluorescence excitation at 313 nm and emission at 400 nm. However, as seen in the right-hand chromatogram, good response for sterigmatocystin is obtained by scanning in the ultraviolet at 240 nm. Ochratoxin A and the aflatoxins remain close to the origin. Ochratoxin A can be separated from aflatoxins after two more continuous developments (total 15 minutes), and are quantitated at fluorescence measurement excitation 313 and emission 400 nm. Patulin and penicillic acid are only detected by ultraviolet measurement and are not resolved. After three additional continuous developments for a total of 20 minutes, baseline separation between penicillic acid and patulin is obtained (Figure 4). We can improve the sensitivity for patulin by scanning it at 280 nm. Luteoskyrin slightly overlaps with patulin, but patulin and penicillic acid do not show any response at 440 nm, where luteoskyrin is analyzed. In this case, baseline resolution is not necessary for quantitative analysis of these components, by taking advantage of their spectral differences.

Three more continuous developments (3 times, 8 minutes each) with solvent II (toluene/ethyl acetate/formic acid at 30:14:4.5)

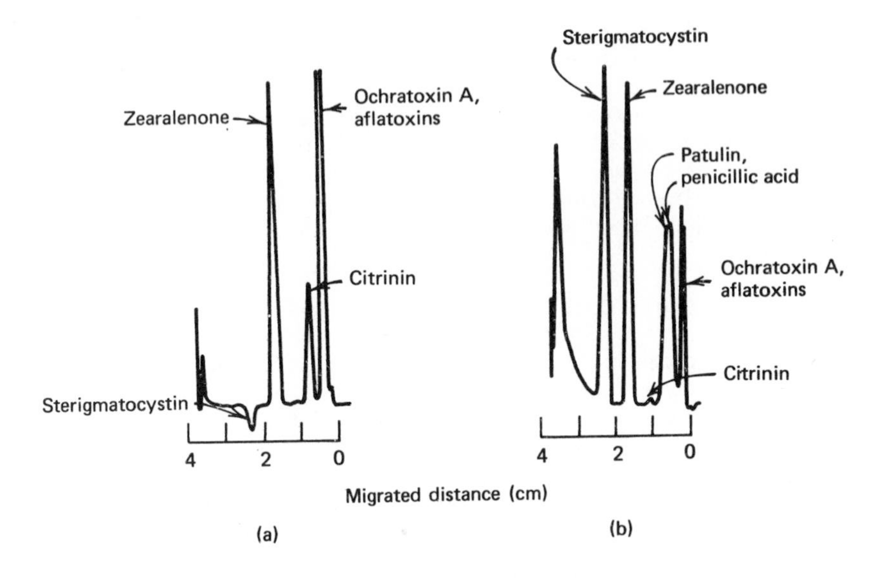

Figure 3. Linear HPTLC chromatogram after single-continuous development for 5 minutes (solvent I = toluene/ethyl acetate/ formic acid 30:6:0.5). (a) Fluorescence, excitation 313 nm, (b) UV 240 nm.

gave a baseline separation of the six aflatoxins, B_1, B_2, G_1, G_2, M_1, and M_2 (Figure 5). All the other mycotoxins have moved beyond these toward the region of the solvent front. Note that these separations take place in only a 3-cm distance by the multiple-continuous development technique. It provides sharper peaks by the reconcentration effect, and reduces the diffusion through fast solvent velocity. It provides higher sensitivity and improved precision for separated components than does linear development alone.

For the fluorescence mode measurement, calibration curves were linear over the range from 200 pg to the upper limit--approximately 2.0 to 3.0 ng. For the UV measurement, 20 to 100 ng was used as a practical working concentration range for quantitation. The precision of the analysis of mycotoxin standards (concentration range 1 to 50 ng) varied between 0.7 and 2.2% relative standard deviation for 12 determinations.

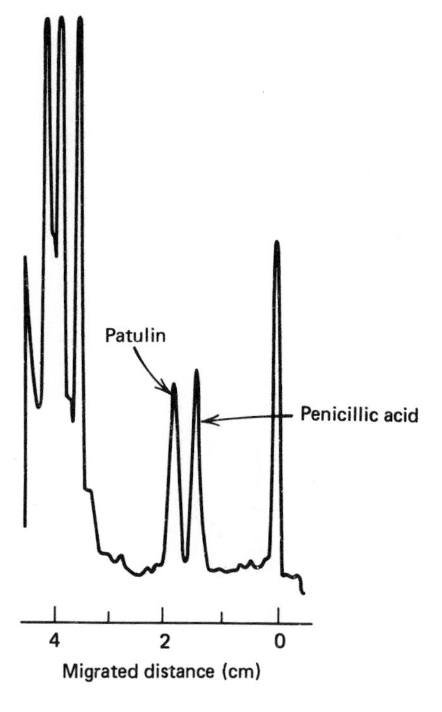

Figure 4. Linear HPTLC chromatogram after the fourth continuous development--4 times, 5 minutes each (solvent I). UV 240 nm.

The total analysis took about 50 minutes for complete separation and quantitation of 13 mycotoxins. If information concerning only one or two mycotoxins was required, then the separation could be terminated at the appropriate development stage.

Figure 6 shows three UV absorption spectra obtained from HPTLC separation of an aflatoxin M_1 standard. By using a scanning spectrophotometer, one can examine the spectra for ultraviolet absorption, fluorescence excitation, and fluorescence emission for each sample of interest, selecting the scanning mode and wavelength that provide the maximum sensitivity and signal-to-noise ratio. Wavelength scanning of each separated spot, while time-consuming, can be used for additional qualitative identification purposes.

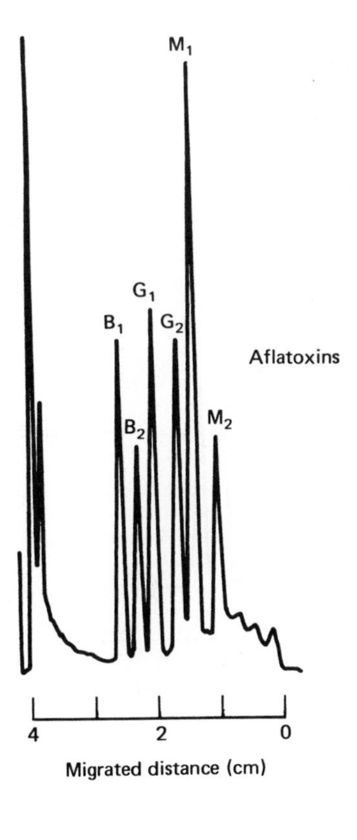

Figure 5. Linear HPTLC chromatogram after the seventh continuous development--4 times, 5 minutes each (solvent I), 3 times, 8 minutes each (solvent II--toluene/ethyl acetate/formic acid 30:14:4.5). Fluorescence, excitation 365 nm, emission 430 nm.

In Table I analytical methods for pure mycotoxin standard mixtures are compared. These data were collected and estimated from published reports that described the analysis of similar mycotoxin mixtures. In the HPLC analysis (4), only UV absorbance measurement was used for detection. If both UV and fluorescence detectors could be used in series, the sensitivity could be similar to the HPTLC system. Usually, the precision of HPTLC and HPLC are similar, but the great difference is analysis speed per sample and amount of solvent usage per sample. In the HPTLC

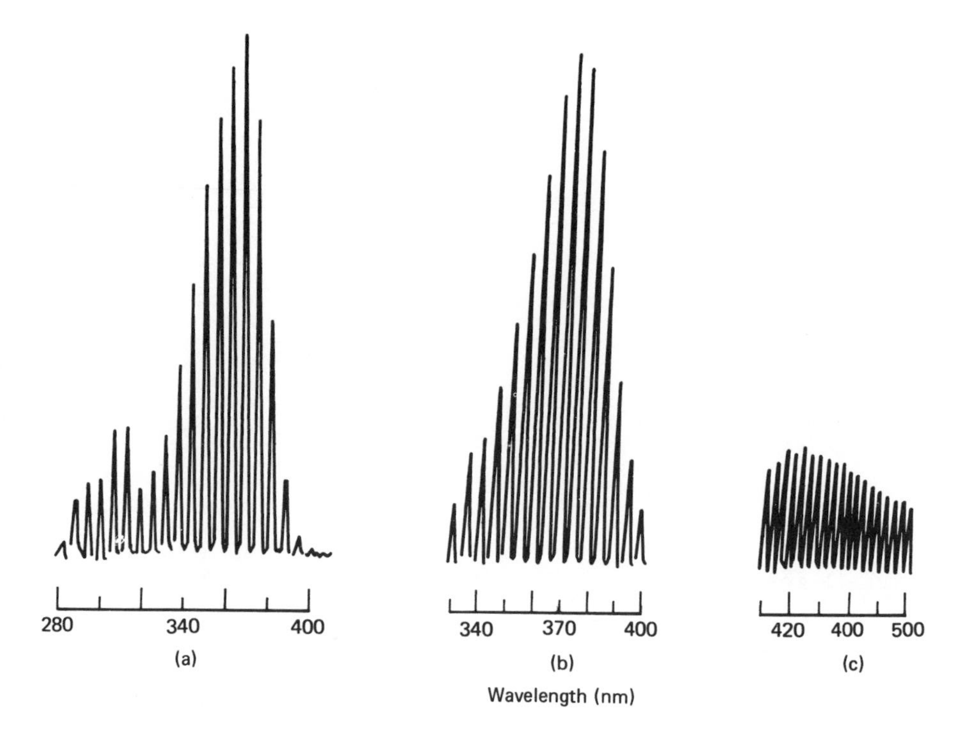

Figure 6. (a) UV absorption, (b) fluorescence excitation, and (c) fluorescence emission (excitation 365 nm) spectra of standard aflatoxin M, separated on HPTLC plate.

case (1), 18 samples were spotted on a plate (10 × 10 cm) for a single, parallel analysis (total time 50 minutes and 50 ml of solvent for 13 mycotoxins). But in the HPLC analysis (4), only one sample at a time could be analyzed (20 minutes and 26 ml solvent for 7 mycotoxins per sample).

To analyze 17 samples, the HPLC analysis must be performed 17 times (total 340 minutes and 442 ml for 7 mycotoxins). Another advantage of the HPTLC system over HPLC is there are no columns to go bad, or degrading resolution by repeated analysis of dirty samples.

Instrumentation cost for HPTLC and HPLC is similar. Conventional TLC is regarded as a low-cost analytical technique because it is generally used in a semiquantitative visual determination mode.

TABLE I. COMPARISON OF ANALYTICAL METHODS FOR STANDARD MYCOTOXIN MIXTURE

	HPTLC (1)	TLC (2, 3)	HPLC (4)
Number of mycotoxins	13	12	>7
Sensitivity	0.1 to 50 ng	>1 μg	25 to 1900 ng
Precision (RSD)	2%	>10%	(Est. 2%)
Analysis speed/sample (n = 17)	3 min	12 min	20 min
Solvent usage/sample (n = 17)	3 ml	24 ml	26 ml
Overall cost/analysis	low	low	medium

The HPTLC system appears to be the preferred analytical approach over conventional TLC and HPLC for multimycotoxin analysis. HPTLC results in fast analysis times per sample, excellent precision, and high sensitivity.

Circular and Anticircular HPTLC Analysis of Mycotoxins

As discussed in the Introduction, circular and anticircular development systems appear to offer some significant advantages over the linear development system. These two approaches were applied to the separation of the 13 mycotoxins.

At the time this work was done, continuous development accessories were not available for circular and anticircular HPTLC development systems. Therefore, the multiple development technique alone was used to separate mycotoxin mixtures using these two systems. For comparison purposes, the same solvent system and EDTA-impregnated HPTLC plates that were used in the linear HPTLC analysis were used in this experiment.

Figure 7 is the chromatogram of the circular HPTLC separation of the mycotoxin mixture. Figure 8 is the result of the anti-

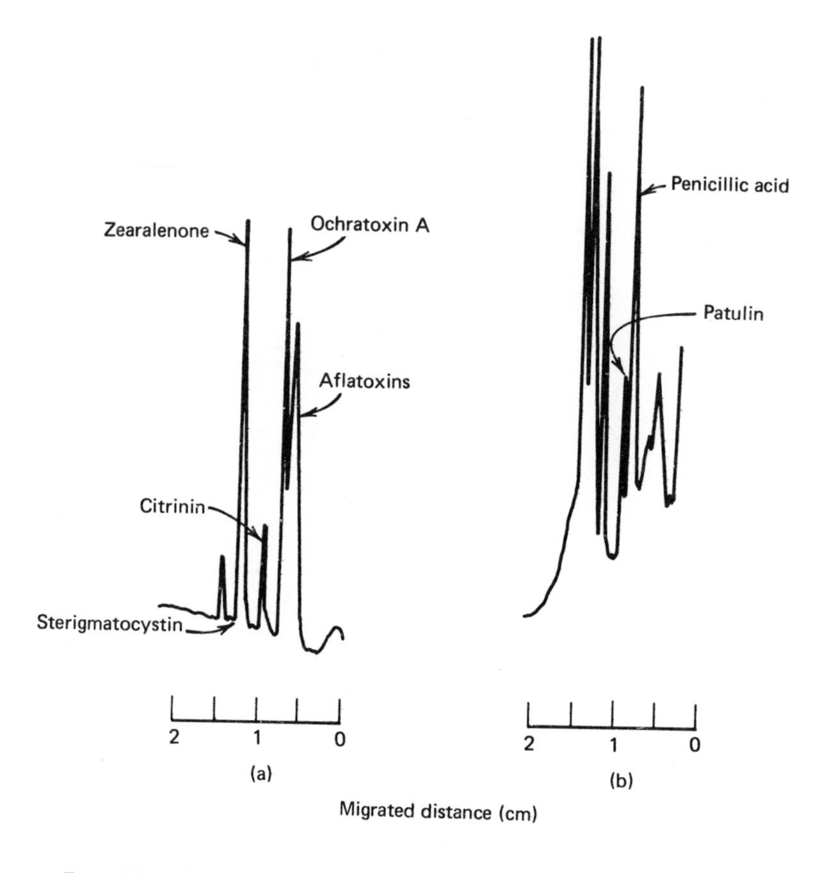

Figure 7. Circular HPTLC chromatogram after the fourth development--4 times, 2.5 minutes each (solvent I). (a) Fluorescence, excitation 313 nm, (b) UV, 240 nm.

circular development. Both cases took only about 10 minutes. For the circular HPTLC, only solvent I was utilized, while solvents I and II were used for the anticircular HPTLC separation. As anticipated, it took a much shorter time to separate the first seven mycotoxins from the aflatoxins. As shown in Figures 7 and 8, the separation pattern is different, and it may be necessary to modify the solvent system for the optimum separation of these mycotoxins. The aflatoxins were not resolved in these analyses.

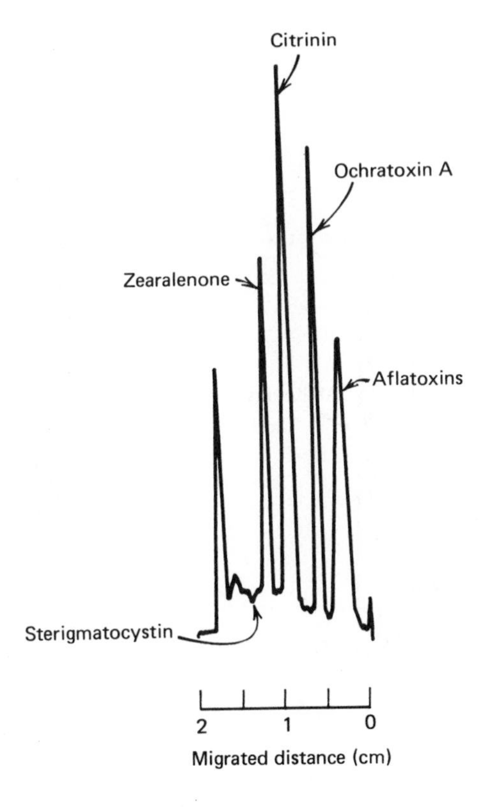

Figure 8. Anticircular HPTLC chromatogram after the sixth development--4 times, 1.5 minutes each (solvent I); 2 times, 2 minutes each (solvent II); fluorescence, excitation 313 nm.

In circular HPTLC, as compounds moved further away from the center, the sensitivity of the compounds was reduced owing to radial broadening of the separated spots. This reduction in sensitivity did not occur using anticircular development.

Aflatoxins usually stay very close to the origin so it should be possible to effectively use the circular HPTLC system for the separation of these compounds (33). In this solvent system, aflatoxins were not well enough resolved for quantitative purposes.

Solvent usage for single development requires only 0.5 ml

for circular, 2 ml for anticircular, and 10 to 20 ml for linear. Thus, solvent consumption is less, and disposal problems are minimized with circular and anticircular techniques, particularly when compared against conventional TLC, which nominally requires 100 ml.

Although much more work is needed, these results indicate that future separation schemes for large-scale mycotoxin screening purposes may well benefit from the application of circular and anticircular HPTLC systems.

In summary, some advantages of the HPTLC system for complex mixtures analysis such as with mycotoxins are listed. The analysis time is short compared to conventional TLC and HPLC. HPTLC also provides one with a flexible developing and detection system. The higher sensitivity and excellent reproducibility enable one to develop an analytical method very rapidly.

Large numbers of samples per HPTLC plate make it applicable to large-scale screening or quality-control purposes--qualitatively or quantitatively. The spectrophotometric detection system for quantitation is expensive. However, with the capability it provides for large-scale operation, faster analysis times, and with the large sample capacities per HPTLC plate, it will eventually reduce the cost per sample analysis to a relatively low level.

REFERENCES

1. K. Y. Lee, C. F. Poole, and A. Zlatkis, Anal. Chem., 52, 837 (1980).
2. B. A. Roberts and D. S. P. Patterson, J. Assoc. Offic. Anal. Chem., 58, 1178 (1975).
3. C. P. Gorst-Allman and P. S. Steyn, J. Chromatogr., 175, 325 (1979).
4. G. W. Engstron, J. L. Richard, and S. J. Cysewski, J. Agric. Food Chem., 25, 833 (1977).
5. C. E. Holaday, J. Amer. Oil Chem. Soc., 53, 603 (1976).
6. R. M. Beebe and D. M. Takahashi, J. Agric. Food Chem., 28, 481 (1980).
7. W. A. Pons, Jr., J. Assoc. Offic. Anal. Chem., 59, 101 (1976).
8. D. M. Takahashi, J. Assoc. Offic. Anal. Chem., 60, 799 (1977).
9. A. E. Pohland, K. Sanders, and C. W. Thrope, J. Assoc. Offic. Anal. Chem., 53, 692 (1970).
10. J. P. Snow, G. B. Lucas, D. Harvan, R. W. Pero, and R. G. Owens, Appl. Microbiol., 24, 34 (1972).
11. J. A. Sphon, P. A. Dreifuss, and H. R. Schulten, J. Assoc. Offic. Anal. Chem., 60, 73 (1977).

12. R. M. Eppley, J. Assoc. Offic. Anal. Chem., 51, 74 (1968).
13. L. Fishbein and H. L. Falk, Chromatogr. Rev., 12, 42 (1970).
14. B. G. E. Josefsson and T. E. Möller, J. Assoc. Offic. Anal. Chem., 60, 1369 (1977).
15. L. Stoloff, L. Nesheim, L. Yim, J. V. Rodricks, M. Stack, and A. D. Campbell, J. Assoc. Offic. Anal. Chem., 54, 91 (1971)
16. P. S. Steyn, J. Chromatogr., 45, 473 (1969).
17. P. M. Scott, J. W. Lawrence, and W. van Walbeek, Appl. Microbiol., 20, 839 (1970).
18. A. Durackova, V. Betina, and P. Nemec, J. Chromatogr., 116, 141 (1976).
19. J. Ripphahn and H. Halpaap, J. Chromatogr., 112, 81 (1975).
20. A. Zlatkis and R. E. Kaiser (Eds.), High Performance Thin-Layer Chromatography (HPTLC), Elsevier, Amsterdam, 1977.
21. W. Bertsch, S. Hara, R. E. Kaiser, and A. Zlatkis (Eds.), Instrumental HPTLC, Dr. Alfred Huthig Verlag, New York, 1980.
22. T. H. Jupille, J. Amer. Oil Chem. Soc., 54, 179 (1977).
23. J. A. Perry, T. H. Jupille, and A. Curtice, Sep. Sci., 10, 571 (1975).
24. J. A. Perry, J. Chromatogr., 113, 267 (1975).
25. J. A. Perry, T. H. Jupille, and L. J. Glunz, in Separation and Purification Method, E. S. Perry, C. J. Van Oss, and E. Grushka (Eds.), Marcel Dekker, New York, 1976, p. 119.
26. J. A. Perry, J. Chromatogr., 165, 117 (1979).
27. U. B. Hezel and D. C. Fenimore, HPTLC Short Courses, Lausanne, Switzerland, 1979.
28. K. Y. Lee, D. Nurok, A. Zlatkis, and A. Karman, J. Chromatogr., 158, 403 (1978).
29. L. Zhou, C. F. Poole, J. Triska, and A. Zlatkis, J. HPC & CC, 3, 440 (1980).
30. J. C. Touchstone and M. F. Dobbins, Practice of Thin-Layer Chromatography, Wiley-Interscience, New York, 1978, p. 143.
31. D. E. Janchen, in Instrumental HPTLC, W. Bertch, S. Hara, R. E. Kaiser, and A. Zlatkis (Eds.), Dr. Alfred Huthig Verlag, New York, 1980 p. 133.
32. R. E. Kaiser, J. HPC & CC, 1, 164 (1978).
33. R. K. Vitek, A. E. Waltkins, and D. M. Kent, Chemo. Abstr., 177, AGFD 48 (1979).
34. D. Nurok and A. Zlatkis, Carbohydr. Res., 81, 167 (1980).
35. D. B. Faber, J. Chromatogr., 142, 421 (1977).
36. G. J. Fischesser and M. D. Seymour, J. Chromatogr., 135, 165 (1977).
37. D. C. Fenimore and C. J. Meyer, J. Chromatogr., 186, 555 (1979)

38. Y. Takeda, E. Isohata, R. Amano, and M. Uchiyama, J. Assoc. Offic. Anal. Chem., 62, 573 (1979).
39. A. Gimeno, J. Assoc. Offic. Anal. Chem., 63, 182 (1980).

CHAPTER 28

Two-Dimensional TLC of Aflatoxins

Anant V. Jain and Roger C. Hatch

INTRODUCTION

Aflatoxins are a group of hepatotoxic mycotoxins produced by the mold *Aspergillus flavus*, from which the name was derived, and the closely related *A. parasiticus*. The toxins were first isolated from contaminated peanut meal. The toxic factor was separated chromatographically in four distinct spots (1). The four components were designated aflatoxin B_1, B_2, G_1, and G_2 after their blue and greenish-blue fluorescent colors and their chromatographic positions. The toxins are potent carcinogens and capable of causing acute (2, 3) and chronic poisoning in animals (4, 5). Aflatoxin B_1 is the most potent and prevalent of the 18 known aflatoxins (6). In many instances, aflatoxins B_2, G_1, and G_2 are also found along with B_1 (7). Aflatoxin B_1 in dairy feeds is metabolized by cows in part to aflatoxin M_1, some of which is excreted in milk (8). Aflatoxin M_1 is also a potential hepatotoxin. When other animals are fed contaminated rations, a small fraction of aflatoxin B_1 and its metabolite M_1 is found in edible tissues, particularly liver (8). Thus, it is important to control aflatoxin contamination of feeds and foods.

Many methods are described in the literature (9, 10) for the determination of aflatoxins. Several of these methods use thin layer chromatography (TLC) for resolving the various aflatoxins, followed by visual or fluorodensitometric quantitation of developed spots. One-dimensional TLC is usually used for the analysis of aflatoxins in simple commodities, whereas two-dimensional TLC is very useful for the quantitative determination of aflatoxins in complex samples.

In two-dimensional TLC a mixture is applied to the TLC plate about 3 cm from one edge and is developed with one solvent system, and after evaporation of the solvent the plate is turned 90° and developed again using another solvent system (11).

Two-dimensional TLC for the determination of aflatoxins was first introduced by Peterson and Ciegler (12). Since then it has been used by several workers (7, 12-40) and has been found useful in the analysis of feeds (13-16), compound feedstuffs containing citrus pulp (7, 14, 17), spices (18-21), peanut butter (22-24), toasted cottonseed (25), tissues (26-29), milk (24, 30-32), cheese (32-35), eggs (35, 36), and figs (37). Two-dimensional TLC has also been used in confirming the identity of aflatoxins (7, 18, 23, 38-40, 44).

In many instances, two-dimensional TLC provides considerable saving of time over the use of column chromatography for extract cleanup (24, 41). Pons (42) described a method for the determination of aflatoxins in mixed feeds. The method involves extraction of the feed with methylene chloride, evaporation of the methylene chloride extract, precipitation of impurities with lead acetate in aqueous acetone, partitioning of aqueous acetone with methylene chloride, and finally column chromatography. We found (14) that for the determination of aflatoxins in feeds, column chromatography is not required for the extract cleanup if two-dimensional TLC is used. With two-dimensional TLC it has been possible to determine aflatoxins even in compound feedstuffs containing citrus pulp (7) without column chromatographic cleanup of the sample extract. It is noteworthy that even with column chromatographic cleanup of the extract it is not possible to determine aflatoxins in compound feedstuffs containing citrus pulp with one-dimensional TLC owing to the presence of excessive fluorescent impurities in the aflatoxin R_f area. Trucksess et al. (36) found two-dimensional TLC more practical for the determination of aflatoxins in eggs than the cellulose column chromatographic cleanup of extracts and one-dimensional TLC.

With two-dimensional TLC we were able to develop an all-purpose procedure for the determination of aflatoxins B_1, B_2, G_1, and G_2 in a wide range of products such as different types of grains, feedstuffs, mixed feeds, silages, chicken litter, mixture of chicken litter with grains or feeds, mixture of silages with grains or feeds, compound feedstuffs containing citrus pulp, goat rumen contents, and pig stomach contents.

THIN LAYER CHROMATOGRAPHY

Precoated or home-made silica gel TLC plates that meet the criteria outlined in Section 26.002 of reference 35 are recommended. An aflatoxin mixture containing 0.5 μg each of aflatoxin B_1 and G_1 per milliliter and 0.15 μg each of aflatoxin B_2 and G_2 per milliliter is used as a resolution and quantitation standard.

The sample extract is prepared according to the procedure of Jain and Hatch (7, 14). The sample extract is screened for gross aflatoxin contamination with one-dimensional TLC. Suspected positive samples are then subjected to two-dimensional TLC for quantitation and confirmation.

One-Dimensional TLC

Routinely, two 10-μl sample spots per sample are applied adjacent to each other at least 1 cm apart on the TLC plate on an imaginary line 3 cm from the bottom edge of the plate. One of the sample spots is superimposed (overspotted) with 10 μl aflatoxin standard mixture ("internal standard"). Three aflatoxin standard spots containing 5, 10, and 15 μl of aflatoxin standard mixture are applied in the middle of the TLC plate on the imaginary line.

The plate is developed in a chloroform-acetone-water system in a tank containing two V-shaped metal solvent troughs. The rear trough contains 50 ml chloroform/acetone (44:6), and the front trough contains 50 ml water. The plate is placed in the rear trough such that the gel layer faces the water trough. The plate is developed for 12 cm in the dark in an unequilibrated tank. The solvent is evaporated from the plate with hot air from a hair dryer.

The TLC plate is examined under long-wave UV light. The expected position of each aflatoxin in the sample is located by comparison with R_f values of the corresponding standards. Figure 1 shows a one-dimensional thin layer chromatogram of six different feedstuffs. The four fluorescent spots originating from D, E, and F in order of decreasing R_f are aflatoxins B_1, B_2, G_1, and G_2. For determining the presence of aflatoxins, the fluorescent sample patterns are compared with the sample patterns containing internal standard. Fluorescent spots in the sample that are thought to be aflatoxins must have R_f values identical to and color similar to aflatoxin spots from samples containing internal standard.

Spraying the plate with 25% sulfuric acid has been recommended

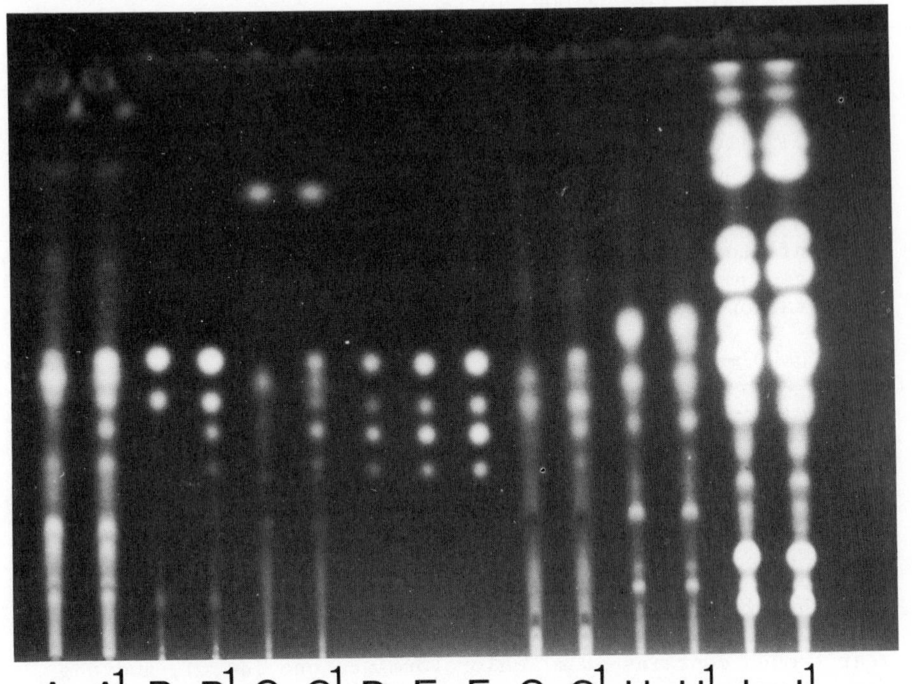

Figure 1. One-dimensional thin layer chromatogram of the extracts from the following feedstuffs: A, cottonseed meal; B, corn; C, mixed grain equine feed; G, rabbit chow; H, pelleted equine feed; I, compound feedstuffs containing citrus pulp. A^1 to I^1 are respective extracts overspotted with aflatoxin standard mixture. D, E, and F are spots of aflatoxin mixture. Photographed under long-wave UV light. The four spots originating from D, E, and F in order of decreasing R_f are aflatoxins B_1, B_2, G_1, and G_2. Developed in dark in unlined, unequilibrated tank ith chloroform/acetone/water (44:6:free). Silica gel layer.

as a confirmation test (43). The bluish fluorescence of aflatoxin B_1 and B_2 and the greenish-blue fluorescence of G_1 and G_2 turns to a yellow fluorescence after acid spray. Twenty-five percent sulfuric acid in methanol could also be used. The color change after spraying with 25% sulfuric acid is indicative but not conclusive identification of aflatoxin contamination, since many nonaflatoxin spots also show this color change. However, failure of color change of the sample spots while the standard aflatoxin spots do show color change rules out aflatoxin contamination.

With the above criterion it is possible to identify many of the negative samples. In some simple commodities such as corn (Figure 1, B), quantitation of aflatoxins is possible from one-dimensional thin layer chromatograms. The quantitation is carried out by visually comparing the fluorescence intensity of the sample aflatoxin spot with corresponding standard spots. Aflatoxin concentration is calculated according to the formula given in reference 7. Densitometric quantitation can be carried out as outlined in Section 26.A07(c) of reference 35. In samples where interferences preclude identification or quantitation, two-dimensional TLC is used.

Two-Dimensional Thin Layer Chromatography

The TLC plate is scored and spotted as shown in Figure 2. Standard spots A, B, C, D, and E contain 4, 6, 8, 10, and 4 μl of aflatoxin standard mixture. S is a sample spot containing 10 μl of sample extract. If necessary, the sample extract is diluted such that the concentration of the suspected aflatoxin B_1 is approximately 20 ppb. The plate is developed in the first direction in chloroform/acetone/water (44:6:free) system as described under one-dimensional TLC. After development in the first direction, the solvent is evaporated from the plate with hot air from a hair dryer. The TLC plate is cooled to room temperature, turned 90°, and developed in the dark in the second direction up to the score line in an unequilibrated tank containing 50 ml toluene/ethyl acetate/formic acid (30:15:5) in a metal trough. For compound feedstuffs containing citrus pulp, the plate is developed in the second direction with toluene/ethyl acetate/formic acid (24:20:6) up to 10 cm (7). The solvents are evaporated by hot air from a hair dryer. All the formic acid must be removed from the plate before quantitation is carried out. Formic acid has a negative influence on the fluorescence of aflatoxin B_1 and G_1, and a positive one on B_2 and G_2 (7).

The TLC plate is examined under long-wave UV light, and a pattern of four fluorescent spots originating from standard spots A, B, C, D, and E is observed. The four spots originating from each standard spot in order of decreasing R_f are aflatoxins B_1 ,

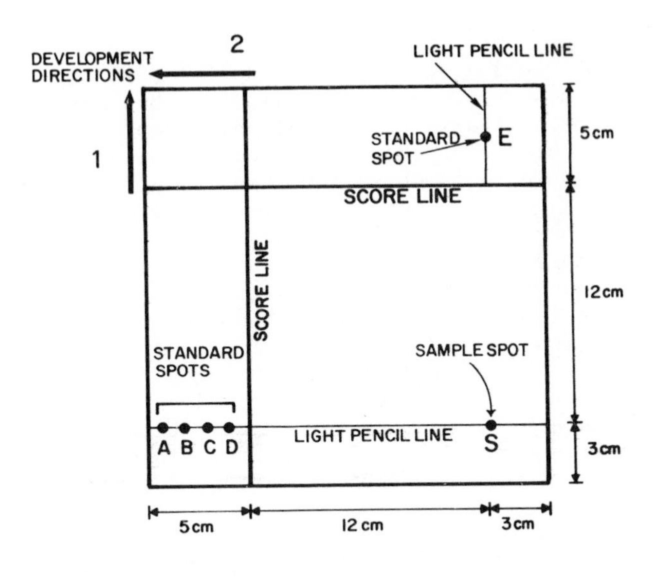

Figure 2. Schematics of the preparation of a TLC plate for two-dimensional development. A, B, C, D, and E are spots containing 4, 6, 8, 10, and 4 µl of standard aflatoxin mixture, respectively. S is a 10-µl sample spot.

B_2, G_1, and G_2. The position of each aflatoxin spot from the sample is located as follows: An imaginary line is projected perpendicular to the direction of first development through aflatoxin B_1 originating from A, B, C, and D. Another imaginary line is projected perpendicular to the direction of second development through the standard aflatoxin B_1 spot originating from E. The junction of these two imaginary lines is the location in which to expect any aflatoxin B_1 from the sample. The process is repeated to locate other aflatoxin spots. The center of the expected aflatoxin B_1 spot from the sample may not be exactly on the intersection of the imaginary lines. In practice the sample spots often run somewhat slower than the standards in the first direction, and somewhat faster in the second direction. This is because impurities retard the sample spots in the first direction, and the sample is on an area of the plate that has been deactivated by the first solvent when it is developed in the second direction. This behavior is solvent dependent (M. J. Nagler, Tropical Products Institute, London, personal communication). With the solvent systems described above, some part of each sample spot falls on the junction of perpendicular projections.

As an additional check on the location and appearance of the aflatoxins spots, it is recommended (23) that a second TLC plate be prepared in which the 10-μl sample spot is overspotted with 10-μl aflatoxin standard mixture and developed in two dimensions and compared with the former plate. Furthermore, the TLC plate overspotted with aflatoxins provides additional information about the presence or absence of aflatoxins in the sample. Comigration of aflatoxin spots with internal standard spots has been recommended as additional evidence of aflatoxin contamination (7).

Figure 3 shows two-dimensional thin layer chromatograms of the following feedstuffs: Figure 3a, pelleted equine feed containing no aflatoxins; Figure 3b, pelleted equine feed with added aflatoxins; Figure 3c, compound feedstuffs containing citrus pulp with no aflatoxins; and Figure 3d, compound feedstuffs containing citrus pulp with added aflatoxins. These two-dimensional thin layer chromatograms were prepared from the same extracts of pelleted equine feed and compound feedstuffs containing citrus pulp as used for one-dimensional TLC (refer to Figure 1; H, H^1, I, and I^1). Comparison of Figures 3a, 3b, 3c, and 3d with Figure 1 H, H^1, I, and I^1, respectively, shows that two-dimensional TLC provides excellent separation of aflatoxins from other impurities. All four aflatoxins are very well resolved and separated from other impurities. There is a faint spot between aflatoxin B_1 and B_2 spots from compound feedstuffs containing citrus pulp (Figures 3c and 3d). However, this spot does not interfere with the identification and quantitation of aflatoxins, since it exhibits a yellow fluorescence under long-wave UV light and is resolved from aflatoxin B_1 and B_2 spots. Using this procedure aflatoxin levels of 2.5 μg B_1 or G_1/kg and 0.8 μg B_2 or G_2/kg of feedstuffs are easily detected. Quantitation is carried out as described earlier under one-dimensional TLC.

Recoveries of aflatoxins B_1, B_2, G_1, and G_2 from compound feedstuffs containing citrus pulp have been reported (7) as 96, 104, 98, and 102% respectively. Our preliminary studies indicate that the recovery of aflatoxins from other agricultural products range between 75 and 100%.

The Smalley Committee of the American Oil Chemist's Society conducts a Check Sample Program for the analysis of aflatoxins. In this program each participating laboratory is assigned an identifying number. The participating laboratories are sent a

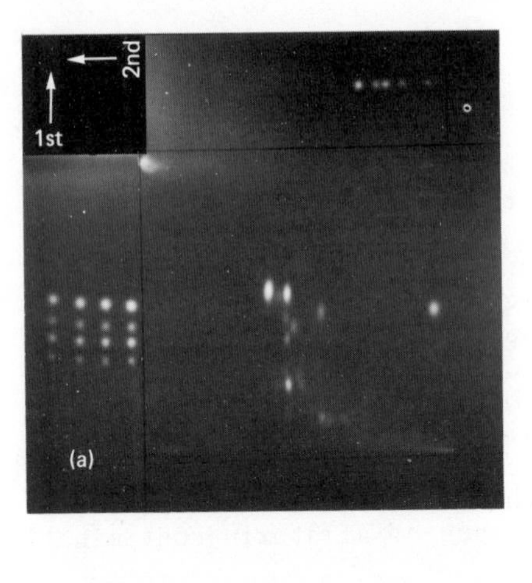

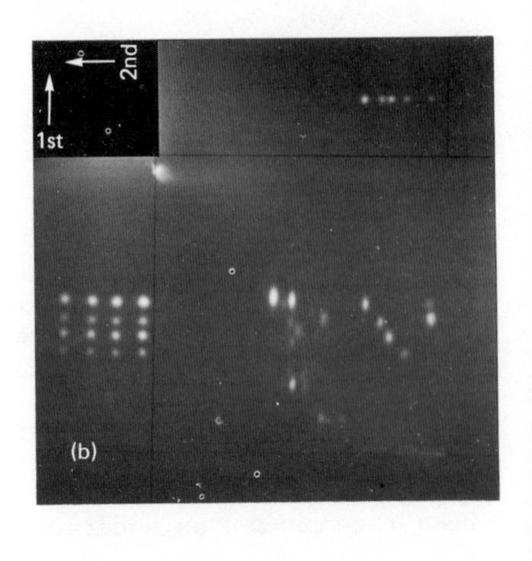

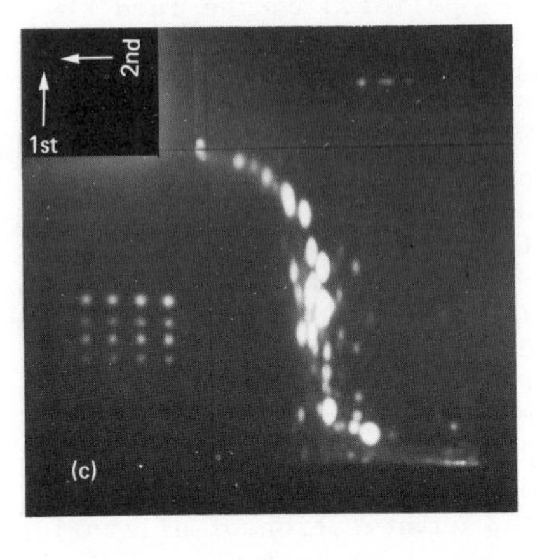

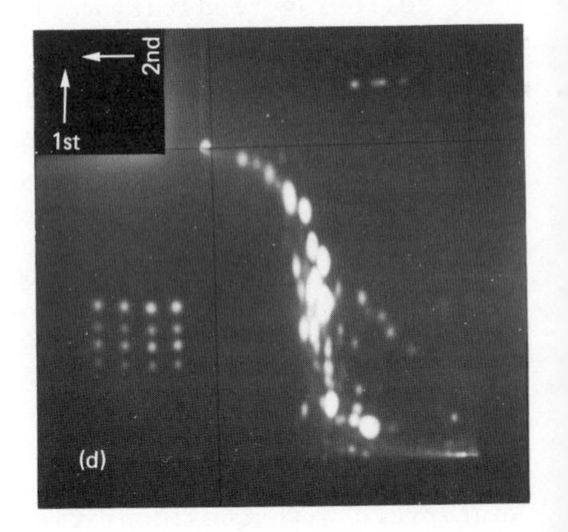

Figure 3. Two-dimensional thin layer chromatogram of extracts from the following foodstuffs: (a) pelleted equine feed containing no aflatoxins, (b) pelleted equine feed overspotted with aflatoxin standard mixture, (c) compound feedstuff containing citrus pulp and no aflatoxins, and (d) compound feedstuff containing citrus pulp overspotted with aflatoxin mixture. Photographed under long-wave UV light, developed in dark in unlined, unequilibrated tank. Direction 1, chloroform/acetone/water (44:6:free). Direction 2, toluene/ethyl acetate/formic acid (30:15:5; a and b), (24:20:6; c and d). Silica gel layer.

series of seven unknown samples approximately 1 month apart. The results of the determination are reported to the Smalley Committee. The mean and standard deviation of each sample is determined, and a report is sent to all participating laboratories. At the end of the series the proficiency index of each laboratory is determined. Proficiency index is computed by a summation of the square of deviation relative to standard deviation values for all samples, dividing by (n - 1) samples and extracting the square root of the result (J. D. McKinney, Ranchers Cotton Oil, Fresno, California, personal communication). The proficiency index is a measure of the deviation relative to standard deviation for the series. During the years 1978-1979 and 1979-1980 we used two-dimensional TLC for the determination of aflatoxins in cottonseed meal series. We were placed in second place among the participating laboratories with proficiency indexes of 0.526 and 0.500, respectively. Thus, our results were within one half of a standard deviation from the mean for both years. These results indicate that the method is reliable and precise.

Aflatoxins G_1 and G_2 are usually not found in cottonseed products, but some extracts (even after column cleanup) may show bluish fluorescent nonaflatoxin spots at or near the R_f of G_1 and G_2 on a one-dimensional thin layer chromatogram (35). On two-dimensional thin layer chromatograms these nonaflatoxin spots are completely resolved from aflatoxins G_1 and G_2.

CONFIRMATION

If the TLC plate after two-dimensional development shows suspected aflatoxin spots, it is compared with the plate on which the sample spot is overspotted with aflatoxin standard mixture. If aflatoxins are present, the sample aflatoxin spots would have comigrated with aflatoxin standard spots (7). Both plates are sprayed with a fine mist of 25% sulfuric acid and observed for the color change as described under one-dimensional TLC. If there is a need for further confirmation, the aflatoxins B_1 and G_1 spots after two-dimensional development are derivatized with trifuloroacetic acid on the TLC plate as described by Trucksess and Stoloff (27) for aflatoxin M_1, and the TLC plate is developed in the first direction with chloroform/acetone/water (85:15:free). The solvent is evaporated from the TLC plate, and the TLC plate is examined under long-wave UV light for blue fluorescent spots of aflatoxins B_{2a} and G_{2a}.

REFERENCES

1. R. D. Hartley, B. F. Nesbit, and J. O'Kelly, Nature London, 198, 1056 (1963).
2. G. T. Edds, J. Am. Vet. Med. Assoc., 162, 304 (1973).
3. R. C. Hatch, J. D. Clark, A. V. Jain, and E. A. Mahaffey, Am. J. Vet. Res., 40, 505 (1979).
4. P. M. Newberne, J. Am. Vet. Med. Assoc., 163, 1262 (1973).
5. J. D. Clark, A. V. Jain, and R. C. Hatch, Am. J. Vet. Res., 41, 1841 (1980).
6. G. Osweiler, Vet. Human Toxicol., 22, 61 (1980).
7. A. V. Jain and R. C. Hatch, J. Assoc. Offic. Anal. Chem., 63, 626 (1980).
8. J. V. Rodricks and L. Stoloff, in Mycotoxins in Human and Animal Health, J. V. Rodricks, C. W. Hesseltine, and M. A. Mehlman (Eds.), Pathotox Publishers, Park Forest South, Ill., 1977 p. 68.
9. S. Nesheim, in NBS Spec. Publ. (U.S.) No. 519, Trace Org. Anal: New Front. Anal. Chem., U.S. Government Printing Office, Washington, D. C., 1979, p. 355.
10. J. G. Heathcote and J. R. Heathcote, Aflatoxins: Chemical and Biological Aspects, Elsevier, New York, 1978, Chapter 4.
11. E. Stahl, in Thin Layer Chromatography, E. Stahl (Ed.), Springer-Verlag, New York, 1969, p. 86.
12. R. E. Peterson and A. Ciegler, J. Chromatogr., 31, 250 (1967).
13. C. A. H. Verhülsdonk and P. L. Schuller, EEC Document No. 2689/VI/73, Rev., 1 (1973).
14. A. V. Jain and R. C. Hatch, in Analytical Toxicology Manual, Kansas State University, Manhattan, Kansas, 1980.
15. L. Petit and G. Sediq, Collect. Med. Leg. Toxicol. Med. (Publ. 1978), 107, 125 (1977); Chem. Abstr., 92, 35575 (1978).
16. H. Arnold, Gesunderheitsgefaerhrdung Aflatoxine, Arbeitstag, 215 (1978); Chem. Abstr., 93, 166182 (1978).
17. L. G. M. Th. Tuinstra, C. A. H. Verhülsdonk, J. M. Bronsgeest, and W. E. Paulsch, Neth. J. Agric. Sci., 23, 10 (1975).
18. P. M. Scott and B. P. C. Kennedy, J. Assoc. Offic. Anal. Chem., 56, 1452 (1973).
19. P. M. Scott and B. P. C. Kennedy, Can. Inst. Food Sci. Technol. J., 8, 124 (1975).
20. P. R. Beljaars, J. C. M. H. Schumans, and P. M. Koken, J. Assoc. Offic. Anal. Chem., 58, 263 (1975).
21. J. I. Suzuki, B. Dainius, and J. H. Kilbuck, J. Food Sci., 38, 949 (1973).
22. P. L. Schuller, C. A. H. Verhülsdonk, and W. E. Paulsch, Arzneim.-Forsch., 20, 1517 (1970).

23. P. R. Beljaars, C. A. H. Verhülsdonk, W. E. Paulsch, and D. H. Leim, J. Assoc. Offic. Anal. Chem., 56, 1444 (1973).
24. B. Altenkirk, J. Chromatogr., 65, 456 (1972).
25. L. Yin, A. D. Campbell, and L. Stoloff, J. Assoc. Offic. Anal. Chem., 59, 102 (1971).
26. C. M. Rossi, A. R. Borgatti, P. Cortesi, and G. Criesetug, Atti. Soc. Ital. Sci. Vet., 25, 437 (1971); Chem. Abstr., 77, 84077 (1971).
27. M. W. Trucksess and L. Stoloff, J. Assoc. Offic. Anal. Chem., 62, 1080 (1979).
28. J. Bartos and Z. Matyas, Vet. Med. (Prague), 25, 495 (1980); Chem. Abstr., 93, 219370 (1980).
29. R. D. Stubblefield and O. E. Shotwell, Abstracts, 94th Annual Meeting, AOAC, Abstract No. 149, 1980.
30. L. C. M. Th. Tuinstra and J. M. Bronsgeest, J. Chromatogr., 111, 448 (1975).
31. D. S. P. Patterson, E. M. Glancy, and B. A. Roberts, Fd. Cosmet. Toxicol., 16, 49 (1978).
32. K. Lemieszek-Chodorowska, Rocz. Parrstw. Zakl. Hig., 30, 141 (1979); Chem. Abstr., 91, 106674 (1979).
33. F. Kiermeier and D. Groll, Z. Lebensmitt.-Untersuch.-Forsch., 142, 120 (1970).
34. F. Kiermeier, Z. Lebensmitt.-Untersuch.-Forsch., 144, 293 (1970).
35. Mycotoxin Methodology, AOAC, Arlington, Va., 1980.
36. M. W. Trucksess, L. Stoloff, W. A. Pons, A. F. Cucullu, L. S. Lee, and A. O. Franz, J. Assoc. Offic. Anal. Chem., 60, 795 (1977).
37. L. Allen, J. Assoc. Offic. Anal. Chem., 57, 1398 (1974).
38. C. A. H. Verhülsdonk, P. L. Schuller, and W. E. Paulsch, Zeszyty Problemowe Postepow Nauk Rolniczych., 189, 277 (1977).
39. M. W. Trucksess, J. Assoc. Offic. Anal. Chem., 59, 722 (1976).
40. H. P. Van Egmond, W. E. Paulsch, and P. L. Schuller, J. Assoc. Offic. Anal. Chem., 61, 809 (1978).
41. P. L. Schuller, W. Horwitz, and L. Stoloff, J. Assoc. Offic. Anal. Chem., 59, 1312 (1976).
42. W. A. Pons, A. F. Cucullu, and L. S. Lee, Proc. Third Intern. Congr. Food Sci. Technol. (505/70) Washington, D. C., 1970, p. 705.
43. P. L. Schuller, Th. Ockhuizen, J. Werringloer, and P. Marquardt, Arzneim.-Forsch., 17, 888 (1967).
44. H. P. Van Egmond and R. D. Stubblefield, J. Assoc. Offic. Anal. Chem., 64, 152 (1981).

CHAPTER 29

Application of TLC in the Study of Aflatoxon Biosynthesis

Dennis P.H. Hsieh and Nick C. Wan

INTRODUCTION

From the various advanced and sophisticated TLC techniques described in the previous papers, there is no doubt that TLC is a major analytical tool in chemical research and monitoring. In this paper some basic, simple TLC techniques will be described as they have been applied to our own research in the biosynthesis of aflatoxins, to further illustrate the usefulness of TLC.

Aflatoxins are a class of potent carcinogenic mycotoxins produced by *Aspergillus flavus* and *A. parasiticus*, frequently detected in peanuts, corn, cottonseed, and other foodstuffs (1). The structures of the four major aflatoxins are shown in Figure 1.

These compounds all contain a characteristic bisfuran ring structure fused to a substituted coumarin moiety that makes them strongly fluorescent under ultraviolet light. The vinyl ether double bond in the bisfuran ring structure is essential for the mutagenic and carcinogenic properties of aflatoxins (2). Thus aflatoxins B_2 and G_2 are nontoxic, whereas B_1 and G_1 are the most potent carcinogenic mycotoxins known so far. Because of the unusual chemical structure of aflatoxins and their food safety significance, the biosynthesis of aflatoxins in *A. parasiticus* has been a subject of active investigation during the last 15 years.

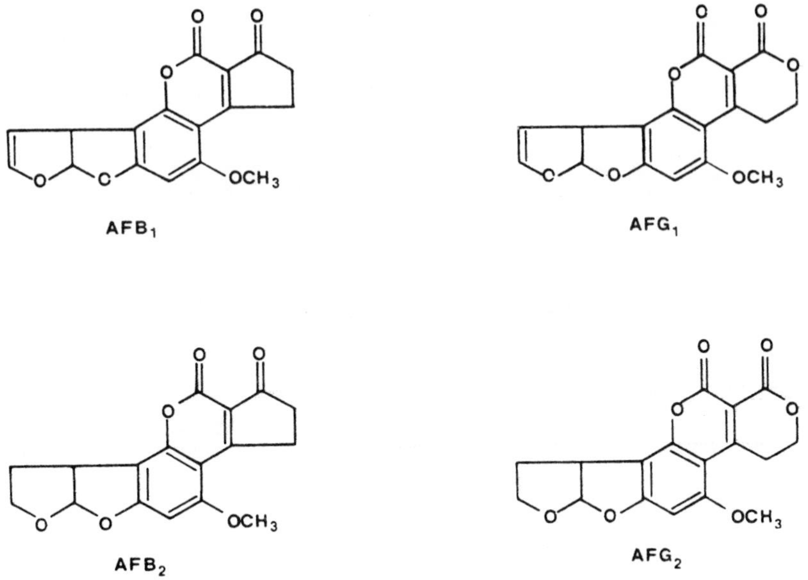

AFB$_1$ AFG$_1$

AFB$_2$ AFG$_2$

Figure 1. The structures of aflatoxins B_1, B_2, G_1, and G_2.

THE PATHWAY OF AFLATOXIN BIOSYNTHESIS

As a result of active investigation, a pathway of aflatoxin biosynthesis in *A. parasiticus* has been established by Singh and Hsieh (3), which is shown in Figure 2.

Ten acetate units in the form of one acetyl-CoA and nine malonyl-CoA are linked in a head-to-tail fashion to form a 20-carbon polyacetate or polyketide intermediate that cyclizes to form a series of C_{20}-anthraguinones such as norsolorinic acid, averufin, and versiconal hemiacetal acetate (VHA). The loss of an acetate unit in the enclosure of the aliphatic side chain of VHA results in the formation of the first bisfuranoid compound, versicolorin A (VA), a C_{18}-bisfuranoanthraguinone in this pathway. The anthraguinone moiety of VA undergoes an oxidative ring cleavage to form a bisfurano xanthone, sterigmatocystin. The xanthone undergoes another oxidative ring cleavage to form aflatoxin B_1 (AFB_1), which is the major and also the most potent member of the family. Aflatoxins B_2, G_1, and G_2 are derived from B_1.

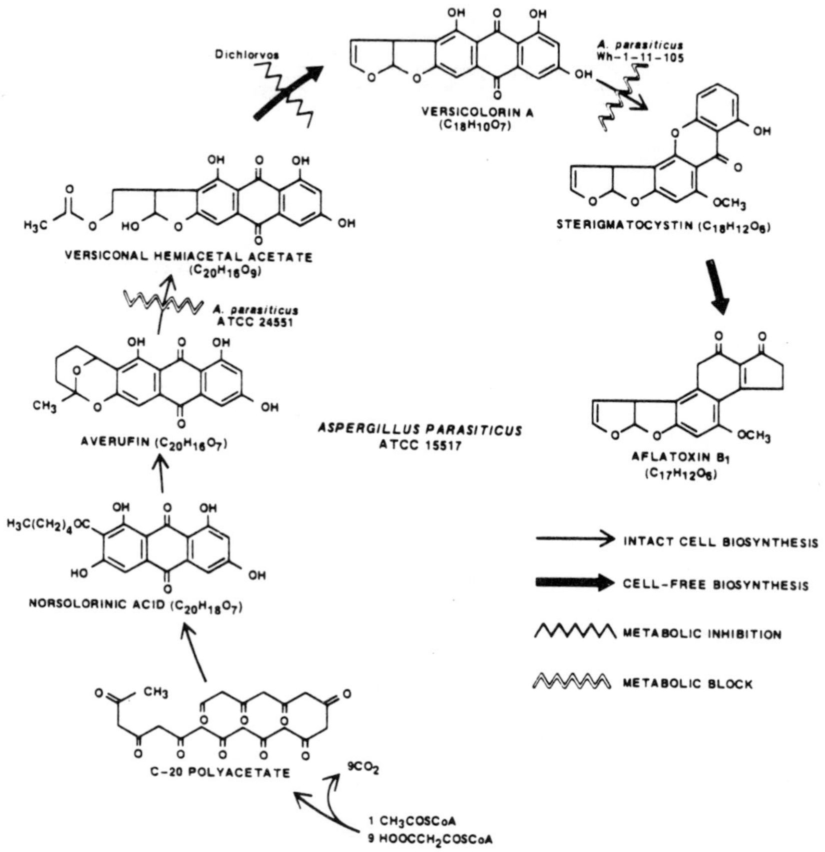

Figure 2. The pathway of aflatoxin biosynthesis.

This scheme was mainly established by conversion of one compound to another by the myceluim of A. parasiticus using isotope-labeled compounds. The purification of compounds has been accomplished mainly by the use of TLC.

THE VERSICOLORIN A SYNTHASE

In the biosynthetic scheme (Figure 2), the characteristic bisfuran ring structure of aflatoxins is formed in the step where VHA is

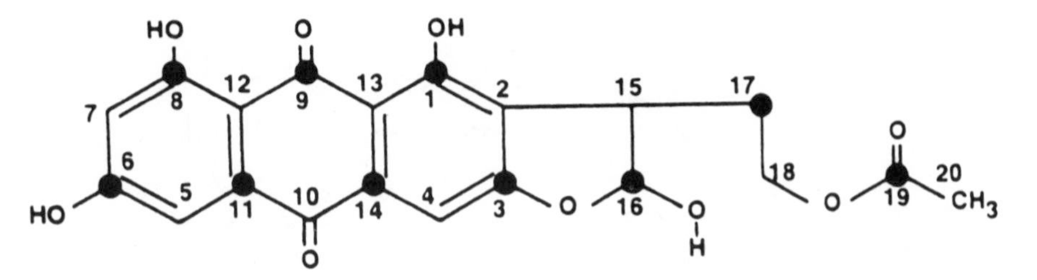

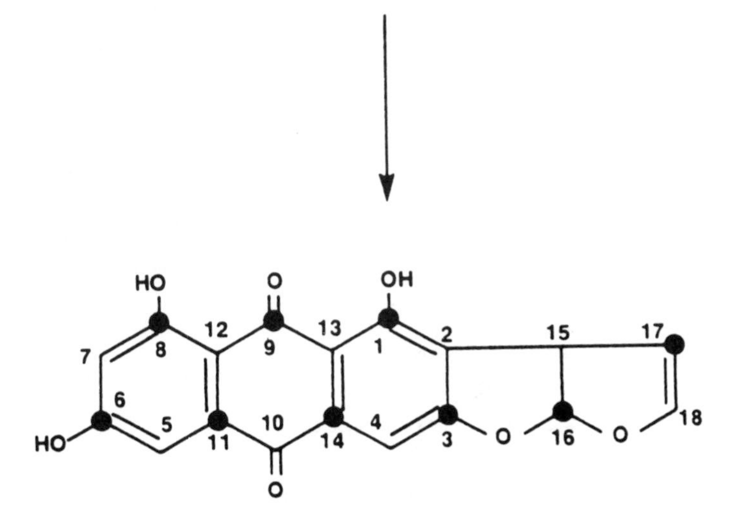

Figure 3. The precursor-product relationship between versiconal hemiacetal acetate and versicolorin A. The labels are carbons derived from the carboxyl carbons of acetate molecules.

converted to VA. A close-up view of this step is shown in Figure 3. In this paper, the discussion of TLC and other methodology is focused on the enzymatic reaction of this step for two reasons:

1. This is a key step in the aflatoxin pathway where the crucial bisfuran ring structure is formed. As mentioned earlier, this functional group confers the mutagenic and carcinogenic property upon aflatoxin molecules and also upon others such as sterigmatocystin and versicolorin A (4). Fungal products with a bisfuran ring structure are now known as bisfuranoid mycotoxins. Characterization of the enzyme involved in this biosynthetic step may shed light on the control of the biosynthesis of all the bisfuranoid mycotoxins. Since this enzyme catalyzes the formation of VA, it is tentatively named the VA synthase.
2. Aflatoxins are secondary metabolites of *A. parasiticus*, and the enzymes involved in any fungal secondary biosynthesis have been experimentally difficult to isolate and characterize. The VA synthase, therefore, serves as a model system for the development of methodology in the enzymatic studies of fungal secondary biosynthesis.

PREPARATION OF ^{14}C-LABELED SUBSTRATE

The first task in the study of this conversion step is the preparation of ^{14}C-labeled VHA as the substrate for enzyme transformations. This is accomplished by adding ^{14}C-labeled acetate to the submerged culture of *A. parasiticus* in the presence of 10 ppm of dichlorvos, an organophosphorus insecticide (5). Dichlorvos is a specific inhibitor of aflatoxin biosynthesis, blocking this particular step to result in accumulation of VHA, instead of AFB_1, so that the ^{14}C-acetate added to the culture is incorporated into VHA.

The ^{14}C-VHA extracted from the mycelium of the cultures was purified by repeated preparative TLC developed in different solvent systems. Thus the concentrated acetone extract from the fungal mycelium was applied to a preparative silica gel GTLC plate (EM Laboratory, Inc., Emsford, New York) of 2 mm thickness as a straight band by a streaker (Applied Science, State College, Pennsylvania). The TLC plate was developed in benzene/ethyl acetate/isopropanol/water (25:10:2:1). After development, the VHA band of absorbent was collected and VHA was extracted, condensed, and applied on another plate and developed in chloroform/acetone/hexane/acetic acid (85:15:20:4) (CAHAA).

The VHA in the isolated band of the second TLC plate was considered chromatographically pure and was extracted for use in the conversion experiments. The purified VHA was checked with the authentic compound by cochromatography in several solvent systems. Cochromatography was always done by applying three spots on an analytical plate: sample, sample plus standard, and standard. For a sample to cochromatograph with the analytical standard, the sample plus standard spot must appear as a single spot, which may or may not have the same R_f values as the other two spots, depending upon whether there are variations in the local conditions on the TLC plates.

EXPERIMENTAL CONVERSION OF VHA TO VA

In a typical conversion experiment, the ^{14}C-VHA is incubated with the enzyme system extracted from the mycelium of _A. parasiticus_ when the culture reached biosynthetic phase. The enzyme system was extracted by grinding the mycelium with glass beads in a neutral phosphate buffer, followed by centrifugation to remove all the cellular debris (6).

After incubation for a period of time, the compounds in the reaction mixture were extracted with chloroform and separated on TLC to assay for the enzyme activity. The latter is defined as the amount of radioactivity from ^{14}C-VHA incorporated into VA. Since the enzyme system was prepared from the aflatoxin producing mycelium, the extract contained a number of biosynthetic intermediates that were not completely separable by a single one-dimensional TLC system. However, VHA and VA were completely isolated on a two-dimensional TLC plate, developed in CAHAA followed by development in hexane/acetone/ethyl acetate/acetic acid (70:30:20:4). The result of one- and two-dimensional TLC is diagrammed in Figure 4. The absorbent in the isolated spots was collected into scintillation vials, and the radioactivity was measured with a scintillation spectrometer (Packard Tri-Carb). The activity of VA synthase of two strains of aflatoxin producer is compared by using the quantities of radioactivity in microcuries incorporated into the product VA in a unit length of time (hours). See Table I for results of such a study.

DETECTION OF RADIOACTIVITY ON TLC PLATES

Two techniques were used in addition to scintillation counting to detect and measure radioactive compounds on TLC plates: autoradiography and spark chamber photography. The autoradiography

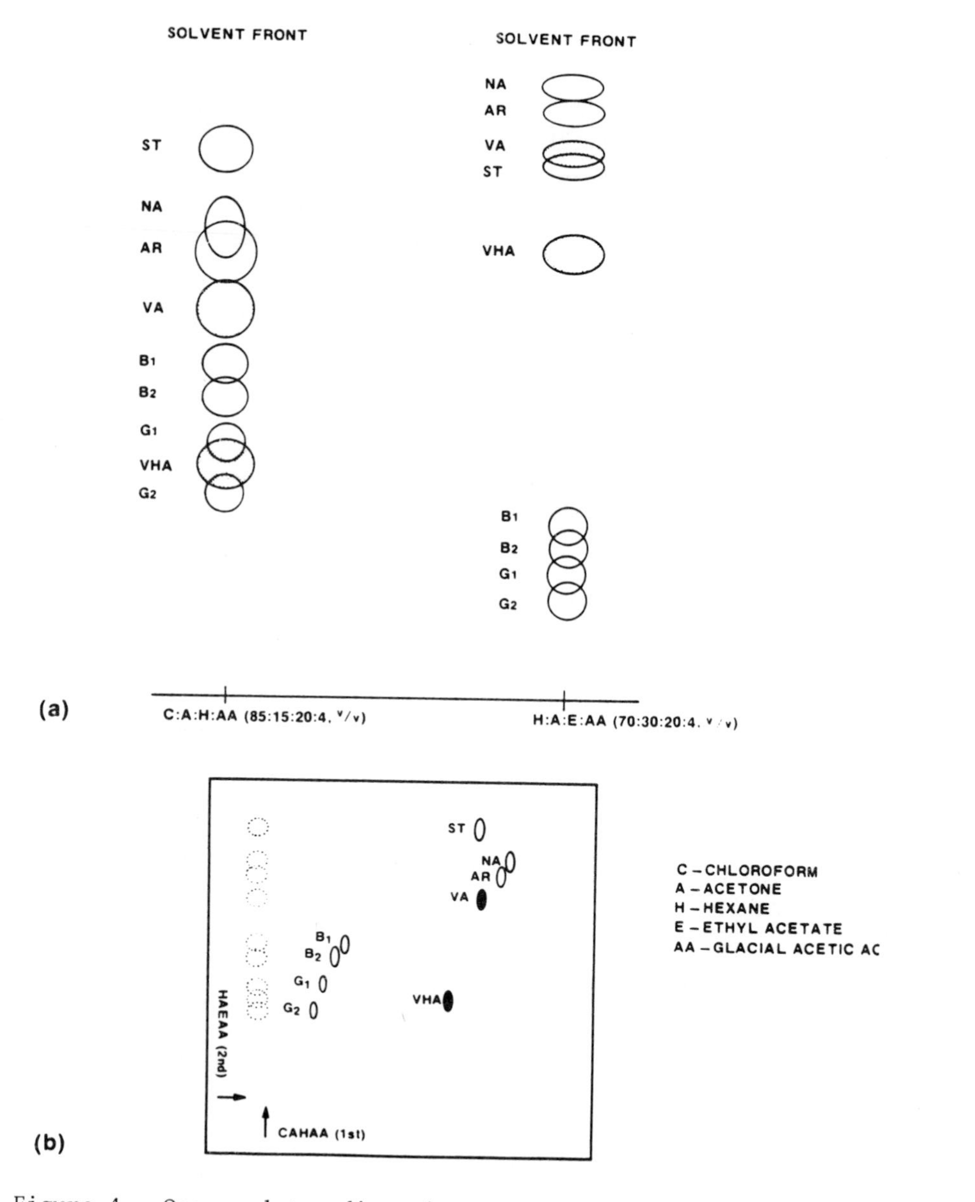

Figure 4. One- and two-dimensional thin layer chromatography of aflatoxins and their biosynthetic intermediates. Abbreviations: ST, sterigmatocystin; NA, norsolorinic acid; AR, averufin; VA, versicolorin A; B_1, aflatoxin B_1; B_2, aflatoxin B_2; G_1, aflatoxin G_1; VHA, versiconal hemiacetal acetate; G_2, aflatoxin G_2; C, chloroform; A, acetone; H, hexane; E, ethyl acetate; AA, acetic acid.

TABLE I. COMPARISON OF THE ACTIVITY OF VA SYNTHASE ISOLATED FROM *A. PARASITICUS* ATCC 36537 and ATCC 15517

	Enzyme Activity[a] (μ Ci/hr) $\times 10^4$
A. parasiticus ATCC 36537	2.77 ± 0.088
A. parasiticus ATCC 15517	3.50 ± 0.081

[a]Enzyme activity is measured as the amount of radioactivity recovered in VA. Average of duplicate samples.

is a classic method that is done by placing an X-ray film (Kodak, no-screen film NS-ST) over a developed TLC plate and exposing the film to the radioactive emission from the spots on the plate in a dark room for a week or two. The X-ray film was then developed and fixed to exhibit radioactive compounds on the TL plate as dark spots. The TLC-autoradiographical technique is especially useful in a time-course study. An autoradiograph of samples taken at different time intervals from an enzyme reaction system is shown in Figure 5. It is apparent that during the 10-hr incubation period, VHA disappeared with time while VA increased in amount, and there were at least four other compounds involved in the conversion, which may be transient intermediates of this conversion step.

The usefulness of the TLC-autoradiographic technique in biosynthetic studies is obvious. The only problem is that it usually takes days or weeks to expose the X-ray film. This problem can be solved now by the use of a spark chamber (Berthold, Postfach, West Germany), which is capable of taking pictures of radioactive spots on a TLC plate with exposure time in only a few minutes. The detailed features of a spark chamber have been described in an earlier article in this book, dealing with radio-scanning of TLC.

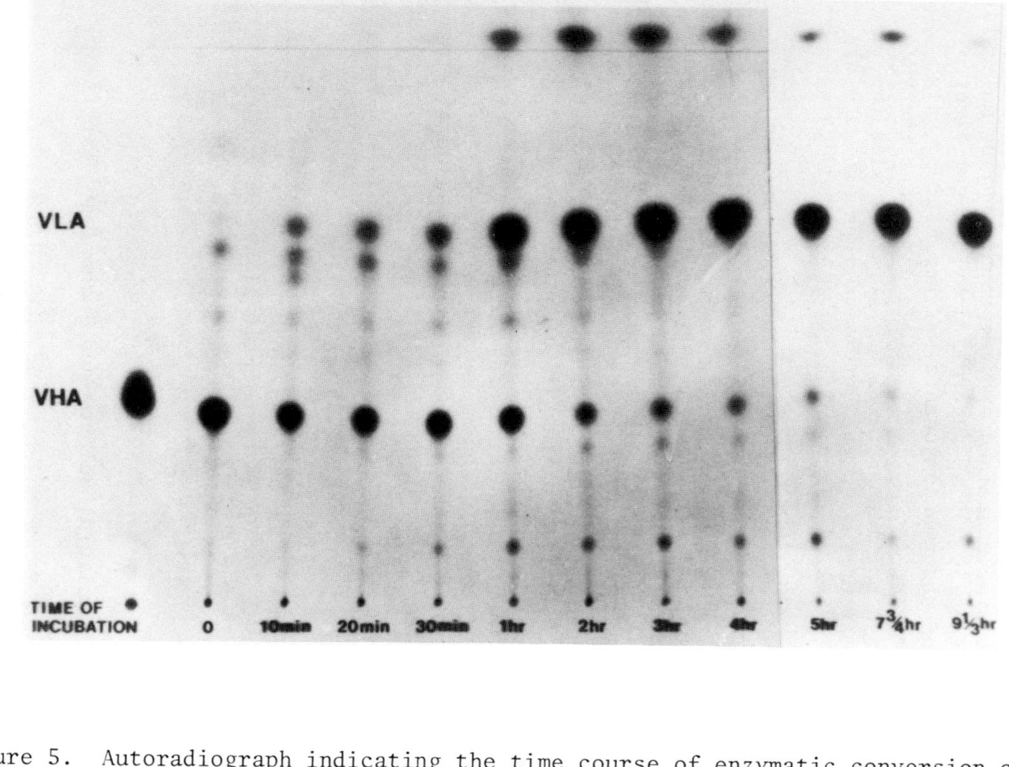

Figure 5. Autoradiograph indicating the time course of enzymatic conversion of VHA to VA.

PURIFICATION AND CHARACTERIZATION OF THE VA SYNTHASE

Using the quantitative TLC techniques described above to assay for enzyme activity, the VA synthase system was purified and characterized (6). A 123-fold purification was achieved through centrifugation, ammonium sulfate precipitation, and Sephadex gel filtration, as summarized in Table II.

The purified enzyme contained two components having molecular weights of 92,000 and 130,000 as determined by SDS gel electrophoresis. Either of the two components was not active; only the combined system gave the activity, confirming that the VHA-to-VA conversion is a multistep reaction. Together with the intermediate spots appearing on the autoradiograph (Figure 5), a scheme for the conversion reactions between VHA and VA was put forward as shown in Figure 6.

The VA synthase was located in the 105,000*g* supernatant fraction of the mycelial homogenate of *A. parasiticus*. The enzyme was stable for 3 hr at pH 5.8 to 7.5 at 22°C; it did not require any cofactors for activity except for oxygen. As expected, the enzyme is very sensitive to dichlorvos inhibition. During aflatoxin biosynthesis in *A. parasiticus*, the enzyme is subject to feedback inhibition and repression regulations by VA and aflatoxin B_1.

In conclusion, the basic TLC techniques have played a vital role in the detailed studies of the VA synthase system in the aflatoxin pathway. This enzyme system is the first one of this biosynthetic pathway to be purified and characterized. The data on the property of this enzyme system may shed some light on the control of aflatoxin formation and the formation of other bisfuranoid mycotoxins in food commodities.

ACKNOWLEDGMENT

Our biosynthetic studies were in part supported by Public Health Service grant ES00612 and by Western Regional Research Project W-122.

REFERENCES

1. Council for Agricultural Science and Technology, Report No. 80, 1979, p. 5.
2. J. J. Wong and D. P. H. Hsieh, Proc. NAS, U.S.A., 73, 2241 (1976).

TABLE II. PURIFICATION OF THE VA SYNTHASE SYSTEM IN *A. PARASITICUS* ATCC 15517[a]

Step	Description	Volume (ml)	Total Activity (μCi)	Protein (μg/ml)	Specific Activity (μCi/μg) × 10^7	Yield (%)	Purification (-fold)
1	Filtered, crude ext.	80	0.129	1778	9.05	100	1
2	Supernat., centrifug. 105,000*g*	78	0.106	1160	11.8	82.6	1.3
3	Redissolved ppt. 50-75% $(NH_4)_2SO_4$	12	0.043	70.8	50.5	33.3	5.59
4	Void, Sephadex G-75	18	0.004	43.1	51.0	3.07	5.64
5	Fract. 26, 27, & 31 Sephadex G-200	5.7	0.0026	11.2	406	2.01	44.87
6	Fract. (26 + 27) A, 31B Sephadex G-200	1.97	0.0015	6.8	1110	1.17	122.96

[a]Results are averages of duplicate samples.

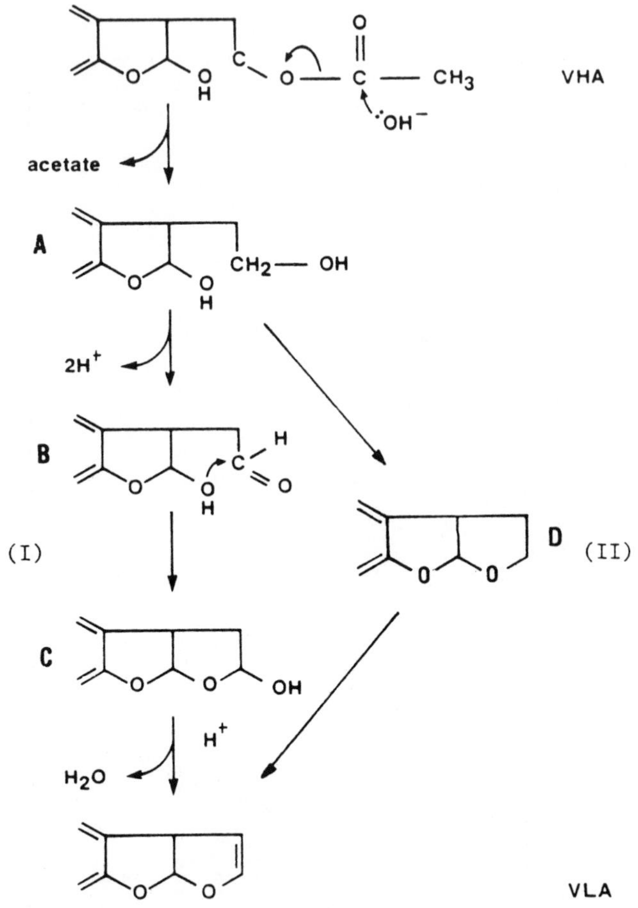

Figure 6. Proposed mechanism for the conversion of VHA to VA.

3. R. Singh and D. P. H. Hsieh, Arch. Biochem. Biophys., 178, 285 (1977).
4. J. J. Wong, R. Singh, and D. P. H. Hsieh, Mutation Res., 44, 447 (1977).
5. R. C. Yao and D. P. H. Hsieh, Appl. Microbiol., 28, 52 (1974).
6. N. C. Wan and D. P. H. Hsieh, Appl. Environ. Microbiol., 39, 109 (1980).

CHAPTER 30

Sample Purification of Mycotoxins by Gel Permeation

Steven R. Tonsager, David A. Maltby,
Robert J. Schock, and W. Emmett Braselton

INTRODUCTION

Mycotoxic fungi have been implicated as a threat to feed and foodstuffs since the fifteenth century, when ergot-infected grains were known to have affected animals and man, but only through extensive research efforts in the past 20 years have we become aware of the diversity of mycotoxins and their associated disease syndromes, or mycotoxicoses (1, 2). Confirmation of a mycotoxicosis presents a difficult challenge to diagnosticians because the manifestations may vary from acute disease and death to more subtle effects on growth, production, disease resistance, and immunity. Final confirmation rests on the identification of the responsible mycotoxin in sufficient amounts in a representative sample of feed or foodstuff implicated in the intoxication. To this end, numerous procedures have been described for the analysis of individual mycotoxins, or closely related species (1-5). However, since subacute mycotoxin poisonings often lead to nonspecific effects such that a specific mycotoxicosis cannot be identified from the symptomology, toxicologists are in need of screening procedures for simultaneous detection of multiple mycotoxins.

Although methods for simultaneous detection of multiple mycotoxins on TLC have been reported by Steyn (6) and Durackova et al. (7), these generally have been limited to analysis of pure compounds. TLC procedures to screen simultaneously for aflatoxins, ochratoxin, and zearalenone have been described (8, 9), and a multiple mycotoxin-detection procedure for aflatoxins, ochratoxins, zearalenone, sterigmatocystin, and patulin in agricultural commodities has been reported by Stoloff et al. (10).

For TLC detection of mycotoxins in feed and foodstuffs to be successful, extensive purification from the sample matrix is required. Methods described to date have utilized removal of colored interferences with a lead acetate (11), zinc acetate (12), or ferric hydroxide gel (13) precipitate and removal of lipid with solvent partition (10, 13) or silica gel absorption columns according to Eppley (8), or combinations of the above (3). Although these cleanup procedures are sufficient to permit TLC analysis of a number of mycotoxins, they are not adequate for TLC detection of the trichothecenes including T-2 toxin, diacetoxyscirpenol (DAS), and deoxynivalenol (vomitoxin), which are of particular importance in the midwestern United States. Stahr (personal communication) has achieved additional cleanup of colored interferences from hay samples using activated charcoal, at the sacrifice of sensitivity. Since we had been using gel-permeation chromatography to purify a number of organic toxicants from biological matrices, we began to investigate the use of gel permeation to purify trichothecenes and other mycotoxins prior to TLC.

MATERIALS AND METHODS

Reagents

Aflatoxins B_1, B_2, G_1, and G_2 were obtained from Applied Science Laboratories (College Park, Pennsylvania); diacetoxyscirpenol (DAS), T-2 toxin, and zearalenone were from Sigma Chemical Company (St. Louis, Missouri); and deoxynivalenol (vomitoxin) was from the Myco-Lab Company (Chesterfield, Missouri). Bio-Beads S-X3 for gel-permeation chromatography were from BioRad Laboratories (Richmond, California). Prepared TLC plates were LHP-K, 10 × 10 cm from Whatman, Inc. (Clifton, New Jersey). Trimethylsilylimidazole (TMSI) was from Supelco, Inc. (Bellefonte, Pennsylvania) and bis(trimethylsilyl)trifluoroacetamide (BSTFA) was obtained from Regis Chemical Company (Morton Grove, Illinois). Solvents were "Distilled in Glass" (Burdick and Jackson) or "Resi-Analyzed" (Baker)

Extraction of Grains and Feed

Grain or feed samples (100 g) were weighed into a Waring blender and homogenized for 5 minutes in 400 ml of 50% methanol/water (14). Hay or silage samples were allowed to soak in the solvent for 5 minutes before blending, and then homogenized for 2 to 3 minutes only. The sample was filtered through Whatman No. 4 filter paper in a Buchner funnel, and the volume of recovered solvent was recorded. The aqueous filtrate was shaken in a separatory funnel with 100 ml of $CHCl_3$, and the layers were allowed to separate. The $CHCl_3$ fraction was passed into a round-bottom flask through a small funnel containing anhydrous Na_2SO_4 and evaporated to dryness under vacuum on a rotary evaporator. The residue was redissolved in toluene/CH_2Cl_2 (85:15) at a final concentration of 100 g equivalents feed per 1 ml and filtered through a small plug of silanized glass wool in a disposable Pasteur pipette to remove the nonsoluble portion of the sample.

Column Chromatography

Gel-permeation chromatography (GPC) was carried out on a 1.5 × 52 cm bed of Bio-Beads S-X3 equilibrated in toluene/CH_2Cl_2 (85:15). Fifty-gram equivalents of sample (25 g for hay or silage) in 0.5 ml was placed onto the top of the column and allowed to drain into the column bed before starting to collect the waste fraction. The column was then eluted with toluene/CH_2Cl_2 (85:15), and the first 50 ml was discarded. Mycotoxins were then collected in the following 5-ml fractions: F1, 50 to 55 ml, contained T-2 toxin; F2, 55 to 60 ml, contained DAS, the majority of any aflatoxins and part of the zearalenone present; F3, 60 to 65 ml, contained aflatoxins, zearalenone, and most of any vomitoxin present; F4, 65 to 70 ml, contained the remainder of the vomitoxin. Figure 1 depicts the overall procedure for purification of mycotoxins by GPC.

TLC Analysis for Mycotoxins

The fractions from GPC were evaporated just to dryness under a stream of N_2 and redissolved in benzene/acetonitrile (98:2) so that 1 g equivalent of sample was contained in 25 µl. Samples (25 µl) were spotted on the adsorbent portion of the LHP-K TLC plates in thin vertical lines. Samples were applied to the plate in the sequence GPC F1; F1 overlaid at the origin with a standard

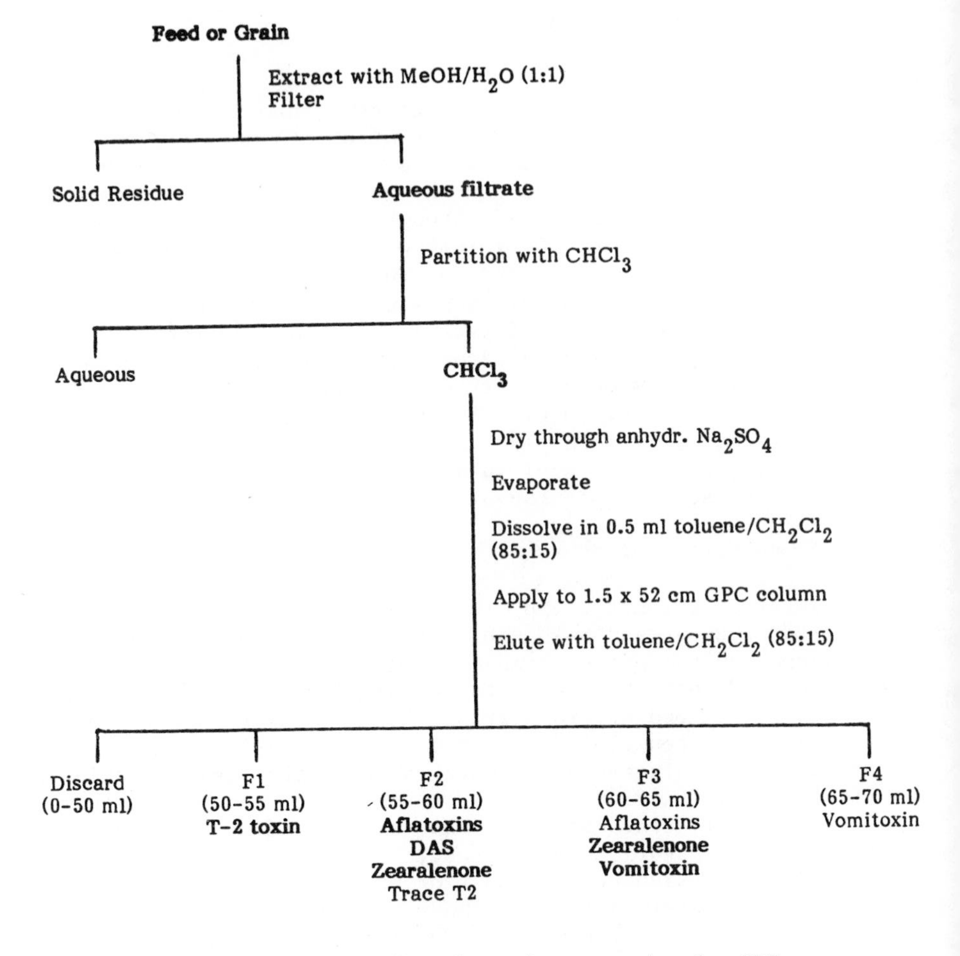

Figure 1. Purification of mycotoxins by GPC.

mixture containing 2 ng each of aflatoxins B_1, B_2, G_1, and G_2, 200 ng of T-2 toxin, DAS, and vomitoxin and 100 ng zearalenone; F2 overlaid with standard mix; F2; and so on. In this manner, two GPC fractions could be spotted per 10 × 10 cm plate, and the presence of interferences could be more easily determined by the overlaid standards.

The plates were developed two times in the solvent system toluene/ethyl acetate/acetone (3:2:1) described by Stahr (13).

Mycotoxins were identifed on the thin layer plates by a specific sequence of visualization procedures: (1) fluorescence under short-wavelength (254-nm) UV light; (2) fluorescence under long UV (365 nm); (3) fluorescence under long UV after charring with H_2SO_4/MeOH; (4) fluorescence under long UV after charring with H_2SO_4/MeOH and heating; (5) observation under visible light after charring and heating (Table I). Visualizing the plates with short-wavelength UV light indicated the presence of zearalenone as a yellow fluorescence. The plates were then observed under long UV, which indicated zearalenone (blue fluorescence) and aflatoxins (blue-green fluorescence). The plates were then sprayed lightly with H_2SO_4/methanol (1:1) and again observed under long UV. The aflatoxins changed from a blue-green fluorescence to a bright yellow. The trichothecenes could be observed after the plates were charred and heated briefly on a hot plate. Under visible light T-2 toxin, DAS, and vomitoxin all appeared as gray-black. Zearalenone appeared as a dull yellow. The charred and heated plates were then observed under long UV, and zearalenone DAS, and T-2 toxin appeared yellow, brown, and gray-white, respectively. Vomitoxin appeared a dull orange-red.

Confirmation and Quantification of TLC Positives

Aflatoxin

The presence of aflatoxins in a sample was confirmed, and amounts were quantified by high-performance liquid chromatography (HPLC) of aliquots of the GPC fractions F2 and F3. HPLC was carried out on a Waters 6000 pump, U6-K injector, and model 450 UV absorbance detector at 365 nm. Samples were chromatographed on a µ Porisil column in $CHCl_3$/cyclohexane/acetonitrile (25:7.5:1), 2% isopropanol, according to the procedure of Pons (15). If sufficient compound was present, HPLC fractions were collected and analyzed by electron-impact mass spectrometry using a solid insertion probe. Mass spectra were taken at 70 eV on a Finnegan 3200 gas chromatograph-mass spectrometer interfaced to a Riber SADR data acquisition and control system.

DAS and T-2 Toxin

DAS and T-2 toxin were confirmed and quantified by GC/MS of the trimethylsilyl ether (TMS) derivatives using selected ion monitoring (SIM). Aliquots of GPC F1 or F2 were placed

TABLE I. THIN LAYER CHROMATOGRAPHY OF MYCOTOXIN FRACTIONS FROM GEL-PERMEATION CHROMATOGRAPHY AND SEQUENTIAL VISUALIZATION PROCEDURES[a]

			TLC Spot Color				
Compound	GPC Fraction	TLC R_f	a. Short UV	b. Long UV	c. H_2SO_4/MeOH Long UV	d. Charred Long UV	e. Charred Visible
Zearalenone	F2-F3	0.70	Yellow Fluorescence	Blue Fluorescence	Yellow Fluorescence	Yellow	Yellow
T-2 toxin	F1	0.49				Gray-White	Gray-Black
DAS	F2	0.46				Brown	Gray-Black
Aflatoxin B_1	F2-F3	0.32		Blue Fluorescence	Yellow Fluorescence		
Aflatoxin B_2	F2-F3	0.27		Blue Fluorescence	Yellow Fluorescence		
Aflatoxin G_1	F2-F3	0.27		Green Fluorescence	Yellow Fluorescence		
Aflatoxin G_2	F2-F3	0.24		Green Fluorescence	Yellow Fluorescence		
Vomitoxin	F3-F4	0.20				Dull Orange-red	Gray-Black

[a]The plates were developed twice in toluene/ethyl acetate/acetone (3:2:1).

in a 1-ml conical vial and taken to dryness with a stream of N_2. The sample was redissolved in 50 µl of CH_2Cl_2 and allowed to react with 50 µl of BSTFA and 10 µl of pyridine for 30 minutes at 60°. The reaction mixture was taken to dryness under N_2 and redissolved in cyclohexane/CH_2Cl_2 (85 : 15) to give a concentration of 1 g equivalent sample/µl. Samples were chromatographed on a 2 mm × 1.5 m column of 1% OV-17 on Ultra-Bond 20 M (RFR Corp., Hope,Rhode Island) at 255°. Selected ions monitored for DAS were m/z 290, 350, and 378, and ions monitored for T-2 were m/z 350, 378, and 436. Retention times for DAS and T-2 toxin were 2.6 and 7.8 minutes, respectively. Quantification was carried out with the SADR software computer-determined peak areas from the selected ion profiles, and interpolation from a standard curve of peak areas was obtained the same day.

Vomitoxin

Confirmation and quantification of vomitoxin were carried out by SIM of the TMS derivative. Aliquots of GPC F3 and F4 were evaporated to dryness in a 1-ml conical vial and redissolved in 50 µl of CH_2Cl_2. Samples were allowed to react with 10 µl TMSI for 15 minutes at room temperature and evaporated under a stream of N_2. The vomitoxin-TMS derivative was redissolved in cyclohexane/CH_2Cl_2 (85:15) at a volume of 1 g equivalent/µl, and chromatographed on 1% OV-17 on Ultra-Bond 20 M at 220°. Ions monitored were m/z 422, 497, and 512, and the retention time was 2.8 minutes.

Zearalenone

Selected ion monitoring was also used to confirm and quantify zearalenone identified on TLC. Aliquots of GPC F2 and F3 were evaporated to dryness under N_2 in a 1-ml conical vial and redissolved in 50 µl of CH_2Cl_2. Samples were allowed to react with 100 µl BSTFA for 20 minutes at 60°, then evaporated with a stream of N_2. The derivatized sample was redissolved in cyclohexane/CH_2Cl_2 (85:15) to a volume of 1 g equivalent sample/µl. Samples were analyzed by GC/MS using SIM on a 2 mm × 1.5 m column of 1% OV-1 on Gas Chrom Q, at 250°. Ions monitored were m/z 333, 429, and 462, and the zearalenone retention time was 1.7 minutes.

RESULTS AND DISCUSSION

The GPC elution fractions of the mycotoxins studied are given in Table I, along with their TLC R_f sequence. Chromatography on Bio-Beads S-X3 provided an efficient separation from the bulk of extracted material and partially separated the mycotoxins into groups that were clean enough to be examined directly by TLC. Use of the high-performance plates and short migration distances enhanced the sensitivity of the procedure such that detection limits on thin layer were in the ranges of 2 ppb for aflatoxins, 25 to 50 ppb for zearalenone and vomitoxin, and 200 to 800 ppb for T-2 and DAS. In addition, the visualization sequence provided maximum sensitivity for the compounds examined by taking advantage of fluorescent properties wherever possible and provided confirmatory evidence for the compounds in question by bringing about one or more characteristic color changes during the sequence. The GPC column also provided a measure of confirmation by the required appearance of specific mycotoxins in their appropriate GPC fractions.

Although the main mechanism of separation, size exclusion chromatography, provided the major effect of removing large-molecular-weight colored interferences and lipids from the lower-molecular-weight mycotoxins, additional mechanisms of affinity were also operating to contribute to the separation. These effects, which probably include weak interactions between the gel matrix and mycotoxin molecules such as hydrophobic bonding, could be manipulated somewhat by the choice of elution solvent. For instance, use of cyclohexane/CH_2Cl_2 (85:15) in place of toluene/CH_2Cl_2 (85:15) resulted in such high affinity of aflatoxins for the column that they could not be eluted with the former solvent. Vomitoxin also gave poor recovery with cyclohexane/CH_2Cl_2. Zearalenone also eluted much later when cyclohexane replaced toluene in the elution solvent. This allowed an extremely good purification of zearalenone by itself and points out that GPC conditions can be tailored for specific mycotoxins where desired. The conditions described in this paper are compromised to allow for a general screen. When chromatography was carried out in the toluene/CH_2Cl_2 solvent system, recoveries of the aflatoxins, trichothecenes, and zearalenone were quantitative. Preliminary experiments with ochratoxin indicated that this compound could not be eluted from the GPC column under the above conditions, and routine screening for ochratoxin had to be carried out by alternative procedures.

Although some colored or fluorescing material still remained in the mycotoxin fractions following GPC, these were resolved from the mycotoxin areas by the high-performance TLC plates used, and by the sequence of visualization procedures described. The GPC fractions were also clean enough for direct analysis by other confirmatory methods, if a positive was found by TLC. Figure 2 shows the high-performance liquid chromatogram of GPC F2 from a corn sample that was positive for aflatoxins by TLC. The chromatogram is compared with that of a reference mixture of aflatoxins

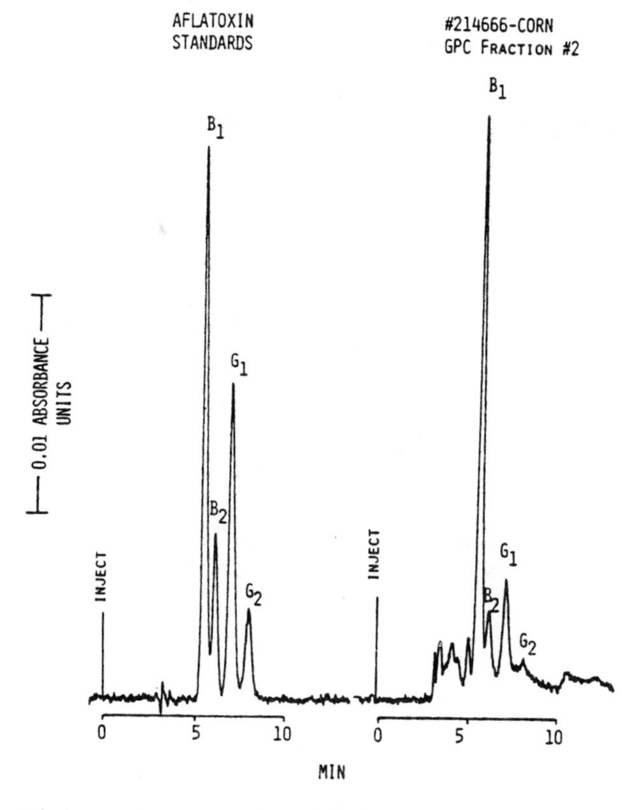

Figure 2. High-performance liquid chromatogram of reference aflatoxins B_1, B_2, G_1, and G_2 and GPC F2 from a corn sample positive for aflatoxins by TLC. Samples were chromatographed on a Waters μ Porasil column in $CHCl_3$/cyclohexane/acetonitrile (25:7.5:1) containing 2% isopropanol (15) at a flow rate of 1.5 ml/min. Compounds were detected by UV absorbance at 365 nm. The reference mixture applied to this column contained 50 ng each of B_1 and G_1 and 15 ng each of B_2 and G_2; the GPC F2 aliquot represented 300 mg equivalents of sample.

B_1, B_2, G_1, and G_2. There was very little background interference in the GPC fraction, and the aflatoxins were easily quantified by measurement of peak heights and interpolation from a standard curve. The identity of the aflatoxin B_1 peak in the sample was further confirmed by collection of the HPLC peak for mass spectral analysis. Figure 3 shows the 70-eV mass spectrum of the aflatoxin B_1 peak from HPLC and a spectrum of the reference standard.

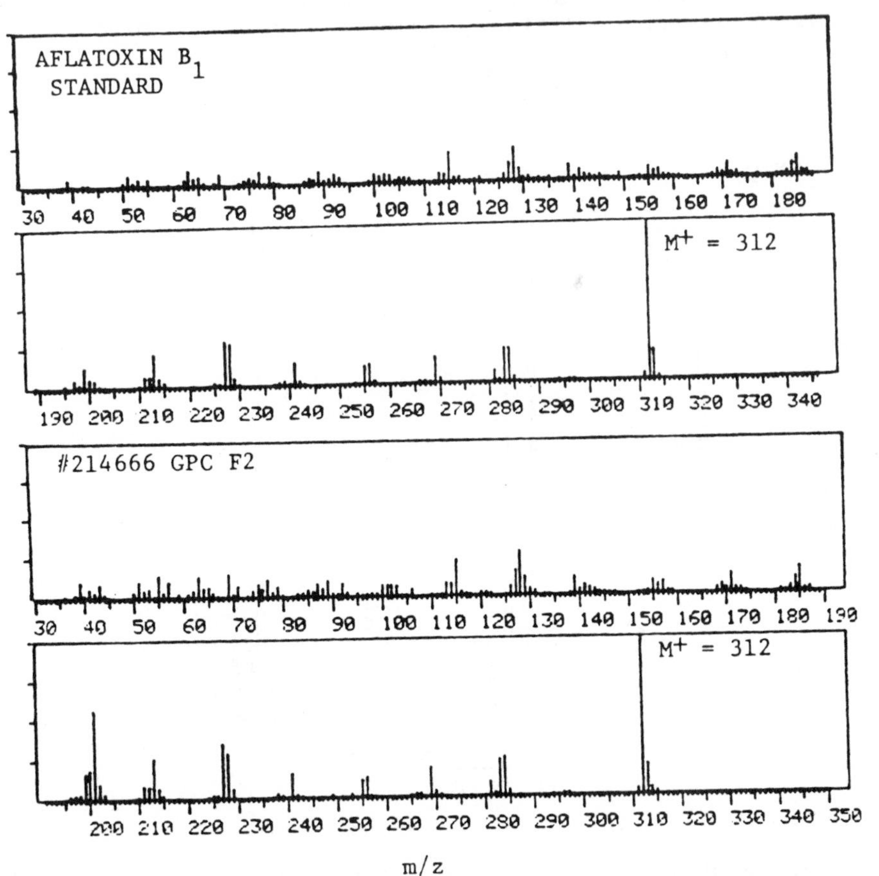

Figure 3. Mass spectra of reference aflatoxin B_1 and the compound isolated at the retention volume of B_1 during high-performance liquid chromatography of GPC F2 of a corn sample positive for aflatoxins by TLC. Spectra were obtained at 70 eV by solid sample direct probe.

GPC fractions containing the trichothecenes or zearalenone were also pure enough for confirmation directly by GC/MS using selected ion monitoring. Figure 4 is the SIM profile of the TMS derivative of an aliquot of GPC F2 from a corn sample that was positive for zearalenone and vomitoxin by TLC. Vomitoxin in GPC F3 was also confirmed by SIM, as shown in Figure 5.

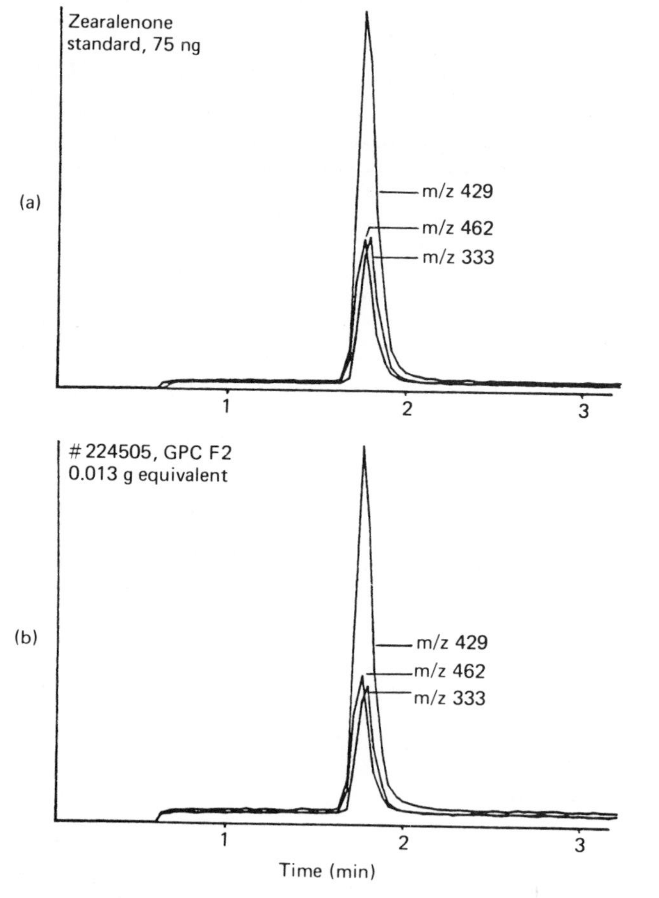

Figure 4. Selected ion-monitoring profile of (a) the TMS derivative of zearalenone and (b) an aliquot of GPC F2 from a corn sample that was positive for zearalenone and vomitoxin by TLC. Compounds were chromatographed on a 2 mm × 1.5 m column of 1% OV-1 on Gas Chrom Q at 250°. Ions monitored (70 eV) were m/z 333, 429, and 462.

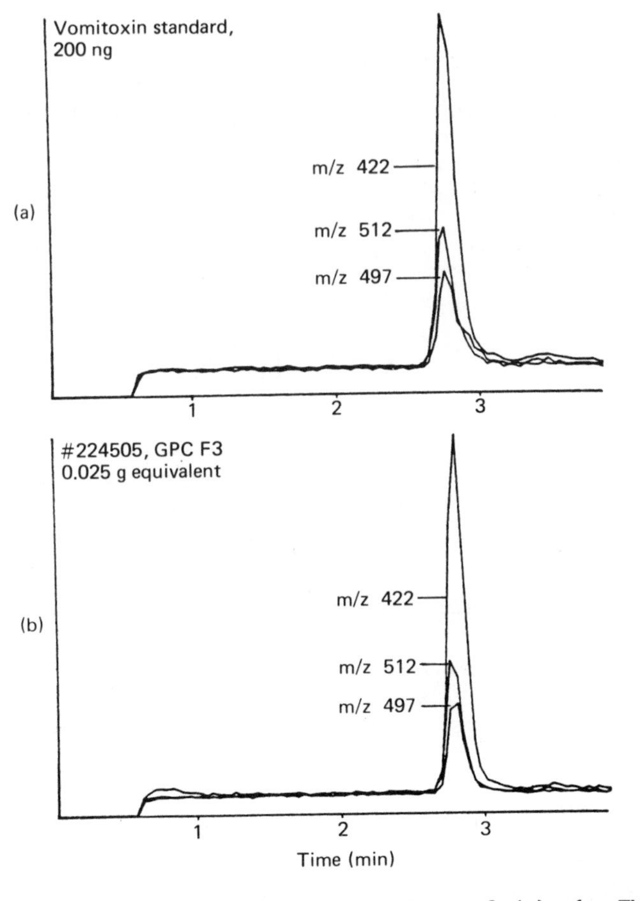

Figure 5. Selected ion-monitoring profile of (a) the TMS derivative of vomitoxin and (b) an aliquot of GPC F3 of a corn sample that was positive for vomitoxin and zearalenone by TLC. Compounds were chromatographed on a 2 mm × 1.5 m column of 1% OV-17 on Ultra-Bond at 220°. Ions monitored (70 eV) were m/z 422, 497, and 512.

Gel-permeation chromatography on Bio-Beads S-X3 thus provided a simple, effective procedure for the purification of aflatoxins, trichothecenes including DAS, T-2 toxin, and vomitoxin, and zearalenone in agricultural commodities. Ochratoxin was not amenable to the procedure as described. The GPC method is currently proving extremely valuable to our analytical toxicology service as a mycotoxin-sreening procedure for animal-health diagnostic purposes. Others have described a GPC method for purification of sterigmatocystin (16), and continuing work is focused on extending the GPC screen to include this and other mycotoxins of proven clinical and economic importance.

ACKNOWLEDGMENTS

The authors wish to express their appreciation to Lisa F. Lepper for technical assistance, Randy A. Cardona for the photography, Debbie Fish for preparation of the figures and Diane K. Hummel and Theresa A. Blasen for preparation of the manuscript.

REFERENCES

1. T. D. Wyllie and L. G. Morehouse (Eds.), Mycotoxic Fungi, Mycotoxins, Mycotoxicoses, An Encyclopedia Handbook, Vol. 1, Marcel Dekker, New York, 1977.
2. W. Shimoda (Ed.), Conference on Mycotoxins in Animal Feeds and Grains Related to Animal Health. Bureau of Veterinary Medicine, Food and Drug Administration, Report No. FDA/BVM 79/139, National Technical Information Service, 1979.
3. L. Stoloff, Clin. Toxicol., 5, 465 (1972).
4. D. L. Park and A. E. Pohland, National Bureau of Standards Special Publication 519, Trace Organic Analysis: A New Frontier in Analytical Chemistry, Proceedings of the 9th Materials Research Symposium, April 10-13, 1978 Gaithersburg, Md., 1979, p. 321.
5. J. L. Richard, J. R. Thurston, and A. C. Pier, Proceedings, 83rd Annual Meeting, U.S. Anim. Health Assoc., 205 (1979).
6. P. S. Steyn, J. Chromatogr., 45, (1969).
7. Z. Durackova, V. Betina, and P. Nemec, J. Chromatogr., 116, 141 (1976).
8. R. M. Eppley, J. Assoc. Offic. Anal. Chem., 51, 74 (1968).
9. I. Balzer, C. Bogdanic, and S. Pepeljnjak, J. Assoc. Offic. Anal. Chem., 61, 584 (1978).
10. L. Stoloff, S. Nesheim, L. Yin, J. V. Rodricks, M. Stack, and A. D. Campbell, J. Assoc. Offic. Anal. Chem., 54, 91 (1971).
11. W. A. Pons, Jr., A. F. Cuculla, A. O. Franz, Jr., L. S. Lee, and L. A. Goldblatt, J. Assoc. Offic. Anal. Chem., 56, 803 (1973).

12. C. E. Holaday and J. Lansden, J. Agric. Food Chem., 23, 1134 (1975).
13. H. M. Stahr, Analytical Toxicology Methods Manual, Iowa State University Press, Ames, Iowa, 1977, p. 164.
14. T. R. Romer, T. M. Boling, and J. L. MacDonald, J. Assoc. Offic. Anal. Chem., 61, 801 (1978).
15. W. A. Pons, Jr., J. Assoc. Offic. Anal. Chem., 59, 101 (1976).
16. A. S. Salhab, G. F. Russel, J. R. Coughlin, and D. P. H. Hsieh, J. Assoc. Offic. Anal. Chem., 59, 1037 (1976).

CHAPTER 31

Analysis of Mycotoxins by HPTLC and RP-TLC

H. Michael Stahr and Marlaine Domoto

INTRODUCTION

Mycotoxins are among the most prevalent biotoxins that foods and feeds contain. They are responsible for numerous animal health problems and are suspected of causing human health problems as well. Ochratoxins cause kidney failure (1). Aflatoxins cause liver tumors (2). Trichothecene mycotoxins can cause symptoms from feed refusal to sudden death (3, 4). Citrinin (5) is a kidney poison and often occurs with ochratoxin.

Field problems involving mycotoxins are common at the Veterinary Diagnostic Laboratory at Iowa State University. The methods we present are the results of our continuing efforts to provide better data for diagnostic toxicology.

EXPERIMENTAL

Reagents and Apparatus

Nanograde* solvents were purchased from Mallinckrodt Chemical Company, or the equivalent quality was made by distillation or laboratory cleanup procedures. Thin layer silica gel chromatography plates were obtained from E. Merck through Brinkmann Instrument Company. Reverse-phase thin layer plates and high-

---

*Trademark, Mallinckrodt, Inc., St Louis, Missouri

performance thin layer plates were obtained from Whatman, Inc. A Kontes fiber optics densitometer, an Aminco Bowman spectro photofluorometer, a Varian 219 UV-visible spectrophotometer, and a Finnigan 4000 GC/MS with a TEK/N VENT data system were used for quantitation and confirmation. Spotting equipment from Becton Dickinson (micropets) and Hamilton syringes were used. Standard aflatoxins were obtained from Cal Biochem Company; T-2 toxin, diacetoxyscirpenol, and vomitoxin were obtained from Mycolab Inc.; and ochratoxin standard was obtained from Cal Biochem Company.

Ochratoxin

Band areas from ochratoxin suspected extracts, made according to Stahr et al. (5), were reapplied on reverse-phase layers and developed in ethanol/water/acetic acid (65:35:1) containing 0.5% NaCl. The bands were visualized with long-wavelength ultraviolet light in the dark.

Mycotoxin extracts containing trichothecene mycotoxins and zearalenone prepared according to Stahr et al. (6) were spotted on normal high-performance TLC plates and reverse-phase TLC plates. Mobile phases toluene/ethyl acetate/acetone (3:2:1) for normal phase and ethanol/water/acetic acid (65:35:1) with 0.5% NaCl were used. Trichothecene spots were visualized with anisaldehyde reagent (6). Zearalenone is visualized by short-wave UV light and anisaldehyde.

The aflatoxin extracts (6) were developed in 5, chloroform/acetone/propanol-2 (85:10:5) for normal silica and water, acetic acid 65:25:1 with 0.5% NaCl, for reverse-phase plates. Aflatoxins are visualized by long-wavelength UV light.

Citrinin extract (6) was developed in methanol/water/acetic acid (60:40:10:1) (solvents are not named) for silica gel and methanol/water/acetic acid with 0.5% NaCl for reverse-phase TLC. Citrinin is visualized by long-wavelength UV after exposure to formic acid vapor. Ochratoxin extracts were cleaned up by a silica gel column when necessary using 0.2 to 0.6 mm coarse silica, 1 to 10 g depending on the sample and dilution fractions of 15 to 150 ml of petroleum ether, 15 to 150 ml of diethyl ether, 15 to 150 ml of 3% methanol chloroform, and 15 to 150 ml of 1% acetic acid/chloroform. The latter portion contained ochratoxin. Methanol/chloroform contained zearalenone and aflatoxin; diethyl ether and chloroform/methanol fractions contained trichothecenes.

TABLE I. TLC DATA FOR MYCOTOXINS ON SILICA GEL

Mycotoxin	R_f (3/2/1)	R_f (3/2/1 + 1% Acid)	Fluorescent Color	Lower Detectability Level On Silica Gel (g hr)[a]	HPTLC[a]
Aflatoxin B_1	0.48	0.88	Blue	1-5 ng[b]	0.1 ng
Aflatoxin B_2	0.45	0.83	Blue	1-5 ng[b]	0.1 ng
Aflatoxin G_1	0.43	0.78	Green	1-5 ng[b]	0.1 ng
Aflatoxin G_2	0.36	0.73	Green	1-5 ng[b]	0.1 ng
Zearalenone	0.69	Solvent Front	Bluegreen	0.2-5 μg	0.1-0.2 μg
Sterigmatocystin	0.80	Solvent Front	Red or Yellow	10-20 ng	0.5-1 ng
Ochratoxin A	Origin	0.64	Blue	10-20 ng	0.1-0.5 ng
Ochratoxin B	Origin	0.48	Blue	10-20 ng	0.1-0.5 ng

[a] Whatman, Inc., Bridewell, New Jersey; Merck, Inc., Brinkmann Instruments, Westbury, New York.

[b] Yellow fluorescence after spraying with trifluoroacetic acid spray.

TABLE II. RP-TLC DATA FOR MYCOTOXINS ON SILICA GEL

Mycotoxin	R_f[a] (65/35/1)	Fluorescent Color	Color After Anisaldehyde	Lower Detectability Level of Silica Gel G (ng)
Ochratoxin	0.68	Blue		Same as normal phase
Aflatoxin B_1	0.47	Blue		
Aflatoxin B_2	0.50	Blue		
Aflatoxin G_1	0.55	Green		
Aflatoxin G_2	0.59	Green		
Zearalenone	0.13	Yellow-blue	Brick red	
T-2 toxin	0.21	No fluorescence	Purple	
Diacetoxy-scirpenol	0.36	No fluorescence	Purple	
Vomitoxin	0.60	No fluorescence	Yellow	

[a] 65 Methanol/35 Water/1 Acetic Acid(0.5% NaCl) developing solvent.

Citrinin samples may be cleaned up by adding 1 mg/ml decolorizing charcoal at the chloroform step in the procedure described in reference 6.

Samples for densitometry are applied in 1/4 to 3/4 in. streaks depending on the sample matrix and the appropriate detecting head is used to measure the fluorescence. Standards bracketing the samples in concentration are used to determine levels of mycotoxins.

RESULTS

Tables I and II show the relative R_f's of the mycotoxins and sensitivity of the analysis.

Figure 1 shows a typical aflatoxin and ochratoxin separated by reverse phase for densitometric analysis. Figure 2 shows the separation of aflatoxins from an International check sample.

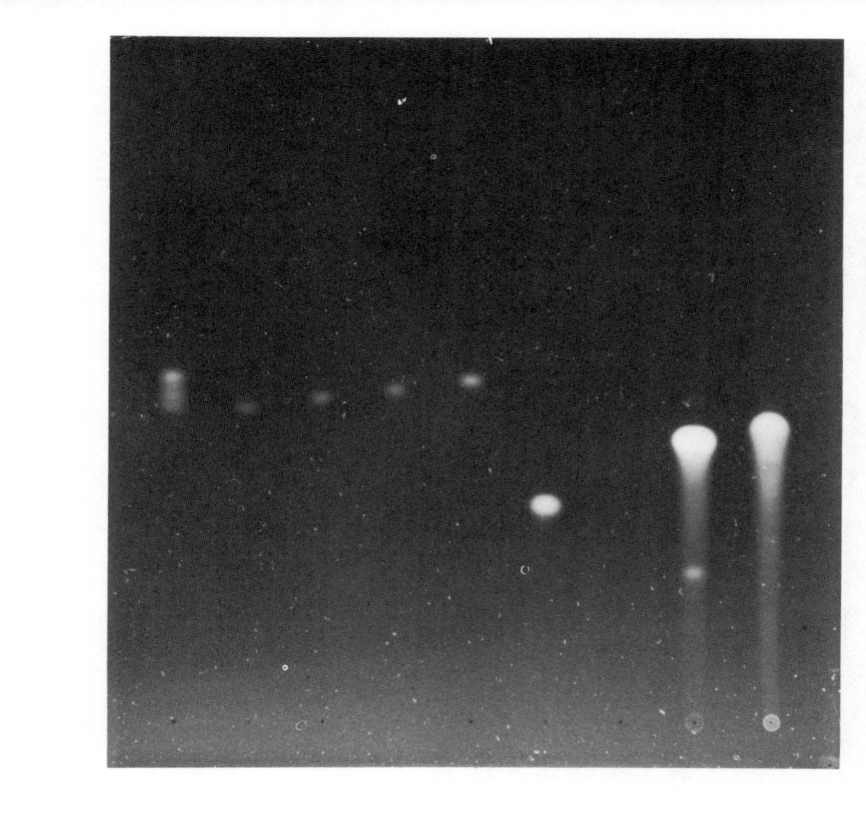

Figure 1. Mycotoxins on reverse-phase TLC. From left to right: channel 1, aflatoxin $B_1B_2G_1G_2$; channel 2, B_1 aflatoxin; channel 3, B_2 aflatoxin; channel 4, G_1 aflatoxin; channel 5, G_2 aflatoxin; channel 6, zearalenone; channel 7, ochratoxin A; channel 8, ochratoxin B.

Figure 2. Aflatoxin bands from International check sample, ready for densitometry.

Figure 3 gives a typical standard curve--that of aflatoxin B, in concentrations from 10 to 120 ng. Figure 4 shows the separation of aflatoxins and zearalenone by normal-phase TLC. Figure 5 gives the chromatogram of ochratoxin A from a field case along with the bracketing standards.

DISCUSSION AND CONCLUSION

The use of reverse- and normal-phase high-performance TLC allows more sensitive analyses to be made for mycotoxins. Reversing the order of elation allows detection of mycotoxins in difficult matrices and in some cases greater sensitivity for the analysis.

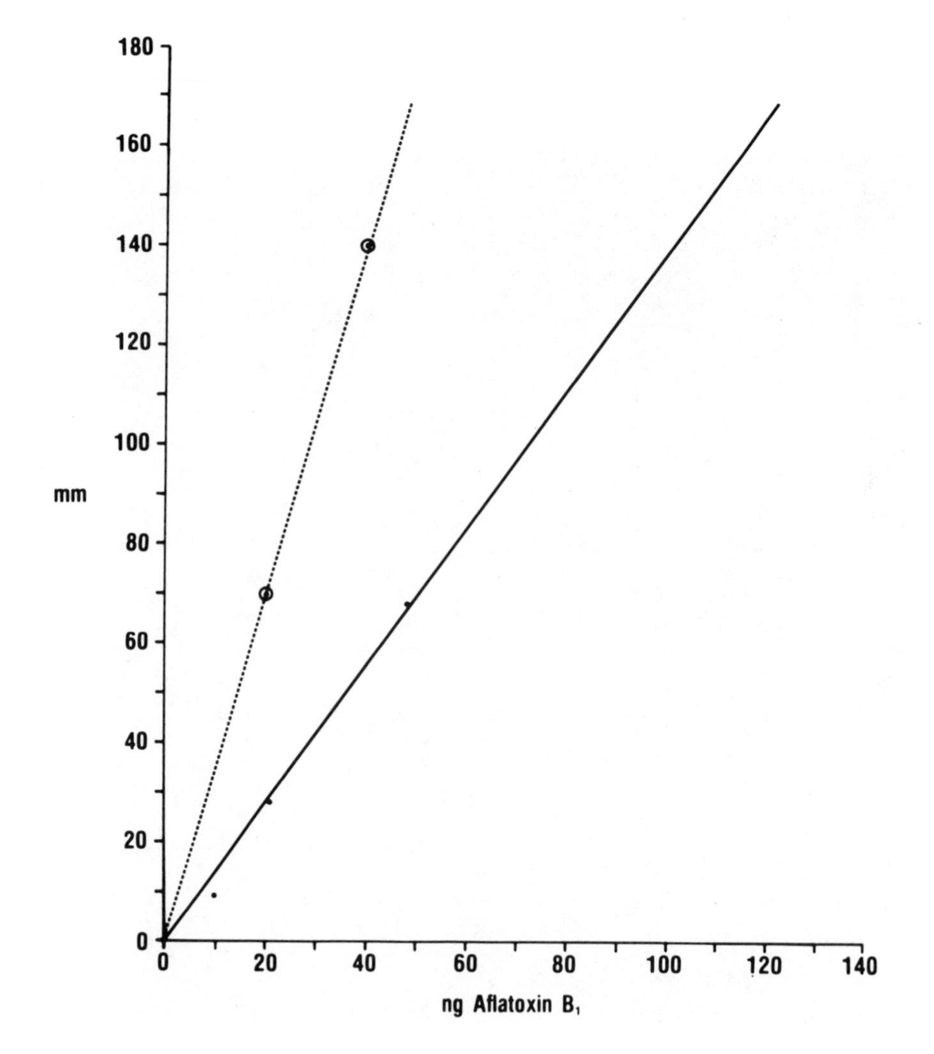

Figure 3. Aflatoxin B concentration curve from Kontes Densitometer.

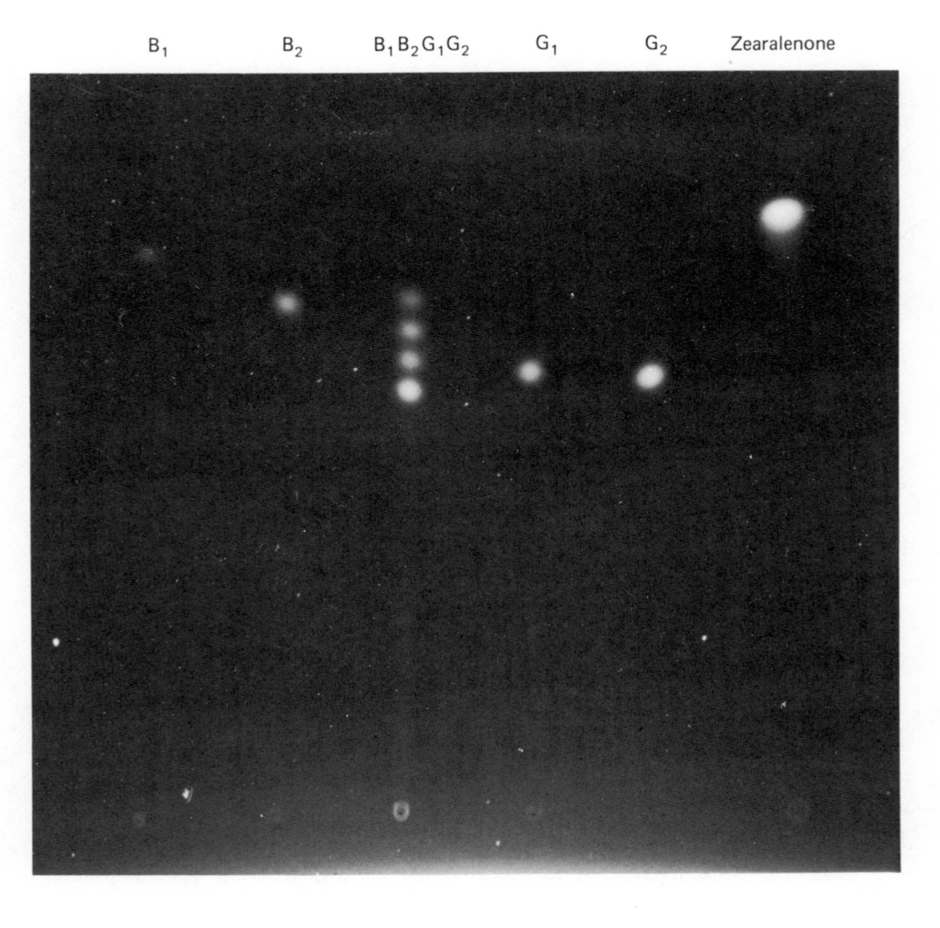

Figure 4. Normal phase thin layer separation mycotoxins.

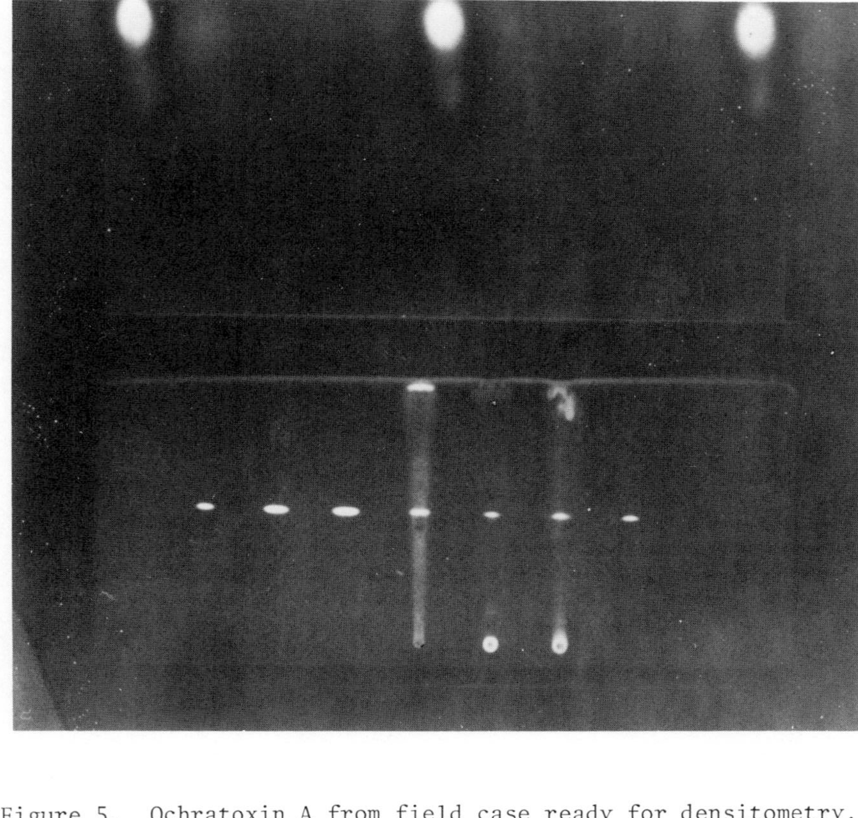

Figure 5. Ochratoxin A from field case ready for densitometry.

REFERENCES

1. P. Krogh, Ochratoxins in Mycotoxins and Human and Animal Health, M. V. Rodricks and C. W. Hesseltine (Eds.), Path. Tox, Park Forest, Ill., 1977, p. 490.
2. R. B. A. Carnaghan, Nature, 298, 308 (1965).
3. W. Shimoda, (Ed.), Proceedings Conference on Mycotoxins in Animal Feeds and Grains Related to Animal Health, Rockville, MD., 1979. Publication #PB300 300 - Bureau Vet. Med. FDA, H.W. Dept. Rockville, MD.
4. H. M. Stahr, A. A. Kraft, and M. Schuh, Applied Spectroscopy, 33, 294 (1979).
5. R. V. Chalam and H. M. Stahr, J. AOAC, 62, 570 (1979).
6. H. M. Stahr (Ed.), Analytical Toxicology Methods Manual, Iowa State Press, Ames, Ia., 500, 1977.

CHAPTER 32

The Analysis of Sulfur in Drinking Water

Ernest H. Wake and Roger A. Ward

INTRODUCTION

Hydrogen sulfide has long been known as an offensive taste and odor constituent in drinking water, occurring naturally in ground water and developing under anaerobic conditions in storage reservoirs or in water-distribution pipelines. It has been shown (1) that the threshold odor level for this substance is in the range 0.1 to 0.01 ppb.

Oxidation by aeration or chlorination or both is commonly the preferred method for sulfide removal in the water works industry. Typically, well water containing sulfide is pumped into a pond or tank where it is aerated by spraying or diffusion, chlorinated, and repumped into the distribution system.

Oxidative treatment of this type effectively reduces the odor level of sulfide-bearing water, but usually the odor is not eliminated. Normally, a sulfurous-type taint persists for a long time after treatment, even in the presence of a high chlorine residual. Often this remaining odor can be detected in a heated sample after considerable dilution with odor-free water.

Monscvitz and Ainsworth (2) proposed that this residual odor is due to traces of "hydrogen polysulfide" formed in the oxidation step. This component was not isolated, however, and these workers did not propose a method for doing so.

Others have closely examined the sulfur oxidation process (3-6). These workers have shown that under ordinary conditions only elemental sulfur and sulfate result from the aqueous oxidation of sulfide. The ultimate oxidation product is sulfate, but sulfur is produced as a durable intermediate, which can persist for days in the presence of a chlorine residual. This effect is illustrated by Figures 1 and 2 (4).

The elimination of sulfide by the action of chlorine was shown to be practically instantaneous by Choppin and Faulkenberry (3).

Polysulfide species have been shown to be temporary intermediates in the chlorine-sulfide reaction. The fact that these components cannot persist in the presence of chlorine was elucidated by the author (7). An equilibrium diagram utilizing the dissociation constants of the polysulfide acids and complex ions was prepared (Figure 3), illustrating the relationships between them, elemental sulfur, and the sulfide species.

Under neutral or acidic conditions, the polysulfides are unstable without the presence of a gross excess of sulfide. The polysulfide anions are predominant under alkaline conditions, but these too depend on the presence of sulfide for their stability.

Since the polysulfide anions have an intense (yellow) color, it is easy to demonstrate their removal upon chlorination by the decoloration, which (typically) occurs in less than 5 seconds in the presence of excess chlorine.

In the authors' experience, whenever a sulfide solution is chlorinated to a residual, a characteristic odor remains. Other oxidants have the same effect. If polysulfides are eliminated, what then could cause the odor?

The authors experimented with common sulfur compounds and discovered that other aqueous reactions that result in the formation of colloidal sulfur also cause the offending odor, which cannot be eliminated with chlorine. A good example of this is the acidification of sodium thiosulfate, which produces free sulfur by a nonoxidative route. The characteristic odor is found to be present before and after chlorination of the acidified solution.

Elemental sulfur is moderately soluble in a number of organic solvent, including the water-miscible ethyl and methyl alcohols. If a small amount of an alcoholic sulfur solution is added to water near the boiling point, the alcohol will flash into vapor, providing a colloidal suspension of pure free sulfur.

The performance of this experiment with sulfur dissolved in methanol results in a solution exhibiting the same characteristic sulfurous-type odor, which persists after sulfide chlorination. In fact, a bit of powdered sulfur added to a flask of hot water causes the odor, but it is much less intense.

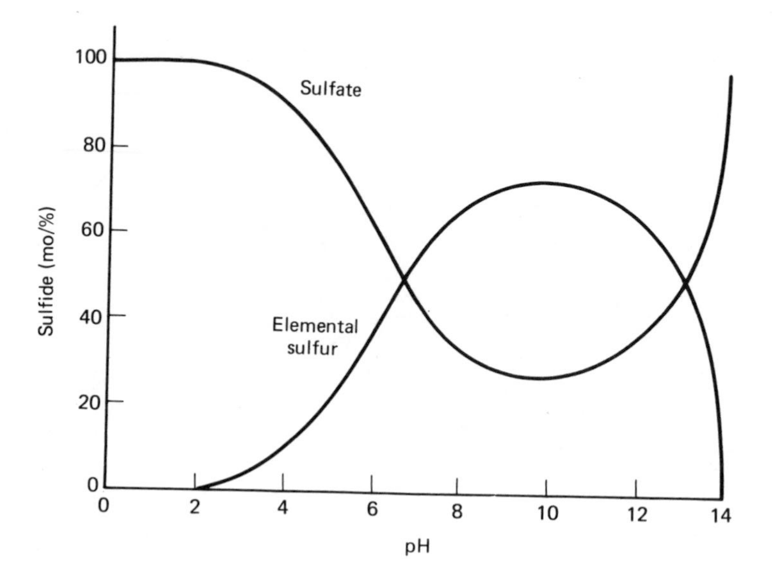

Figure 1. Reaction products resulting from chlorination of sulfide in water (10 minutes reaction time).

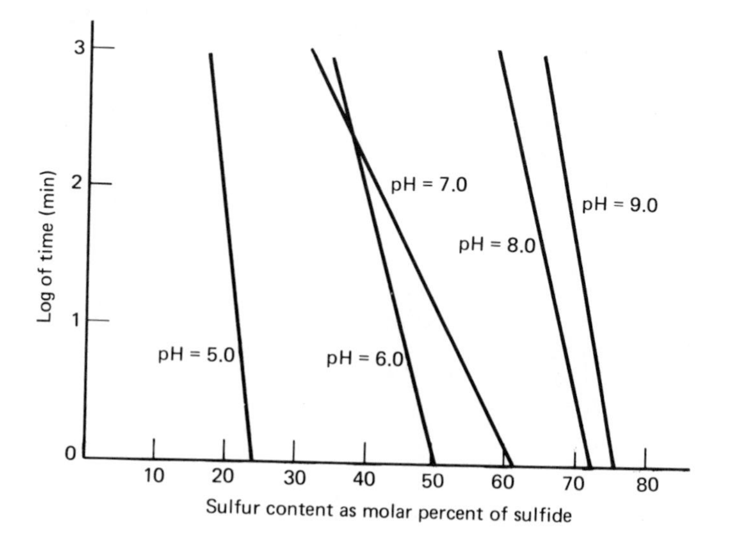

Figure 2. Elemental sulfur resulting from the chlorination of sulfide.

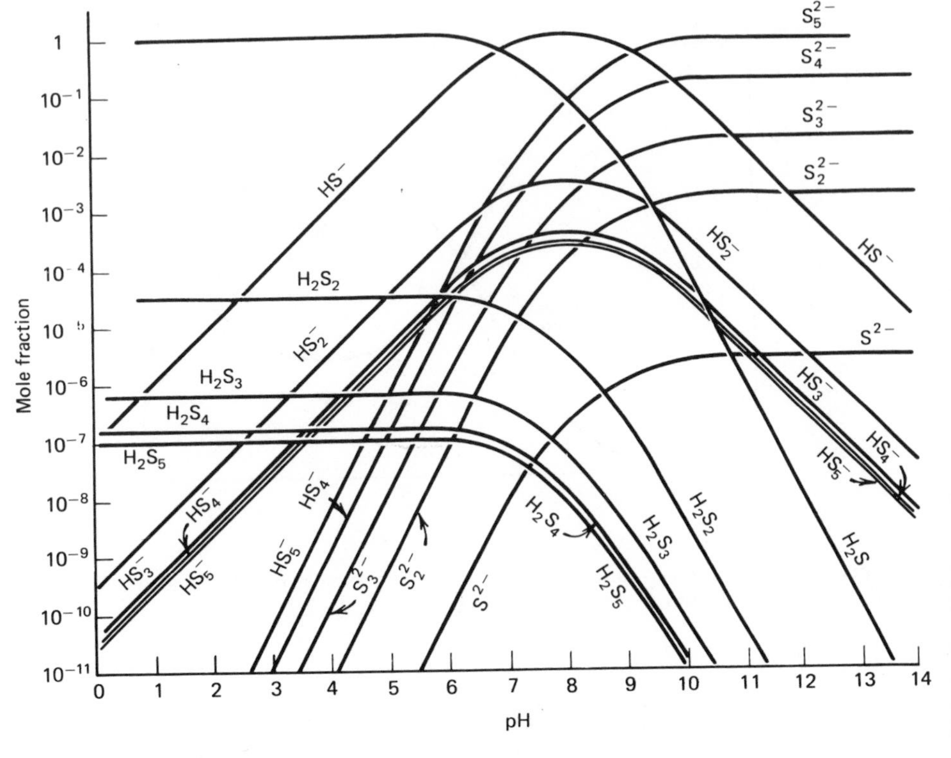

Figure 3. Equilibrium diagram for sulfide species in water, saturated with elemental sulfur. Sulfide molarity was constant through pH range.

Considerable dilution (of the sulfur colloid made from an alcohol solution) with odor-free water is necessary before a sample is obtained that is odorless (when heated to 60°C). An aliquot from a methanolic solution containing 100 ppm sulfur must be diluted more than 10,000 times before the odor is eliminated, indicating a threshold odor level of less than 10 ppb.

Could it be possible that free pure elemental sulfur as such can evoke the perception of odor, and at such a low concentration level? Most first-year chemistry textbooks say of the element "it is yellow, insoluble, and has no odor." Experienced chemists (including the authors) find the idea hard to accept. Pure flowers of sulfur in its reagent bottle has a barely detectable scent--how could a solution diluted to 10 ppb still have an odor? Other substances that can cause odors at trace levels in water are volatile (the mercaptans, for example), and a whiff of the pure material in a bottle would "knock your head off."

One might assume that the odor of pure elemental sulfur is due to a trace contaminant, or is a decomposition product, but sulfur does not decompose readily (It is one of the few elements that occur in the earth's crust in abundance in the elemental form.) Dilute laboratory solutions are stable for weeks, and sulfur resists the action of acids, oxidants, and reducing agents. The idea of a trace contaminant or decomposition product retaining its odor after a 10,000-fold dilution is even more unbelievable.

In the performance of a standard threshold odor test, dilutions of the test substance are prepared with odor-free water and heated to 60°C in covered containers. Sufficient time is allowed for vapor-liquid equilibration, and the highest dilution level exhibiting odor is observed.

It can be reasoned that the threshold odor concentration for a substance depends on its vapor pressure and solubility at 60°C, its "intrinsic" odor level, and on the amount of water solution in its heated container. It has been proposed (7) that for elemental sulfur this can be expressed mathematically by the expression:

$$C_{\ell 1} = C_{g2}\left(\frac{C^*_{\ell 2}}{C^*_{g2}} + 1\right)$$

Where $C_{\ell 1}$ is the threshold odor level for sulfur in water (in micrograms per liter), C_{g2} is the "intrinsic" odor level for sulfur vapor (in micrograms per liter), $C^*_{\ell 2}$ is the solubility of sulfur in water at 60°C (in micrograms per liter), and C^*_{g2} is the sulfur vapor concentration above pure elemental sulfur in the supercooled liquid form at 60°C. This assumes the volume of the liquid solution is equal to the volume of the air space in the container

Substituting known and estimated values into this expression gives:

$$C_{\ell 1} = \left(0.013\ \mu g/l\right)\left(\frac{487\ \mu g/l}{0.6\ \mu g/l} + 1\right) = 10.6\ \mu g/l$$

This value is in fair agreement with experience and indicates that the hypothesis that elemental sulfur as such is an odor problem in drinking water is not an unreasonable one.

In the water works industry, sulfur has generally been considered to be totally insoluble in water, and it has been assumed that it can be completely removed by filtration, as determined by effluent turbidity. A finite solubility for sulfur has been accurately determined (by others) at 178 ppb (for 25°C), however. This is far above the threshold odor concentration, which appears to be below 10 ppb. Clearly, simple filtration cannot be expected to remove sulfur odor totally.

If sulfur at the parts-per-billion level in water is an odor problem, a sensitive analytical method is of great interest. Many methods are described in the literature, but few are available with the required sensitivity.

One unique characteristic of elemental sulfur that is useful analytically is its UV absorbance spectrum. This is depicted for an 8.64-mg/l methanolic standard sulfur solution (Figure 4), done with an ordinary spectrophotometer with a UV light source. Quartz sample cells must be used in this wavelength range. In this case the light path was 1 cm in length.

Since sulfur is much more soluble in organic solvents than it is in water, it can be extracted and detected in the solvent phase simply by measuring its UV absorbance. This method is useful for laboratory solutions and is quite sensitive. Since most organics absorb UV light, however, this method can only tell you what the maximum sulfur level might be in a sample of natural water.

High-pressure liquid chromatography is presently the most sensitive method available for the determination of sulfur in water. Cassidy (11) reported a sensitivity of below 10 ng for elemental sulfur extracted in chloroform. This technique depends on the elements' UV absorbance for its detection.

Elemental sulfur is an inorganic substance that in many respects acts like an organic compound. Because of this behavior, some techniques normally only useful to organic chemists can be used for sulfur. A good example of this is thin layer chromatography (TLC).

To be a useful tool enabling the water chemist to foresee an odor problem due to sulfur, the method must be quantitative in the submicrogram range. If a TLC method employing spot area measurement is to be suitable, sensitivity must be such that a visible spot is caused by a few tens of nanograms since much more than the least visible amount must be present to provide reasonably accurate spot-area measurements.

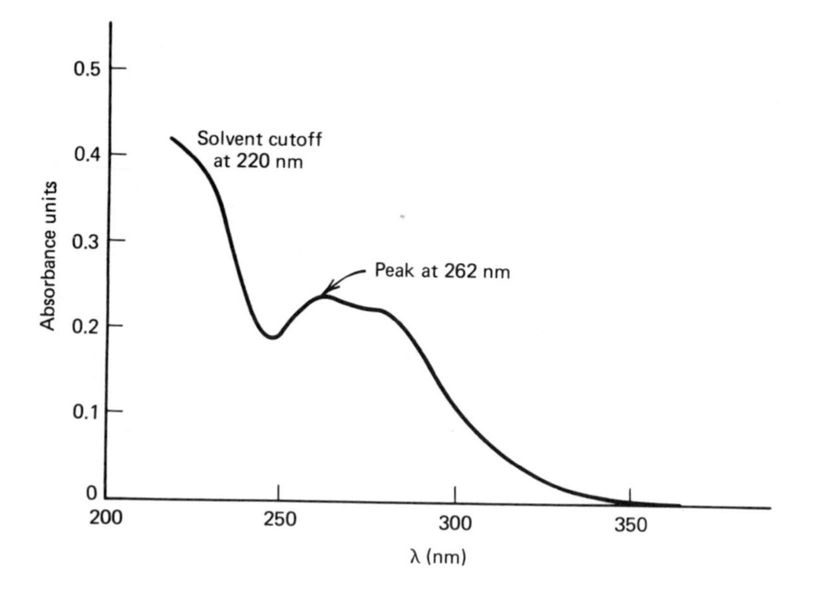

Figure 4. Elemental sulfur in methanol UV absorbance spectrum (8.64 mg/l S°).

Davies and Thuraisingham (8) developed a TLC method for elemental sulfur in rubber utilizing an iodine-azide visualization technique. This technique was adapted for the determination of free sulfur in chemical reagents by Banszkiewicz. Quantitation was by means of spot-area measurement.

Paster and Kabacoff (9) employed TLC spectrodensitometry to determine sulfur in aqueous emulsions. With this method, water is eliminated by reacting a small aqueous sample with 2,2-dimethoxypropane and acid. Water reacts with the reagent to form acetone and methanol, which are good solvents for sulfur.

The method of Davies and Thuraisingham employing iodine-azide visualization was adapted for drinking-water analysis by the authors and co-workers with limited success. 500-ml water samples containing sulfur were extracted by shaking with methylene chloride. The solvent fraction was filtered and evaporated to ∿20 μl under vacuum and spotted onto 20 × 20 cm silica gel plates that were developed to 10 cm in heptane.

The spray reagent for this test consists of 0.1N elemental iodine in a mixture of water and ethanol (3:7 v/v) with 3% sodium azide. Sulfur spots are white on a yellow-brown background, while organic material appears as brownish areas, making the technique specific for sulfur, which has an R_f value of 0.65. Unfortunately, the sulfur spots are short-lived and indistinct, so that good quantitation is difficult. Less than 1 μg of sulfur can be seen on the plate, but only for a few seconds, and respraying will not cause its reappearance.

In an attempt to attain better sensitivity and quantitation, high-performance silica gel plates containing phosphor were utilized in an adaption of the method described by Paster and Kabacoff (9). Sulfur extraction and solvent removal were done in the same manner as with the iodine-azide method, but development (in hexane) was to only 4 cm, and visualization was by UV quenching. By this technique the minimum detectable level is about 100 ng of sulfur.

Although more sensitive, this latter technique was insufficiently specific for sulfur, the only means of identification being the R_f value and the UV absorbance. Since many organic substances absorb UV light, some, which might be contained in a natural water sample, could have chromatographic properties similar to sulfur. Some UV-absorbing contaminants streak the plate and obscure the sulfur spots.

When a solution of cuprous ammonium sulfate is added to a solution of elemental sulfur in acetone or butyl carbitol, colloidal curpous sulfide is precipitated and can be measured photometrically. Mack and Hamilton (10) have determined the elemental sulfur in plant spray residue by this reaction.

The reagent for this test is an aqueous solution made from cupric sulfate, ammonium hydroxide, and hydroxylamine.

Experiments by Ward have shown this mixture to be highly effective as a spray reagent for sulfur visualization, with a high sensitivity for sulfur. Under some conditions, as little as 10 ng can be seen. The cuprous sulfide spots formed are enduring and have a distinctive "copper" (red-brown) color. The R_f value and characteristic colored-spot development provide a highly selective test for elemental sulfur, and the high sensitivity allows for sulfur quantitation by spot-area measurement in a range low enough to provide determinations at the submicrogram level.

This visualization technique is incorporated in the following method for the determination of elemental sulfur in drinking water.

METHOD AND MATERIALS

A water sample is extracted with methylene chloride, the solvent is evaporated to 20 µl under vacuum and applied to a silica gel TLC plate. The chromatogram is developed to 10 cm in heptane, and elemental sulfur is fixed and visualized with copper-hydrazine spray reagent. Quantitation is by spot size measurement.

Reagents

1. Silica gel ("Kieselgel G," for TLC plates).
2. Methylene chloride, spectrograde.
3. Elemental sulfur, recrystallized.
4. Acetone, reagent grade.
5. *n*-Heptane, spectrograde.
6. Cupric sulfate, pentahydrate, reagent grade.
7. Ammonium hydroxide, reagent (∿28% NH_4OH).
8. Hydroxylamine hydrochloride, reagent grade.
9. Standard sulfur solution. Dissolve 500 mg of crystalline sulfur in acetone and make up to 1 l. Heating the sulfur in an oven at 100°C for 3 hr before weighing will ensure that it is crystalline, free of amorphous sulfur, and completely soluble in acetone.
10. Copper-hydroxylamine reagent:
    a. Stock solution 1--dissolve 4.0 g of cupric sulfate pentahydrate in water, add 48 ml of 28% ammonium hydroxide, and add distilled water to bring the solution volume up to 100 ml.
    b. Stock solution 2--dissolve 20.0 g of hydroxylamine hydrochloride in distilled water, bring the volume up to 100 ml.
    c. Reagent solution--mix 10 ml each of solutions 1 and 2, then add water to make 100 ml and mix. Prepare this solution fresh each day, since it is unstable. Light, heat, and oxygen increase the rate of decomposition.

Standard Curve Preparation

Prepare silica gel plates, and activate them with heat in the usual manner, utilizing the kieselgel G. The plates should be 20 × 20 cm in size and the film 250 µl in thickness (or purchase ready-made plates).

Transfer aliquots of the 500 mg/1 sulfur standard solution in the range 1 to 20 μl to a TLC plate, to provide 0.5 to 10 μg of sulfur at nine levels on the plate, using suitably sized capillary pipettes (1 μl = 0.5 μg). Locate the spots 2 cm apart and 1 cm from the bottom of the plate. Develop the standard chromatogram in a conventional glass tank lined with filter paper, containing *n*-heptane. Remove the plate and air dry when the solvent front reaches exactly 10 cm.

Support the plate on its bottom edge in a fume hood, and spray liberally with the freshly prepared copper-hydroxylamine reagent. Allow 5 minutes for color development, then respray, this time directly on the visible Cu_2S spots. Allow an additional 5 minutes for the reaction to go to completion, then measure the spots with a caliper or rule calibrated in millimeters. Calculate the spot areas in square millimeters.

Prepare a calibration graph by plotting the square roots of the spot areas versus the log of the sulfur quantities applied. This plot yields a straight line in the useful range.

Sample Extraction and Chromatography

Add 50 ml of methylene chloride to 500 ml of freshly collected water sample in a 500-ml separatory funnel, and extract by shaking for 10 minutes. Remove and filter the solvent layer (bottom layer) by drawing through glass-fiber filter paper into a 125-ml round-bottom extraction flask. Immerse the flask in lukewarm water, and remove the solvent by applying vacuum by means of glass fittings (avoid rubber or neoprene). Reduce the solvent to 5 to 10 ml, and transfer it to a 15-ml conical centrifuge tube, rinsing the flask with a little methylene chloride. Continue solvent removal by applying vacuum to the centrifuge tube immersed in the warm water bath. Use a polyethylene stopper to connect the vacuum to the tube. Reduce the volume to about 20 μl, carefully avoiding evaporation to dryness.

Transfer the sample to the activated TLC plate using a 5-μl capillary pipette. Rinse with one-half drop of methylene chloride, and transfer this to the plate. Other samples, extracted in the same manner, can be spotted on the same plate. Locate the spotting points at least 1 cm above the bottom of the plate and 2 cm apart. Always include two or more levels of sulfur standard on the same plate. Develop the plate with *n*-heptane to 10 cm as with the standard plate. Visualize with the copper-hydroxylamine spray reagent, and measure the Cu_2S spots as described before.

Some loss of sulfur in the extraction and solvent-removal process is inevitable and must be taken into account if a reasonable degree of accuracy is to be achieved. For this purpose a percent sulfur recovery curve should be prepared by extracting aqueous standard sulfur solutions containing amounts of sulfur across the range of the standard curve. This is done by adding microliter amounts of the 500-mg/l sulfur standard in acetone to 500-ml volumes of distilled water, say at four levels, 0.5 μg, 2 μg, 5 μg, and 10 μg, using suitably sized capillary pipettes. These aqueous standards are extracted, concentrated, and spotted as if they were water samples, along with a distilled water blank.

Calculation of Results

The standard curve and percent recovery curve should be documented for all future sulfur determinations done by this method, but several standard sulfur levels should be included on every sulfur TLC plate to compensate for differences in development.

Utilizing the areas of the standard spots developed on the sample plate, construct a corrected curve, completed by paralleling the length of the standard curve.

Measure the areas of the sample spots, compute the square roots, and determine the sample quantities on the plate by use of the corrected standard curve. Determine the original amounts in the samples by prorating the results, using the percent recovery curve.

DISCUSSION

Some aspects of the proposed method have not been determined, and further work is needed before a more detailed, optimized technique can be recommended. The precision and accuracy have not been measured, for example, although this would be expected to be better than that reported by Davies and Thuraisingham (8) for the iodine-azide visualization technique, because of the better spot stability and higher visualization sensitivity.

The sensitivity of the method appears to be on the order of 300 ng of sulfur. A much lower level of sulfur is visible on the plate, but losses in the extraction and solvent-removal steps cause the limitation.

Interferences appear not to be a problem with this test. The copper-hydroxylamine spray reagent is somewhat selective. Inorganic sulfides, xanthates, carbon disulfide, and carbon tetrachloride react with it, but the sulfides and xanthates are either not extracted into methylene chloride or will not appear at the sulfur R_f level, while carbon disulfide and carbon tetrachloride are removed in the plate preparation (by drying). Sulfates, sulfides, thiosulfates, thiocyanates, organic sulfides, and disulfides, sulfones, mercaptans, and thiophenols do not interfere (10).

REFERENCES

1. D. Pomeroy and H. Cruse, J. Am. Water Works Assoc., 61, 677 (1969).
2. J. Monscvitz and L. Ainsworth, J. Am. Water Works Assoc., 66, 537 (1974).
3. A. Choppin and L. Faulkenberry, J. Am. Chem. Soc., 59, 2203 (1937).
4. A. P. Black and J. B. Goodson, J. Am. Water Works Assoc., 47, 309 (1952).
5. K. Y. Chen and J. C. Morris, J. Env. Sci. Tech., 6, 529 (1972)
6. M. R. Hoffman, J. Env. Sci. Tech., 11, 61 (1977).
7. E. H. Wake and R. A. Ward, Elemental Sulfur in Drinking Water, (an unpublished paper presented at the 1980 Fall conference of the California-Nevada Section of the American Water Works Assoc., October 21, 1980, available from the authors).
8. J. R. Davies and S. T. Thuraisingham, J. Chromatogr., 35, 513 (1968).
9. A. Paster and B. Kabacoff, J. Chromatogr. Sci., 14, 572 (1976).
10. G. L. Mack and J. M. Hamilton, Ind. Eng. Chem., Anal. Ed., 14, 604 (1942).
11. R. M. Cassidy, J. Chromatogr., 117, 71 (1976).

CHAPTER 33

Determination of Sulfonamides in Animal Tissues and Feeds

Michael H. Thomas and Karen E. Soroka

INTRODUCTION

Sulfonamide drugs have proved to be effective antibacterial agents in many food-producing animals. The U.S. Food and Drug Administration has set the tolerance level for residues in uncooked edible tissues at 0.1 ppm (1). Unfortunately, use of these compounds has resulted in a continuing tissue-residue problem, largely through inadvertant contamination of unmedicated finishing feeds (2). In response to this problem, several researchers in this laboratory (3) have developed a GC/MS procedure for regulatory use to replace the nonspecific colorimetric procedure currently used by USDA monitoring laboratories (4). In the past a qualitative TLC/GLC procedure was employed (5) to screen samples prior to the lengthy determination as well as to provide a degree of specificity. Although GC/MS provides needed structural (confirmatory) data, sample throughput is low to moderate. For this reason an accurate screening procedure is necessary to prevent further analysis of excessive false positive and false negative samples.

For rapid screening, thin layer chromatography (TLC) is extremely attractive as it offers the ability to simultaneously analyze multiple samples. Furthermore, unlike GLC or HPLC the method of detection can be chosen without consideration for carrier gas or mobile-phase incompatibility, thus optimizing both resolution and sensitivity or selectivity. Because detection is off-line, the number of instruments a laboratory needs is reduced.

Sulfonamides of common veterinary and residue interest (Figure 1) possess a primary aromatic amino moiety that is readily derivatized with fluorescamine. Although these derivatives can be formed prior to chromatography (6), postchromatographic derivatization (7) was employed so as to utilize an existing

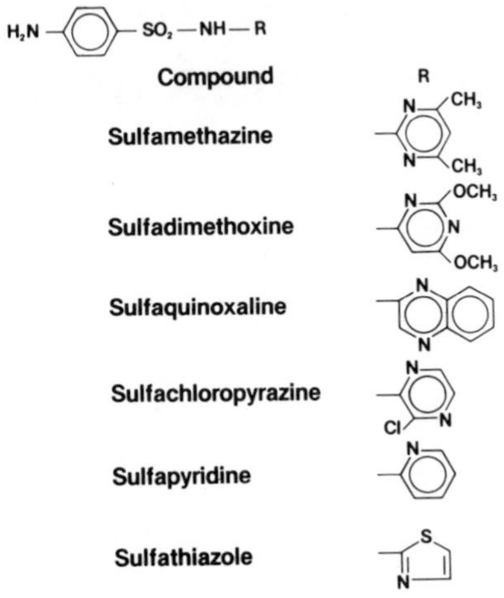

Figure 1. Structures of some sulfonamides of veterinary interest.

separation of the underivatized drugs. The technique has been applied to sulfadiazine residues in tissues (8) and for sulfamethazine in an antemortem blood screening procedure in swine (9). A method was previously reported for sulfamethazine in swine tissues suitable for regulatory use (10). In this paper an adaptation of the method to simultaneous measurement of sulfamethazine, sulfadimethoxine, and sulfaquinoxaline residues is reported.

Improvements in accuracy, reproducibility, and speed over earlier TLC methods have been made through the use of an internal standard and preadsorbent spotting zones.

EXPERIMENTAL

Apparatus

A Camag (Applied Analytical Industries, Wilmington, North Carolina) TLC/HPTLC scanner was used for all quantitation. A medium-pressure mercury lamp was used to excite fluorescence at 400 nm, and total fluorescence was measured by reflectance. Slit dimensions were 5.0 × 0.3 mm and scan speeds were 1 to 2 mm/sec.

Materials and Reagents

Precoated 20 × 20 cm TLC plates containing a preadsorbent spotting area (LK6D, Whatman, Inc., Clifton, New Jersey) were used after predevelopment in methanol.

All solvents except tert-butanol were distilled in glass (Burdick & Jackson Laboratories, Inc., Muskegon, Michigan). The tert-butanol was reagent grade (Fisher Scientific Company, Springfield, New Jersey).

Fluram* (fluorescamine) was obtained from Pierce Chemical Company (Rockford, Illinois). The derivatizing solution was prepared by dissolving approximately 30 mg of fluorescamine in 250 ml of acetone. This was replaced after treating eight to nine plates.

Sulfamethazine (SMZ) and sulfaquinoxaline (SQX) were obtained from Pfaltz and Bauer (Stamford, Connecticut). Sulfapyridine (SPY) was obtained from Sigma Chemical Company (St. Louis, Missouri), and sulfadimethoxine (SDM) was obtained from Hoffmann-LaRoche, Inc. (Nutley, New Jersey). Fortification standards containing SMZ, SQX, and SDM, at 1.25, 2.50, and 5.00 μg/ml (equivalent to tissue concentrations of 0.05, 0.10, and 0.20 ppm) were prepared in 0.05*M* phosphate buffer pH 6.5. All solutions contained a constant concentration of SPY, the internal standard, of 2.50 μg/ml (0.10 ppm).

Extraction Procedure

Tissues

A 2.5-g sample of homogenized liver or muscle was weighed into a 50-ml polypropylene centrifuge tube and fortified with 100 μl of the internal standard solution. After allowing

---

*Trademark, Roche Laboratories, Nutley, New Jersey.

15 minutes for the adsorption of the SPY, 25 ml of ethyl acetate was added, and the tubes were tightly capped. Muscle samples were blended with a Tekmar SDT tissumizer (Cinncinati, Ohio), and liver samples were shaken on a horizontal shaker for 20 minutes. After centrifugation (5 minutes at 2500 rpm), the ethyl acetate was transferred to a clean 50-ml polypropylene centrifuge tube, the tissue was discarded, and 10 ml of 0.2M glycine buffer, pH 12.25, was added to the tube containing the ethyl acetate extract. The tube was again mechanically shaken for 5 minutes and the layers were completely separated by centrifugation (5 minutes at 2500 rpm) The organic phase was aspirated and discarded, and the pH of the aqueous phase was adjusted to 5.2 to 5.3 by addition of 2 ml of a 1:1 mixture of 2M phosphate buffer, pH 5.25, and 1.7M HCl. The pH was checked and final adjustment, if necessary, was made with 0.1N HCl or NaOH. The aqueous phase was then washed with 10 ml of n-hexane by mechanical shaking for 5 minutes. The layers were separated by centrifugation, and the hexane was aspirated and discarded. The aqueous phase was then extracted with 10 ml of methylene chloride, shaken for 5 minutes and the layers were completely separated by centrifugation. Occasional emulsions were broken by increasing the centrifuge time and speed to 10 minutes and 3500 rpm, respectively. The aqueous phase was aspirated and discarded. Ten microliters of diethylamine was added to the methylene chloride to minimize adsorption-caused losses on concentration. The methylene chloride was evaporated just to dryness under a stream of nitrogen using an N-Evap (Organomation, Northborough, Massachusetts) with the water-bath temperature at 40°C. During the evaporation, the walls of the tube were periodically rinsed with additional methylene chloride. The residue was reconstituted in 100 µl of methanol The procedure is summarized in Figure 2.

Feeds

The above procedure was used for feeds with the exception that the sample size was increased to 10 g. For feeds containing 1 to 10 ppm of drug, no concentration step is required. For samples in this therapeutic range, the internal standard was added at 5 ppm.

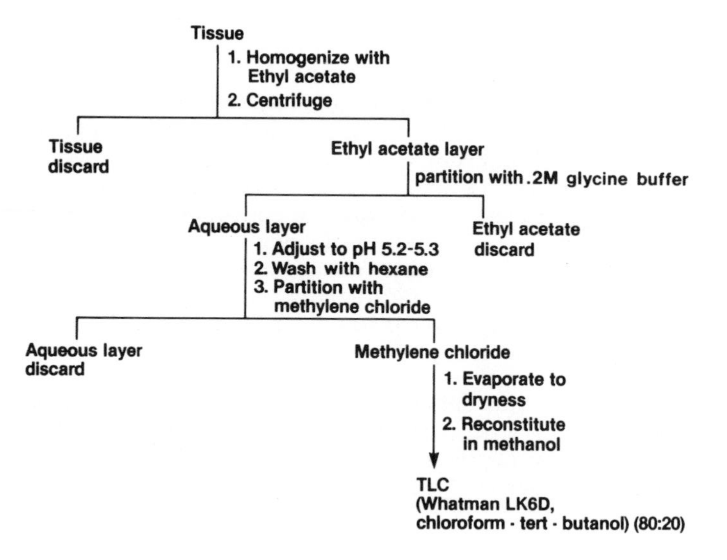

Figure 2. Schematic of sample cleanup procedure.

Thin Layer Chromatography

A 20-μl portion of sample was applied to the preadsorbent layer of the plate using a Corning microcapillary pipette. Use of these pipettes rather than a syringe facilitated a reproducible sample application rate. This was done to maximize uniformity of concentration across the band (11). During spotting, the preadsorbent zone was heated with an AIS spotting device (temperature, 85°C). The plate was first developed 1 cm in methanol followed by two developments in chloroform-_t_-butanol (80:20) for 6-cm and 12-cm distances from the sorbent interface. The plate was dried in an oven at 110°C for 1 minute between each development. Development was done in a twin-trough fully saturated tank. The temperature was maintained at 30 to 33°C by placing the tank in a drying oven with the heating element turned off but the fan on. The derivatization was accomplished by briefly dipping the dried plate in the fluorescamine solution. Derivatization was complete after allowing the plate to air dry for 15 to 30 minutes. No attempts were made at stabilizing the derivatives as reported elsewhere (12, 13), as they were stable for several hours.

Quantitation

With each set of samples, three control samples were fortified with each sulfonamide at 0.05, 0.10, and 0.20 ppm and analyzed concurrently with incurred samples. In this way variations in the extraction procedure and absolute recovery were corrected. Additionally, any effect that matrix components exerted on the chromatography would be present in both samples and standards.

A standard curve was constructed by plotting peak height ratio of each sulfonamide of interest to that of the internal standard versus the concentration of the sulfonamide of interest. The 95% confidence interval about the linearly regressed line was then computed (Figure 3). For screening samples prior to GC/MS analysis, the point on the calibration line corresponding to an upper confidence limit of 0.11 ppm is set as the threshold for confirmation by GC/MS. In this fashion, the threshold varies directly with the quality of the standard curve

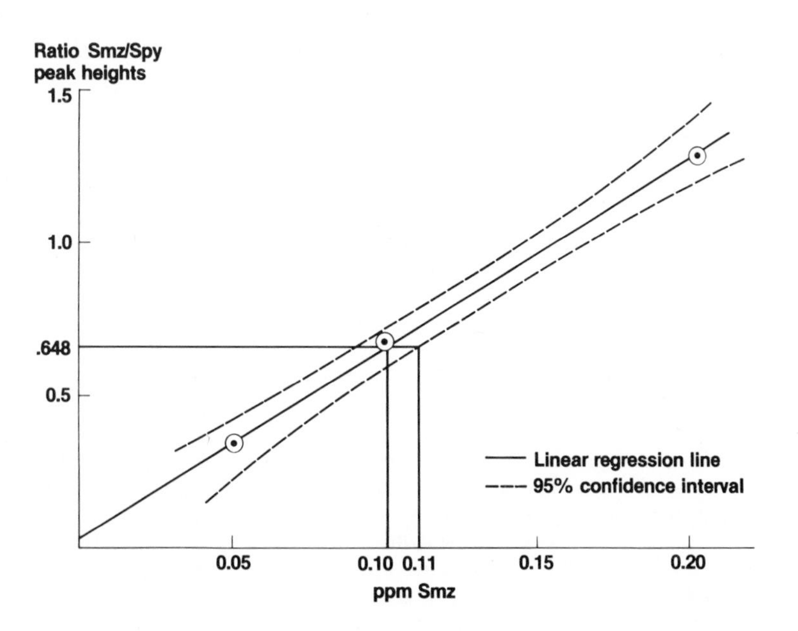

Figure 3. Calibration curve with 95% confidence interval. For this sample the threshold for confirmation is 0.10 ppm.

RESULTS AND DISCUSSION

The method provides the ability to monitor for multiple sulfonamides simultaneously at the tolerance level with a very high degree of accuracy and precision (Table I). The two most significant factors responsible for these characteristics are the internal standard and the preadsorbent spotting zone. In our hands, without the internal standard the CV deteriorates to 10 to 20%, in large part owing to losses on evaporation to dryness. It should be noted that while it is well recognized that an internal standard corrects for volume changes or losses, it is less appreciated that an internal standard will not automatically correct for variations in experimental technique. Therefore, the choice of the internal standard must be made with care such that if a critical step is varied, for example, a pH adjustment, both the internal standard and the analytes behave similarly. Figure 4 indicates that SMZ has a very different solubility from SDM and SQX, the former being amphoteric, while the latter two have very low acid solubility. Sulfachloropyrazine (SCL) would be a very good internal standard for SDM and SQX, whereas SPY has solubility and pKa close to SMZ. Because SMZ is the most common residue and SPY can be resolved from the other sulfonamides of interest chromatographically, SPY was chosen as the internal standard. Nevertheless, the pH profile indicates that if the pH is allowed to exceed 5.5, the rate of change of solubility for SDM and SQX is significantly higher than for SPY. Obviously then, error will be introduced if the method is varied beyond certain well-defined limits.

When dealing with relatively crude biological extracts containing high concentrations of salts or ionic species, preadsorbent spotting zones provide an opportunity to minimize matrix effects in the chromatography. Rather than applying the sample in a concentrated spot, it is dispersed over the width of a 10-mm lane, diluting the nonmigrating components. In our experience a short (0.5 to 1.5 cm) development with a strong solvent such as methanol then "smears" out these strongly retained components, further diluting them with respect to sorbent and lessening their influence on the chromatography. The resulting chromatograms of tissue samples show no evidence of tailing or band distortion that is often characteristic of biological extracts (Figure 5). The only effect that is noticeable is slight edge distortion, a result of the large sample volume (20 μl) applied to the scored lanes. This effect is eliminated by adjusting the densitometer slit width to half-lane width,

TABLE I. ACCURACY AND PRECISION FOR SIMULTANEOUS MULTIPLE SULFONAMIDE DETERMINATION

Tissue[a]	Compound	Mean ppm Found (n = 12)	Mean Within-Day CV (n = 4)	Day-to-Day CV
Liver	SMZ	0.101	4.70	3.28
	SDM	0.101	5.78	3.74
	SQX	0.101	5.60	7.33
Muscle	SMZ	0.103	3.89	1.36
	SDM	0.103	6.27	2.79
	SQX	0.104	4.13	1.72

[a]Samples fortified with 0.10 ppm of each drug.

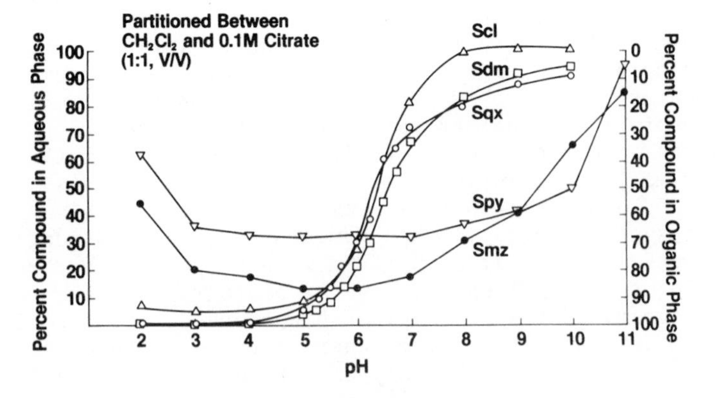

Figure 4. Solubility of certain sulfonamides as a function of pH.

5 mm. Because of the sensitivity inherent in fluorescence, this can be done without compromising the method sensitivity. The analog scanning curve for a 0.1-ppm sample from the plate shown in Figure 5 indicates the signal-to-noise level associated with measurements at the tolerance level (Figure 6). Baseline drift on these plates, while noticeable, is minimal and generally constant from plate to plate.

To achieve scanner-resolved separation of the six sulfonamides shown in Figure 7, multiple development was required. Even more important, and less frequently discussed in the literature, is the need for careful temperature control of the chromatography. For example, resolution of sulfathiazole and sulfapyridine is lost when the separation takes place below 25°C. Between 30 and 35°C baseline separation is achieved. Similarly, selectivity changes for SMZ, SQX, and SDM across this temperature range. Constant temperature throughout the developing tank is also required to obtain reproducible and constant R_f values across the plate. Drafts in the laboratory can cause R_f shifts across the plate that will necessarily affect quantitation. For this reason the tank is placed in a forced-air oven (heating elements off) to maintain a homogeneous environment.

As described, several factors were manipulated to obtain chromatography that approached the near-ideal case for chromatography of dilute standards of analyte in solvent only. Nevertheless, subtle differences in band shape or density may arise between analyte in solvent and analyte in matrix, and error may be introduced from quantitation based on such a comparison. In many procedures comparisons are made between near-perfect standard spots or bands and grossly distorted sample spots. Here the error will be significant. Furthermore, differences in absolute recovery between various analytes and the internal standard can only be corrected by carrying the standards through the procedure. It is thus imperative in trace analysis to use standard curves prepared from fortified control samples as a basis for quantitation.

Quantitative TLC is a sensitive and practical approach to regulatory determinations of sulfonamide tissue residues prior to GC/MS confirmation. For laboratories not having access to a GC/MS instrument or a need for confirmatory information (such as those involved in quality control), the technique is sufficiently accurate and sensitive to function as the determinative procedure. As 18 to 24 samples can be analyzed completely per day, it offers the added benefit of high sample throughput.

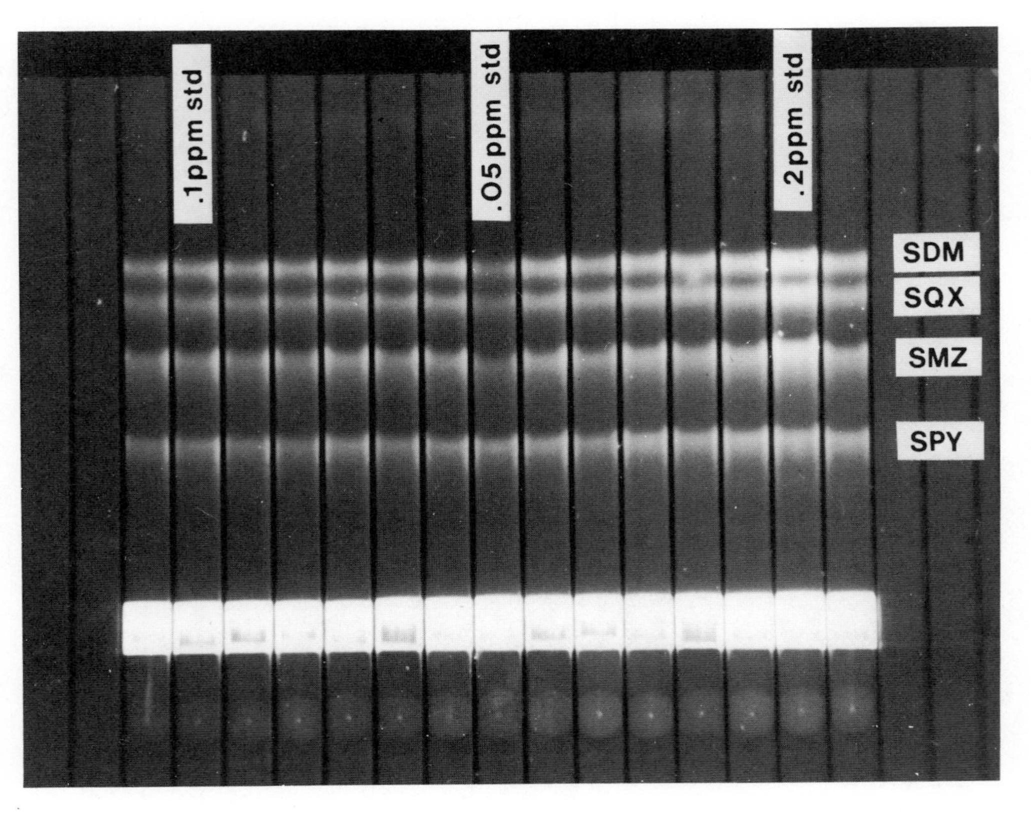

Figure 5. Chromatogram of a set of liver samples. Three fortified control samples (standards) are positioned in lanes indicated. Bright band at sorbent-preadsorbent interface consists of matrix components from the short methanol development. Chromatographic conditions are described in the text.

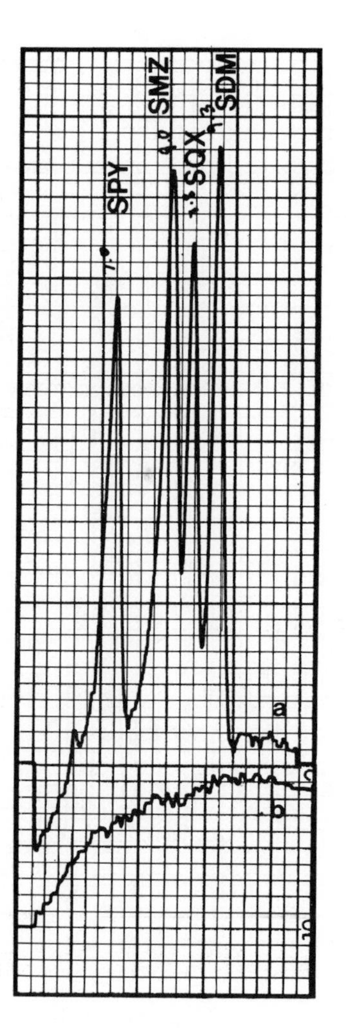

Figure 6. Analog scanning curve for one sample lane in Figure 5. Upper curve (a) is the response from the sample containing 0.10 ppm of SPY, SMZ, SQX, and SDM. Lower curve (b) is a scan of the baseline of an unused lane.

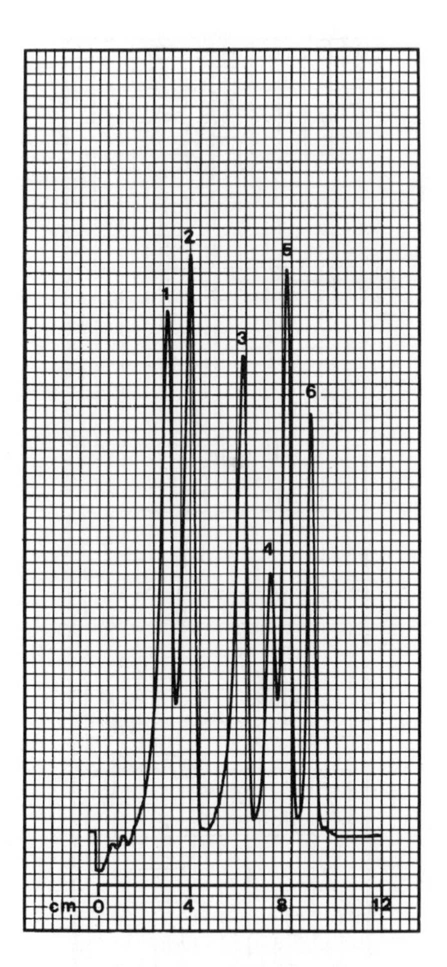

Figure 7. Separation of six sulfonamides, including the internal standard: 1, sulfathiazole; 2, sulfapyridine, internal standard; 3, sulfamethazine; 4, sulfaquinoxaline; 5, sulfadimethoxine; 6, sulfabromomethazine. Approximately 100 ng of each component, same conditions as Figures 5 and 6.

REFERENCES

1. Code of Federal Regulations, 21, 1977. Food Additive Regulation 558.128; New Animal Drug Regulation 546.110 (e), 546.113 (b), and 520.2260, 566.670.
2. G. L. Cromwell and T. S. Stahly, unpublished (paper presented at USDA-FDA-Swine Industry Task Force Report on Sulfonamide Residue "Self-Help" Program. Washington, D.C., January 9-10, 1980).
3. F. B. Suhre, R. M. Simpson, and J. W. Shafer, J. Agric. Food Chem., 29, 727 (1981).
4. F. Tishler, J. L. Sutter, J. N. Bathish, and H. E. Hagman, J. Agric. Food Chem., 16, 50 (1968).
5. D. P. Goodspeed, R. M Simpson, R. B. Ashworth, J. W. Shafer, and H. P. Cook, J. Assoc. Offic. Anal. Chem., 61, 1050 (1978).
6. H. Nakamura and J. J. Pisano, J. Chromatogr., 121, 33 (1976).
7. H. Nakamura and J. J. Pisano, J. Chromatogr., 121, 79 (1976).
8. C. W. Sigel, J. L. Woolley, and C. A. Nichol, J. Pharm. Sci., 64, 973 (1975).
9. R. F. Bevill, K. M. Schemske, H. G. Luther, E. A. Dzierzak, M. Limpuka, and D. R. Felt, J. Agric. Food Chem., 26, 1201 (1976).
10. M. H. Thomas, K. E. Soroka, R. M. Simpson, and R. L. Epstein, J. Agric. Food Chem., 29, 621 (1981).
11. H. Halpaap and K. F. Krebs, J. Chromatogr., 142, 823 (1977).
12. J. C. Touchstone, J. Sherma, M. F. Dobbins, and G. R. Hansen, J. Chromatogr., 124, 111 (1976).
13. K. Imai, P. Bohlen. S. Stern, and S. Undefriend, Arch. Biochem. Biophys., 161, 161 (1974).

CHAPTER 34

Sample Cleanup of Carbamate Pesticides for TLC

Walter Hyde, H. Michael Stahr, Rhonda Moore, Marlaine Domoto, and Richard Pfeiffer

INTRODUCTION

In recent years, carbamate pesticide use has increased, replacing many of the organo-chlorine and organo-phosphate pesticides.

Carbamate pesticides are substituted esters of carbamic acids with substituents on the heteroatoms. Carbamates inhibit cholineserase and are used as insecticides and fungicides. Carbamates are difficult to analyze because of their thermal instability. Breakdown occurs on GLC columns even when heavily silanized or similarly treated. This, coupled with the rapid breakdown of applied carbamates in the field, makes analysis difficult. High-performance liquid chromatography has been used but is expensive and more of an art than is TLC. It also requires derivatization and fluorescence readouts for sensitivity and selectivity.

This paper discusses the need for additional sample cleanup prior to TLC in the analysis of some highly pigmented and colored samples: feed, rumen contents, silages, and haylages prior to TLC analysis for carbamate pesticides.

Figure 1 illustrates the eight carbamates used as representatives for comparison of the cleanup methods and the evaluation of the TLC plates and sprays.

We used bovine rumen contents as a typical example of a highly pigmented sample commonly submitted for carbamate analysis. The extraction procedure (Figure 2) requires blending 20 g of the

A **Carbofuran** Furadan* — FMC

3,3-dihydro-2,2-dimethyl-7-benzofuranyl methylcarbamate

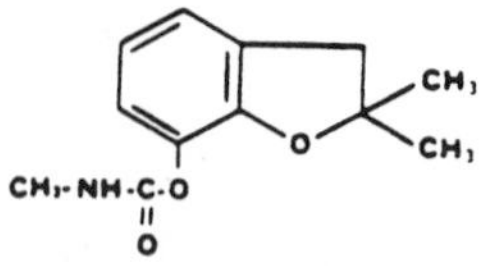

CH_3-NH-C(=O)-O — CH_3, CH_3

B **Carbaryl** Sevin* — Carbide

1-Naphthylmethylcarbamate

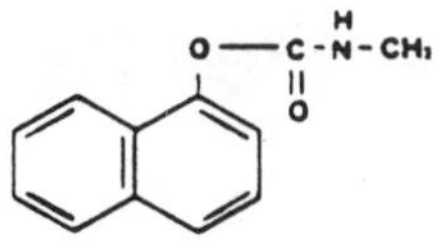

O—C(=O)-N(H)-CH_3

C **Methomyl** Lannate* Nudrin* — Du Pont, Shell

S-Methyl-N-[(methylcarbamoyl)-oxy] thioacetimidate

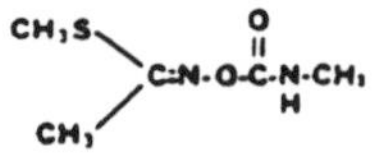

CH_3S, CH_3 C=N-O-C(=O)-N(H)-CH_3

D **Baygon*** Propoxur — Chemagro

2-Isopropoxyphenyl N-methylcarbamate

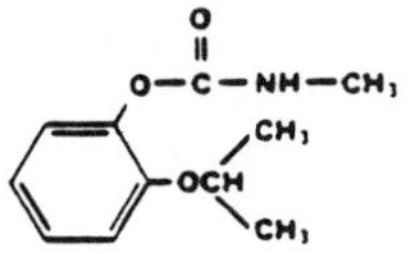

O—C(=O)—NH—CH_3; OCH(CH_3)$_2$

E F **Temik*** Aldicarb — Carbide

2-Methyl-2-(methylthio)propionaldehyde O-(methylcarbamoyl)oxime

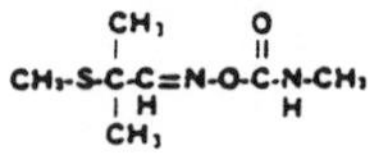

CH_3-S-C(CH_3)$_2$-C(H)=N-O-C(=O)-N(H)-CH_3

G H **Landrin*** — Shell

Mixture of about 20% 2,3,5-Trimethylphenyl methylcarbamate; and 80% 3,4,5-trimethylphenyl methylcarbamate

$(CH_3)_3$—C$_6$H$_2$—O-C(=O)-NH-CH_3

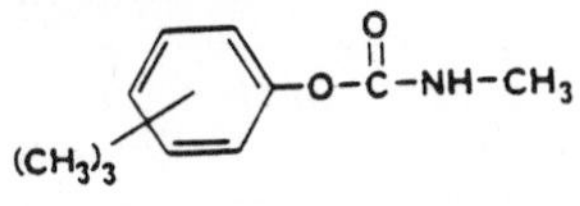

Figure 1. Carbamate standards. All standards were prepared in nanograde methanol at ∿1000 ppm concentration.

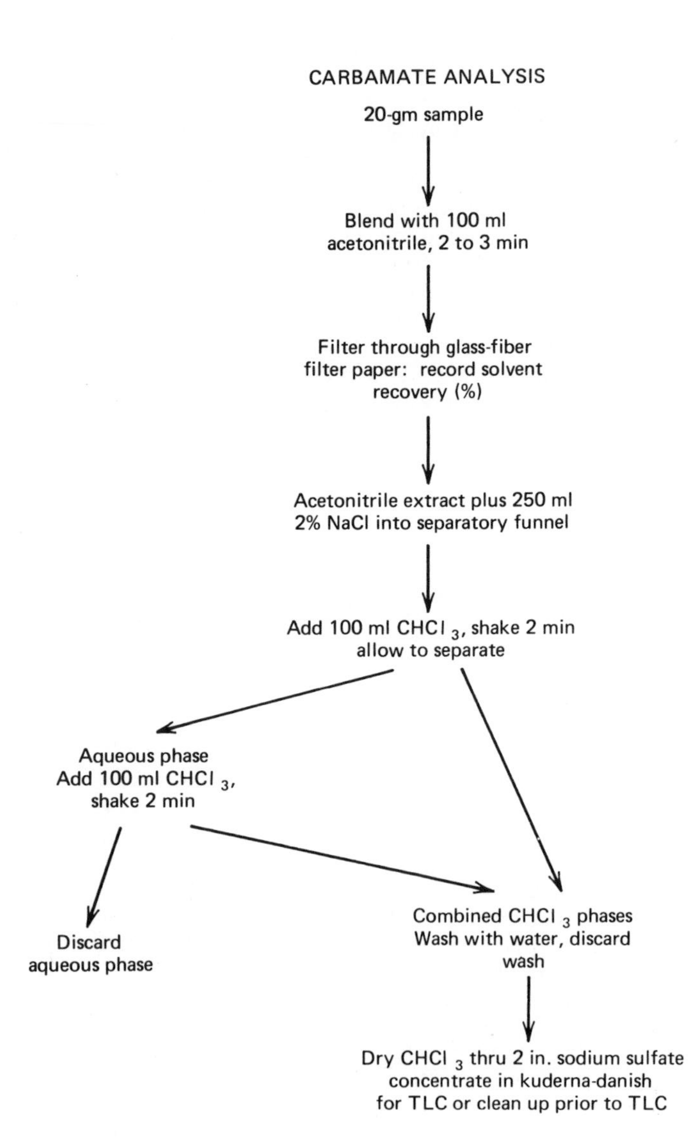

Figure 2. Carbamate analysis.

sample with 100 ml acetonitrile in a waring blender for 2 to 3 minutes. The acetonitrile phase is filtered through glass-fiber filter paper into a graduated cylinder, noting the percent solvent recovery. The acetonitrile phase is added to 250 ml sodium chloride solution (2% in water) and 100 ml chloroform in a 500-ml separatory funnel. The phases are shaken for 2 minutes, allowed to separate, and the $CHCl_3$ layer is drawn off and saved. Another 100 ml $CHCl_3$ is added to the acetonitrile-aqueous phase, shaken, and separated. The two $CHCl_3$ layers are combined, and both are washed with water, discarding the wash. The $CHCl_3$ extract is dried through a 2-inch column of granular, anhydrous sodium sulfate. Finally, the extract is concentrated to 0.5 ml in a kuderna-danish apparatus for further cleanup or direct TLC analysis.

Several commercially available TLC plates were used in this work.

Normal Phase

1. Whatman HP-K HPTLC.
2. Merck Silica Gel 60 Type G.
3. Whatman LK6DF "carbamate" plates.
4. EM Reagents silica gel 60 aluminum-backed plate.

Reverse Phase

1. Whatman KC-18.

All normal-phase developments were accomplished in 80:20 toluene/ethyl acetate. The reverse-phase plates were developed in 65:35 ethanol/water with 1% acetic acid and 0.5 g sodium chloride added.

Thin layer plate visualization was accomplished using either anisaldehyde or trichloro-*p*-benzoquinoneimine (TCBI). Anisaldehyde has been used as a spray to visualize carbamates and results in colors for the standard ranging from pink to brown. Trichloro-*p*-benzoquinoneimine (TCBI available from Aldrich) has been recently suggested for carbamates (1) and results in colors ranging from tan to brown. TCBI seems to be more selective but less sensitive for low μg amounts of the carbamate standards.

Figure 3 shows the attempted TLC of the extraction residue from bovine rumen contents to which carbamate standards had been added. None of the TLC plates were able to sufficiently separate interfering compounds from the added carbamate to standards.

The TLC of rumen contents plus added standard resulted in almost completely unreadable chromatograms, regardless of the

Figure 3. Thin layer chromatogrpahy of rumen content extract with standards added. Note that none of the standards can be differentiated from the colored. Interfering materials that were coextracted with the carbamates. EM reagents silica gel 60 plates were used, developed in 80:20 toluene/ethyl acetate, and were visualized with anis-aldehyde.

plate or visualization spray used. The concentrate direct from extraction had too many interfering compounds that not only masked the carbamates, but also interfered with proper TLC separations. Cleanup was necessary with these highly pigmented samples.

The Kontes sweep codistillation apparatus used silated sand and glass beads to provide a crude chromatographic separation of many pigment- and fat-type interferences from other easily eluted compounds, such as many of the pesticides. Recoveries of carbofuran standard varied with the temperature of the column oven. Table I shows the relationship between temperature of the column oven and the percent recovery of parent carbofuran.

The chromatography occurred in a column packed with a glass-wool plug, sand, glass beads, and finally another glass-wool plug (Figure 4).

The column was placed in an elevated temperature oven (140 to 250°C, depending on the application) and nitrogen was passed through it. The column was extensively deactivated by injections of silyl-8 or other silanizing reagents. The sample extract was then injected and allowed to vaporize through the column followed by three injections of $CHCl_3$ and collection. Finally, the teflon tubes that connect the end of the column to the collecting tubes were flushed $CHCl_3$. The total treated extract was reconcentrated and analyzed by TLC.

After Sweepco treatment, TLC visualization with anisaldehyde resulted in sensitive spike recoveries in the midst of a highly colored background.

TABLE I. CARBOFURAN RECOVERY FROM SWEEP CODISTILLATION

Oven Temperature (°C)	Column Condition	Recovery (%)
160	Unsilated	40- 80
160	Silated	95-100
140	Unsilated	80- 85
140	Silated	95-100
120	Silated	85- 95

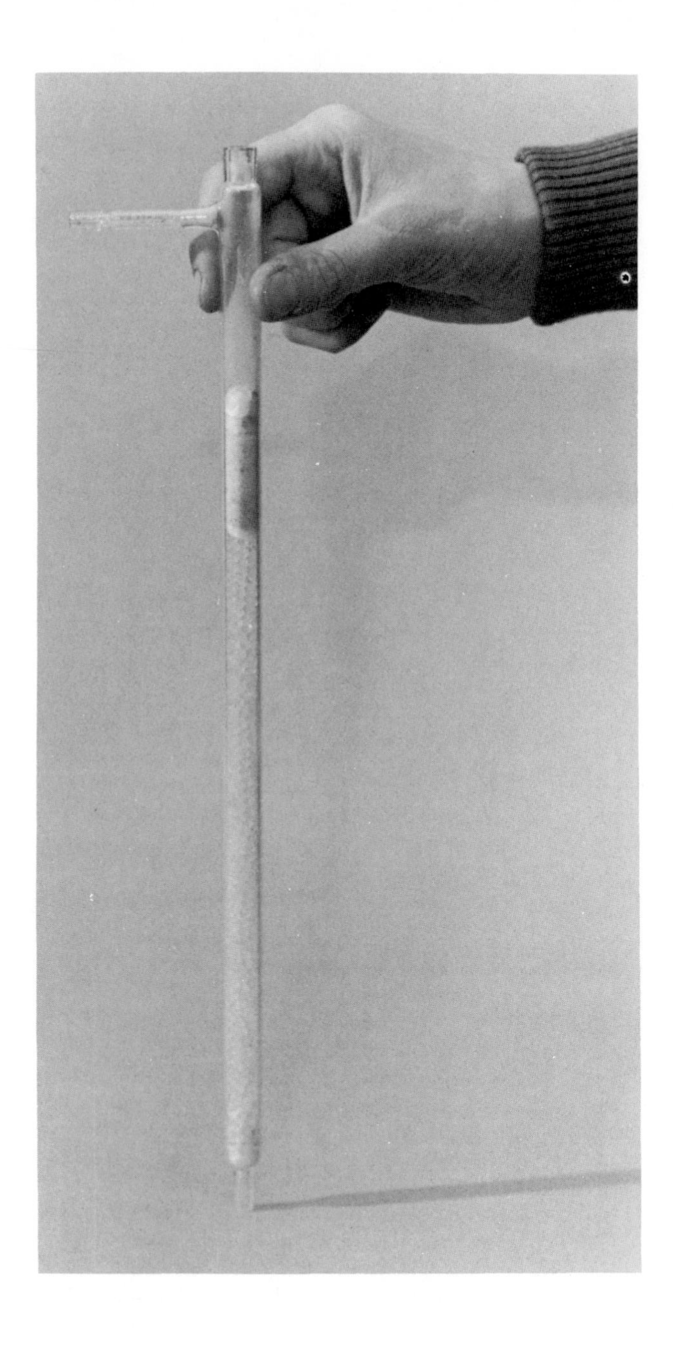

Figure 4. Kontes sweep codistillation column. The column consists of glass-wool plug, sand, glass beads, and glass-wool plug. Carrier gas enters the tube near the top of the tube at the side arm. The sample is injected directly into the top glass plug through a long syringe needle.

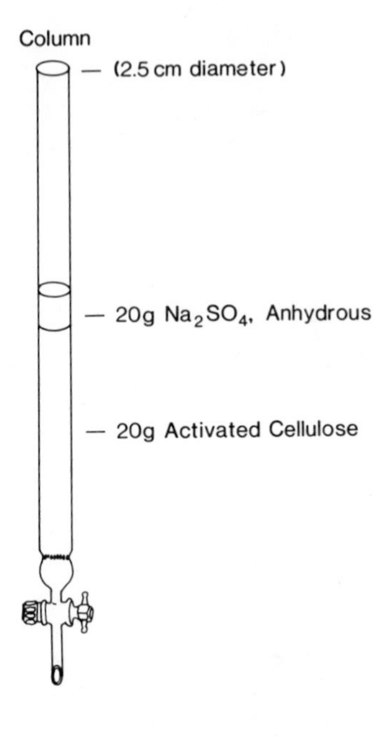

Figure 5. Cellulose cleanup column.

TCBI was selective enough that the carbamate spikes were readily discernible from the RC components, with very little background.

We have previously reported the use of cellulose and cellulose-activated charcoal columns for the cleanup of pesticide extracts as well as sweepco for one carbamate (2) and some organophosphate compounds. We wanted to compare the sweepco cleanup to these cleanup procedures. The cellulose column consisted of 20 g of activated cellulose and 20 g of sodium sulfate in a glass frit column (Figure 5).

The cellulose was activated overnight at 85°C and under 35 in. vacuum. The column was slurry-packed in petroleum ether or $CHCl_3$, and an upper layer of Na_2SO_4 was added. The sample concentrate was added to the top of the column and eluted with 100 ml 8% benzene in petroleum ether and reconcentrated for TLC analysis.

The chromatography of RC plus standard treated in this manner illustrated the superiority of the sweepco cleanup. The plate sprayed with anisaldehyde was still masked by other compounds in the rumen contents and the TCBI sprayed plate indicated that the recovery of the standards was less than that of the sweepco.

The cellulose-carbon consisted of 0.5 in. of sand, 5 g of cellulose, 0.5 in. of sodium sulfate (Figure 6). The cellulose portions of the column were slurry-packed with $CHCl_3$ or petroleum ether. The sample was added and eluted with 100 ml $CHCl_3$ and reconcentrated for TLC analysis.

The TLC of RC plus standard was still lacking in comparison to the sweepco cleanup. In fact, there was little apparent difference between the cellulose and the cellulose-carbon column cleanup results. The standard recoveries were a little more evident with the cellulose-carbon column.

Several commercially available plates were used in this work. The retention times for the different plates were not markedly different from each other. Retention times for the standards with different plates are compared in Table II. The most complete separation of the eight standards was achieved using a Merck silica gel 60 type "G" full-size plate. The cluster of standards in the middle were, upon careful inspection, completely discernible, and all eight standards were visible with both the anisaldehyde and the TCBI sprays. Whatman HP-K high-performance plates nearly equaled this separation but were less sensitive in their color development.

Whatman kindly supplied several samples of a special linear LK-6DF plate for carbamate TLC. These plates and EM reagents silica gel 60 aluminum-backed plates gave similar separations with incomplete resolution of the middle four standards and less sensitivity to some of the standards.

In conclusion, cleanup was found to be necessary prior to TLC in the case of some matrixes. Kontes sweep codistillation cleanup removed interfering pigments and materials resulting in a much cleaner TLC chromatogram. Both anisaldehyde and TCBI exhibited advantages, with TCBI being very selective for the carbamates and anisaldehyde being very sensitive. These two sprays complemented each other. Finally, of several commercially available TLC plates, Merck silica gel 60 gave the most complete separation of the eight carbamates tested. With other applications perhaps another plate would work better.

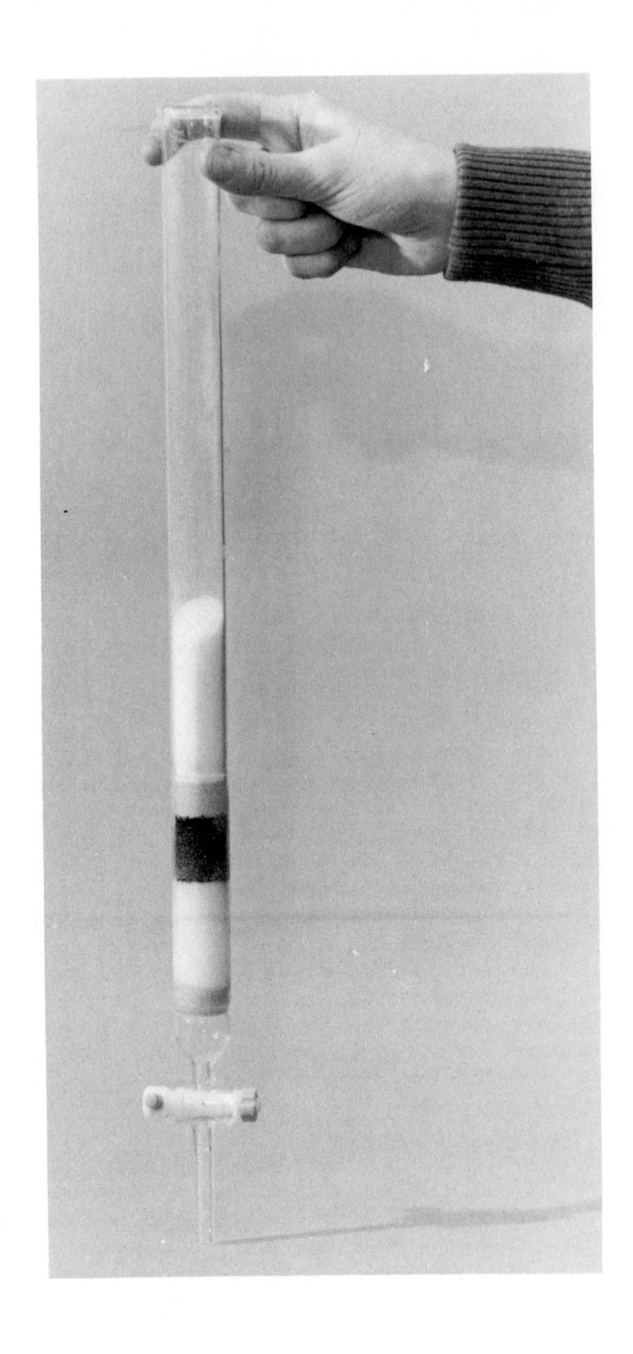

Figure 6. Cellulose-carbon cleanup column. The column consists of 0.5 in. sand, 5 g activated cellulose, 0.5 in. sand, 5 g charcoal 0.5 g sand, 5 g activated cellulose, and 5 g anhydrous sodium sulfate.

TABLE II. THIN LAYER PLATES RETENTION COMPARISONS[a]

Plates Used	Carbamate Standards RRT								Avg. Spot Width
	A	B	C	D	E	F	G	H	
Reverse Phase	43.5	40.9	ND[b]	43.5	ND[b]	ND[b]	33.0	32	4.3%
Whatman LK	27.1	32.4	5.7	30.5	13.3	1.0	36.2	38.0	6.2%
Whatman High Resolution	37.4	45.8	9.7	43.9	20.6	1.3	47.1	51.0	4.5%
Merck Silica 60	30.8	37.5	8.3	35.8	17.9	1.7	40.8	43.3	4.2%

[a]The figures represent the percent of the solvent front each spot moved and was measured at the middle of the spots. The average spot width was a measure of how diffuse or wide the spots were. The Merck 60 plate achieved the most complete separation of the standards from each other, followed closely in performance by the Whatman high resolution HP-K plates. Of course, visualizing the actual colored plates is far superior to trying to measure the retention times of the spots. Average spot width = [(average spot width)/(solvent retention)] × 100. The smaller the number the tighter the bands.

[b]ND = not detectable.

ACKNOWLEDGMENT

We would like to acknowledge Whatman Company for the contribution of samples of the LK-6DF plates.

REFERENCES

1. Whatman Publication 508, 9/79, 1979, p. 16.
2. M. Stahr, D. Lerdahl, and W. Hyde, J. Liq. Chromatogr., 12 (1980).

CHAPTER 35

Screening for Organochlorine and Organophosphorous Pesticides

Richard Pfeiffer and H. Michael Stahr

INTRODUCTION

The analysis of organochlorine and organophosphorous pesticides has been performed largely by gas chromatography employing a variety of specific detectors (1). While gas chromatography has received a great deal of emphasis, several investigators have reported the use of thin layer chromatography for the analysis of organochlorine pesticides (2-5) and organophosphorous pesticides (2, 6, 7) as separate groups. The analysis of both the organochlorine and organophosphorous pesticides together has also been investigated (8).

MATERIAL AND METHODS

Apparatus

Thin layer chromatographic equipment: a developing tank (27 × 20 × 7 cm) with glass top.
Micropipettes (from Becton Dickinson).
Hair dryer.
Spraying bottle.
UV light source.
Merck silica plates (Brinkmann Instrument Company).
Densitometer (Kontes model 1).

Samples and Standards

Standards were supplied by EPA and were 99.9% pure. Samples were prepared by R. Moore, Veterinary Diagnostic Laboratory, Iowa State University.

Chromogenic Reagents

o-Tolidine (1% v/v in acetone).
Silver nitrate reagent. Silver nitrate (1.7 g) treated with 5 ml of H_2O and 10 ml of NH_4OH and diluted to 200 ml with acetone.

Sample Preparation and Cleanup

All samples were extracted using procedures outlined by Stahr et al. (1). Organochlorine pesticides underwent additional cleanup using a florisil column (1, 9), and organophosphorous pesticide cleanup utilized a cellulose column (1, 10). Further cleanup by sweep codistillation was used for both groups of pesticides in feeds, rumen contents, and stomach contents (12, 13).

Thin Layer Chromatography Procedure

The samples are applied using 20 to 50 µl of sample dissolved in isooctane with a micropipette. Plates are dried with a hair dryer and then placed in a developing tank containing the mobile phase *n*-heptane/acetone (91:9). The plates are allowed to develop to within 1 cm of the top, removed, and dried with a hair dryer, after which they are sprayed with *o*-tolidine. They are then dried and exposed to UV light for approximately 10 minutes. The organochlorine pesticides appear as blue spots against a white background. The plates are then sprayed with the silver nitrate reagent and exposed to UV light for 10 minutes. The organophosphorous pesticides appear as white spots against a black background. The organochlorine pesticides will also appear as dark spots against this background.

RESULTS AND DISCUSSION

The R_f's of several organochlorine pesticides in the *n*-heptane/acetone mobile phase are shown in Table I. The chromatograms of brain, liver, and kidney tissues were free from any interferences, and the sensitivity was 100 ng for aldrin. The characteristic

TABLE I. R_f's OF COMMON CHLORINATED PESTICIDES RELATIVE TO ALDRIN

Pesticide	R_f
Chlordane	0.81[a]
op-DDT	0.79
op-DDE	0.79
op-DDD	0.50
Hexachlorobenzene	1.11
2,4,6-TCP	0.19
PCP	0.13
2,4D-Butylester	0.45
2,4D-Acid	0.00
BHC	0.31
Lindane	0.47
Aldrin	1.00
Dieldrin	0.59
PBB	1.00
Arochlor-1254	1.00[a]
Toxaphene	0.75[a]
Heptachlor	0.91
Endrin	0.66

[a] Gives a broad band.

blue spot on a white background was excellent for visualization, and quantitation was easily accomplished by densitometry, as is shown in Figure 1. The chromatograms of feeds analyzed for organochlorine pesticides were not entirely free from background interferences as were the chromatograms of tissue extracts. The interference was on the lower one fourth of the plate and so does not interfere with the majority of the organochlorine pesticides. Sweep codistillation has been used successfully to remove the interfering material from feed extracts.

The R_f's of several organophosphorous pesticides are given in Table II. The resulting chromatograms are not as "neat" as those for the organochlorine pesticides, but the white spot on a dark background is easily visible for microgram amounts. The

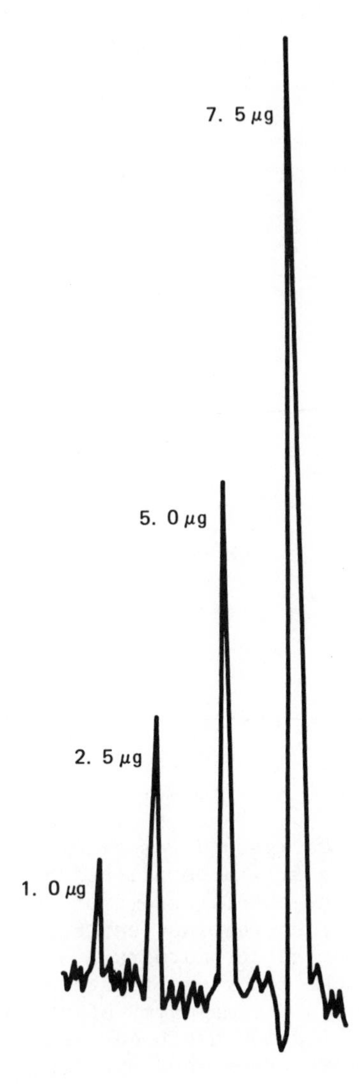

Figure 1. Quantitation of aldrin from TLC by densitometry.

TABLE II. R_f's OF ORGANOPHOSPHOROUS PESTICIDES RELATIVE TO METHYL PARATHION

Pesticide	R_f
Counter	2.83
Thimet	2.67
Dyfonate	2.50
Diazinon	1.33
Malathion	0.67
Methyl Parathion	1.00

organochlorine pesticides appear as dark spots with the silver nitrate, but appear on the upper two thirds of the plate versus the lower one third for the organophosphorous pesticides. Sweep codistillation is a must for the analysis of feed samples for the organophosphorous pesticides. The interfering band is completely removed, and the pesticides are recovered quantitatively from the cleanup.

REFERENCES

1. H. M. Stahr, Analytical Toxicology Methods Manual, ISU Press, 1977, p. 89.
2. K. C. Walker and M. Beroza, J. Assoc. Offic. Agric. Chem., 46, 250 (1963).
3. L. C. Mitchell, J. Assoc. Offic. Agric. Chem., 41, 781 (1958).
4. M. F. Kovacs, Jr., J. Assoc. Offic. Agric. Chem., 49, 635 (1966).
5. E. J. Thomas, J. A. Burje, and J. H. Lawrence, J. Chromatogr., 35, 119 (1968).
6. J. Beck and M. Sherman, Acta Pharmacol. Tox., 26, 35 (1968).
7. M. E. Getz and H. G. Wheeler, J. Assoc. Offic. Anal. Chem., 51, 1101 (1968).
8. M. A. Khan and J. Paul, Microchem. J., 24, 333 (1979).
9. AOAC Methods Manual, 13th ed., 27.015, 1980, p. 471.
10. J. Kovac, V. Batora, A. Hankova, and A. Szokolay, Bull. Envir. Cont. Toxicol., 13, 692 (1975).
11. H. M. Stahr, M. Gaul, W. Hyde, and R. Moore, Microchem. J., 24, 97 (1979).

12. Sweep Co-distillation: A Bibliography, prepared by Kontes, Vineland, New Jersey.
13. H. Stahr, D. Lerdahl, and W. Hyde, Clean-up of Samples for High Performance Liquid Chromatography and Instrumental Analysis, unpublished.

CHAPTER 36

Quantitative TLC of Polycyclic Aromatic Hydrocarbons

Haleem J. Issaq

INTRODUCTION

Polycyclic aromatic hydrocarbons (PAHs) are widely present as contaminants in the atmosphere. They occur in free form or adsorbed to soot and dust particles, in waterways, in the ocean, as well as in the food chain. Major sources of PAHs are combustion or pyrolysis processes such as emissions from transportation systems, heat and power generating plants, refuse burning, and industrial processes (1). A significant number of these PAHs are known to be carcinogenic (2, 3). Despite the fact that several thin layer chromatographic (TLC) methods have been described for the separation of PAHs, the results should be treated with reservation. As early as 1964, Inscoe (4) pointed out that some PAHs are unstable when applied and developed on silica gel plates. Rao and Vohra (5) reported on the fading of benzo(a)-pyrene (BaP) spots on silica gel plates and concluded that any in situ evaluation is impossible. Seifert (6) also found the decomposition of BaP on silica gel plates within minutes. He also reported that impregnating the plate with 5% parafin wax reduced BaP losses.

This study evaluates the possibility of quantitatively analyzing some of those PAHs that decompose on the plate and examines methods that can be used to separate the photooxidation products. BaP was selected for this study because it has been one of the most studied carcinogenic PAHs in the evironment, and because it has been reported that it decomposes on silica gel plates (4-6).

EXPERIMENTAL

BaP and dimethylbenz(a)anthracene were obtained from Aldrich Chemical Company, and 1,6- and 3,6-BaP quinones were received from the NCI Chemical Carcinogen Reference Standard Repository (a function of the Division of Cancer Cause and Prevention, National Cancer Institute, NIH, Bethesda, Maryland). All solvents used were glass-distilled (Burdick and Jackson). Silica gel and reverse-phase silica gel C_{18} plates were from Whatman, Inc. An automatic elution system, the Eluchrom (Camag), was used to elute the separated compounds from the plate. A 15-A Xenon lamp, Zenarch (Nems and Clasic, Silver Spring, Maryland), was used for the photodecomposition of BaP. Fluorescence measurements in solution and in situ on the plate were made on an MPF-2 Spectrofluorimeter (Perkin-Elmer) with a special TLC attachment. Five microliters of a solution of BaP, in benzene (300 μg/ml), was applied under nitrogen on the left corner of a silica gel plate, 2 cm from the bottom and 2 cm from the left side. The plate was developed in benzene/ethyl acetate (9:1, first dimension) and cyclohexane/isopropyl alcohol (5:1, second dimension). When the sample was applied on C_{18} reverse-phase plates, methanol was used for the first dimension and methanol/water (7:3) for the second dimension.

RESULTS AND DISCUSSION

Two basic experiments were carried out to verify the claims that BaP decomposes rapidly on the plate (4-6); BaP cannot be quantitatively evaluated in situ (5).

The first experiment was carried out to find out if it is possible to apply, develop, and elute BaP from the plate without any appreciable loss by decomposition. The plate was applied under nitrogen and developed in a tank that was flushed with nitrogen before and after the solvent had been introduced. After development in benzene/ethyl acetate (9:1), the plate was taken out, dried under nitrogen, and covered with a clear glass plate. The two plates were clamped tightly together to prevent BaP exposure to oxygen. When viewed in a box under long-UV radiation, one blue fluorescent spot could be seen. The spot was mechanicall eluted with 200 µl methanol, and its fluorescent density was measured at an excitation wavelength of 366 nm and an emission wavelength of 450 nm. The results indicated no loss of BaP. However, when the same experiment was carried out in the hood, and the BaP zones were allowed to dry in a stream of air instead of nitrogen, and then developed, several spots were observed after development. This is in agreement with the results of other researchers (4-6), except that not all the BaP was decomposed. When 5 µl of BaP solution was applied to the plate and allowed to dry in air in the hood, the spot turned a light brown color. After development, in the first dimension, nine spots were observed, of which the top spot (R_f=0.95) was BaP. Other spots were observed at R_f=0 to 0.1, 0.24, 0.37, 0.42, 0.48, 0.57, 0.72, and 0.82. Two spots at R_f=0.42 and 0.48 were characterized by their R_f values and micro infrared as the 1,6- and 3,6-BaP quinones, respectively.

When 10 µl of BaP solution was applied to a silica gel plate, exposed to UV radiation for 1 hr in the presence of air, and then developed in two dimensions, 35 spots were observed. One of these was BaP. This indicates that even after BaP has been subjected to intense UV radiation for 1 hr, it is not all destroyed. It is safe to assume that when the BaP solution is spotted on the silica gel layer (250 µm thick), it penetrates into the silica gel and forms a series of microlayers, of which the upper exposed layer (layers) shields the bottom layers from UV radiation and air and slows their decomposition. Similar results were observed by Falk et al. (7) for BaP adsorbed on filter paper, by Lane and Katz (8) for BaP that was thinly dispersed in a petri dish, and by Gibson and Smith (9) for BaP on silica gel powder. It was found (7, 8) that the rate of decomposition of BaP slowed with time, which suggests that the upper exposed layer reacted quickly and then hindered the decomposition of subsurface material. Table I shows the decomposition of BaP with time on silica gel plates exposed to sunlight. When plates loaded with BaP solutions were left in the hood for a month before being developed, some undecomposed BaP was found. Figure 1 shows the decomposition of BaP and the production of new compounds. A time study of the decomposition of DMBA was performed by applying 50 µg of DMBA on several 5 × 20 cm silica gel plates and exposing them to UV light (the sun) and air for variable lengths of time (Figure 2). It appears that stable products are formed within 90 minutes. The plates were developed in benzene/ethyl acetate (9:1) and scanned on a densitometer at 366 nm.

TABLE I. PHOTODECOMPOSITION OF BaP ON SILICA GEL PLATES EXPOSED TO SUNLIGHT WITH TIME

Time of Exposure (min)	Undecomposed BaP (%)
1	100
60	94
240	88
300	83
360	77

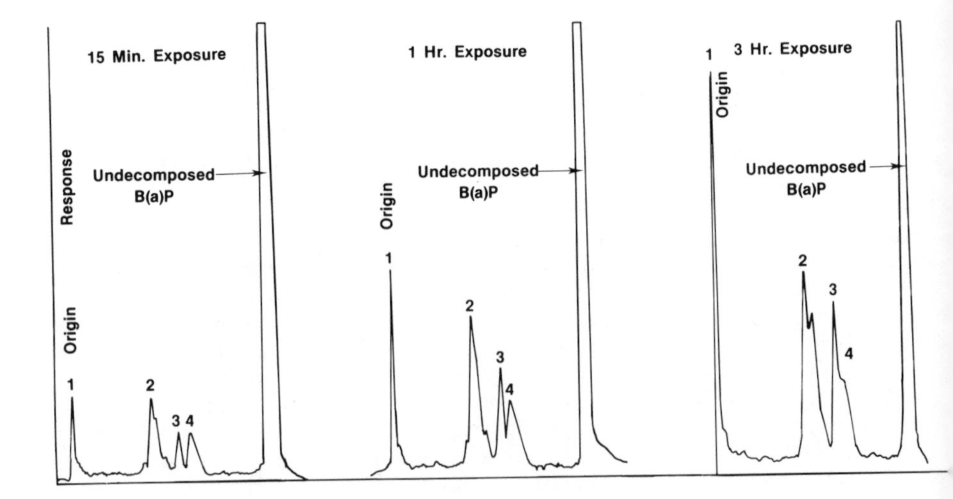

Figure 1. Decomposition of BaP.

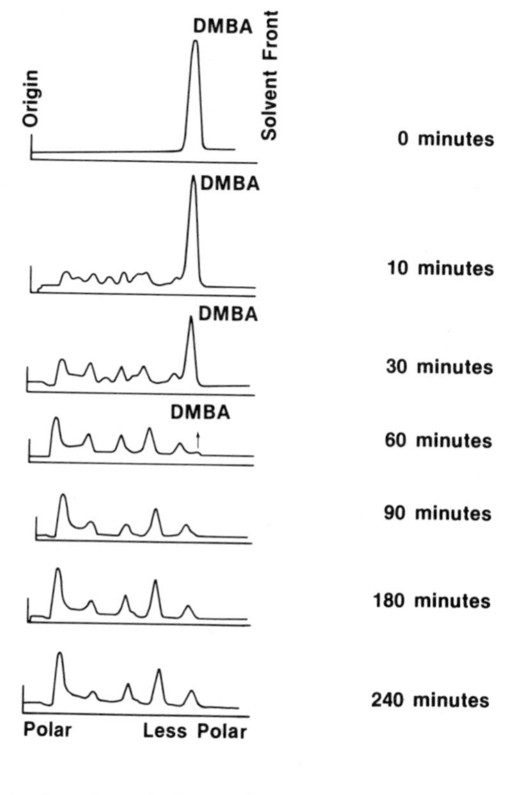

Figure 2. Photodecomposition of DMBA on silica gel TLC plates.

An interesting phenomenon, which could lead to erroneous quantitative results, was observed. When a plate freshly loaded with BaP (spot not completely dry) was placed under long-UV radiation, the spot fluoresced blue. However, after a few minutes, when the spot was completely dry, it fluoresced green. When the spot was wetted with benzene, it again fluoresced blue, and turned green again when it dried. We measured the effect this had on the quantification of BaP in situ on the plate. A plate was spotted and then scanned while the spot was wet. A large peak was observed. When the plate was completely dry, no peak was observed. This clearly indicates that measuring the fluorescence of completely dry spots gives results that are different from those measured when the spots are wet.

Reverse-phase plates were used to separate the polar components of the photooxidation products of BaP, which did not move off the origin on silica gel plates. Most of the photooxidation products were identified by other workers (9).

CONCLUSION

Our results show that it is possible to quantitatively measure BaP in situ on the plate. It was also found that not all the BaP decomposed on the plate after long periods of exposure to air and light. Thirty-five different products of decomposition of BaP were separated by TLC using two-dimensional development in two different solvent systems.

ACKNOWLEDGMENTS

This work was supported by Contract No. NO1-CO-75380 with the National Cancer Institute, NIH, Bethesda, Maryland.

REFERENCES

1. National Academy of Sciences, Committee on Biologic Effects of Atmospheric Pollutants, Particulate Polycyclic Organic Matter, Washington, D.C., 1972.
2. IARC Monographs on the Evaluation of Carcinogen Risk of the Chemical to Man, Certain Polycyclic Aromatic Hydrocarbons and Heterocyclic Compounds, Vol. 3, Lyon, France, 1973.
3. A. Dipple, in Chemical Carcinogens, C. E. Searle (Ed.), ACS Monograph 173, American Chemical Society, Washington, D.C., 1976, p. 245.
4. M. N. Inscoe, Anal. Chem., 36, 2505 (1964).
5. A. M. M. Rao and K. G. Vohrs, Atoms. Environ., 9, 403 (1975).
6. B. Seifert, J. Chromatogr., 131, 417 (1977).
7. H. L. Falk, I. Markul, and P. Kotin, A.M.A. Arch. Ind. Health, 13, 113 (1956).
8. D. A. Lane and M. Katz, in Fate of Pollutants in the Air and Water Environments, Part 2, I. A. Suffet (Ed.), Wiley-Interscience, New York, 1977, p. 137.
9. T. L. Gibson and L. L. Smith, J. Org. Chem., 44, 1842 (1979).

CHAPTER 37

Analysis of Gossypol in Feeds by TLC

Ramasy Ventalchalam and H. Michael Stahr

INTRODUCTION

Gossypol, a polyphenolic, binaphthyl aldehyde, is a poisonous pigment that occurs in the subepidermal glands throughout the cotton plant (Gossypium ssp. of the order Malvacae) (1). In the numerous varieties of the cotton plant cultivated for fiber, gossypol occurs in the seed kernel, which is extensively used in animal feeds. Gossypol has been found to occur in the kernels at levels ranging from 0.4 to 1.7% (2). It is recognized as toxic to both ruminants and nonruminant (1) insects (3) and certain microorganisms (4). In experimental animals it has been found to be a gastrointestinal irritant, and in large doses it causes edema of lungs, shortness of breath, paralysis, and death (1).

Marlewski in 1899 first isolated gossypol and Adams et al. (5) established its structure, $C_{30}H_{30}O_8$ as shown in Figure 1. Withers and Carruth (6) showed that gossypol was readily removed from unprocessed cottonseed kernels by extraction with diethyl ether but not after the cottonseeds were processed by conventional wet cooking (7) prior to oil extraction. It was suggested that this was because of the changes in gossypol brought about by the binding reactions of its aldehyde groups to that of the free amino groups of the proteins in the kernels, and the diethyl ether-extractable gossypol from the unprocessed kernels was free gossypol.

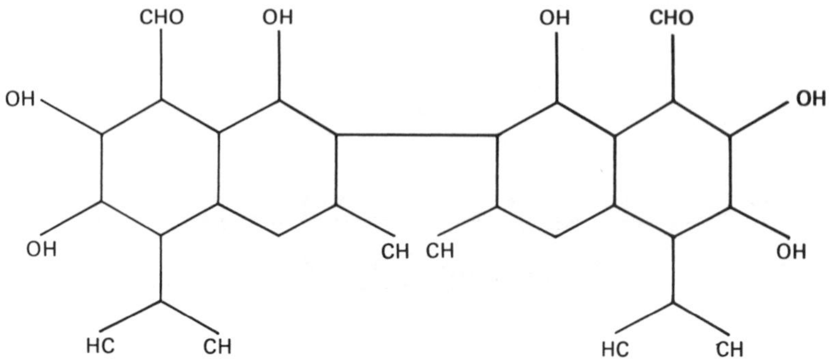

Figure 1. Structure of gossypol, 2,2´-bis(1,6,7-trihydroxy-3-methyl 5-isopropyl-8 aldehydo naphthalene).

Numerous methods have been proposed for the quantitation of gossypol. Carruth (6) extracted it by percolating the ground kernels with diethyl ether for 2 to 3 hours, but this did not give consistent results. To provide more reproducible results Holverson and Smith (8) modified the method by adding water and alcohol to ether and extended the percolation extraction time to 72 hr. Although this procedure increased the sensitivity of the method, the results were still variable. In all these methods estimation of the extracted gossypol was based on gravimetric precipitation of gossypol as the dianilino derivative. These procedures were also tedious and were subject to error due to coprecipitation of extraneous materials in the extract. Pons and Guthrie (9) extracted gossypol by shaking the ground kernels with acetone/water (70:30) for about an hour. The reaction of gossypol with p-anisidine provided a spectrophotometric estimation of the extracted gossypol. The method they suggested extracted the minimal amount of lipids and did not hydrolyze the bound gossypol very much. Although this method slightly increased the sensitivity to 50 to 100 ppm, quantitation was not always satisfactory, and the results were inconsistent. Raju and Carter (10) proposed a gas liquid chromatographic method for determining gossypol in aqueous acetone extracts of cottonseed products. The detection limit of this method was between 1 and 2 ppm, but the method was unable to detect gossypol in cooked and pressed cottonseed meals and glandless seeds. On the contrary, a less sensitive paper chromatographic method proposed by Martin (11) detected low levels of gossypol in the same products. However, the recovery of these methods was about 70%. Other quantitative systems such as the reduction of Fehling's solution by gossypol and titration of the precipitated cuprous oxide (12), and polar-agraphy (13) have been proposed for quantification of gossypol, but none of these methods has been widely used.

METHOD

Apparatus and Reagents

1. Silica gel G, type 60 (Scientific products). Mix 40 g silica gel and 90 ml of 95% ethanol for spreading.
2. Solvent system benzene/dioxane/acetic acid (91:10:4).
3. Gossypol standard solution 0.5 mg/ml diethyl ether.
4. Fluorodensitometer, model N1. K-495000 (Kontes Glass Company, Vineland, New Jersey).
5. TLC plates, 0.5 mm. Prepare using slurry dispenser and plate template. Activate overnight at 65°C before use.

Procedure

Analysis is to be completed in a day; otherwise, store sample residues in a refrigerator. Exposure of sample residues to light or heat results in degradation of gossypol and makes estimation difficult.

Weigh two 10-g ground feed samples and transfer to two blender jars. Spike one sample with 40 µl of 0.5 mg/ml gossypol in ether standard. Proceed as follows for each sample: add 10 ml water and grind samples again for 30 sec. Add 4% sodium hydroxide while constantly mixing samples until pH is 9.0 (range 8.5 to 9.0). Add 100 ml of diethyl ether, and blend again for 30 sec. Transfer contents to a centrifuge bottle, and centrifuge for 2 minutes at 1500 rpm. Remove top ether layer with pipette and discard it. This helps to remove interfering materials while gossypol remains in the aqueous layers. Transfer contents of centrifuge bottle to blender again and add 2N HCl slowly while mixing contents constantly until pH is $3.\overline{5}$ (range 2.5 to 3.5). Add 100 ml of diethyl ether, and blend again for a minute. Centrifuge contents for 2 minutes at 1500 rpm. Transfer ether layer immediately into a conical flask that has been completely wrapped with aluminum foil except for the mouth of the flask.

Evaporate contents of the flask to 1 ml over low or no heat using nitrogen.

Thin Layer Chromatography

Of the 1-ml ether concentrate of the extract, spot 50-μl portions on the TLC. On same plate spot 1, 3, 5, and 10 μg each of the standard gossypol. Spot 5 μg of gossypol on one of sample spots, and use this as an internal standard.

Using benzene/dioxane/acetic acid (91:4:1) as the mobile phase, develop the plate at room temperature (about 25°C) in a TLC tank in a dark room with little or no light until gossypol is resolved sufficiently for quantitative determination. (Developing 90% of the plate will carry with it most of the interfering materials.) Gossypol has an R_f value of 0.40 when this mobile phase is used (i.e., on 8 8-in. plate gossypol band is seen between 2.8 and 3.2 in.). Quantitatively determine gossypol in the sample visually using ultraviolet (UV) fluorescence or by fluorodensitometry. If the intensity of the sample spot exceeds that of the highest standard, adjust the concentration of the sample extract so that the sample spot intensity will fall in the range of standard concentration spotted (14).

RESULTS AND DISCUSSIONS

A variety of feeds were examined by the method. The recoveries of the added standard (2 μg/g) ranged between 85.4 and 90.9% (Table I). The proposed method should prove applicable to most types of feeds, including cotton seeds, corn, ready mixed feeds, and feed pellets, and it is capable of detecting gossypol at the 0.25-μg/g level. Among the feeds examined by the proposed method, though the extracts of the ready mixed feeds and feed pellets produced the greatest amount of interfering substances before TLC, the sensitivity of the proposed method was not affected.

The pH manipulation of the sample and extraction of gossypol with diethyl ether was selected after using a combination of various extracting solvent systems, such as ethanol/water (70:30), acetone/water (60:40), ethanol/water/ether (60:30:10) and acetone/water/ether (60:30:10). The extraction of gossypol with diethyl ether after selective pH manipulation of the sample also gave the lowest level of interfering materials.

Among the various mobile phases investigated to isolate the gossypol spots cleanly from the other residual interfering

TABLE I. RECOVERIES OF GOSSYPOL FROM FEED SAMPLES[a]

Sample	Number of Analyses	Mean Recovery (%)	Standard Development
Cotton seeds	6	90.9	±3.3
Corn	7	90.1	±1.5
Mixed feeds	12	87.1	±1.1
Feed pellets	5	85.4	±1.3

[a]Overall mean recovery when all samples were taken into consideration was 88.3% ± 3.89.

materials present in the samples, benzene/dioxane/acetic acid (91:10:4) gave the best results. Though the chloroform/acetone/ethanol/water (60:40:10:1) mobile phase proposed by Chalam and Stahr (15) for the determination of citrinin gave equally good results, all quantitation of gossypol in the various feed samples was done using the former. The various other mobile phases such as methanol/water/formic acid (90:5:5) and ethanol/diethyl ether/acetic acid (90:10:5) either did not resolve the gossypol from the interfering materials or made the gossypol standards streak on the plate. The spiked gossypol bands when the benzene/dioxane/acetic acid mobile phase was used were compact and deep bluish and were at R_f 35 to 40 on the plate.

The time taken to develop the plate was about 1 hr, and the whole analytical procedure for gossypol can be completed in 2 hr.

REFERENCES

1. L. C. Berardi and L. A. Goldblatt, in Toxic Constituents of Plant Foodstuffs, Academic Press, New York, 1969, p. 211.
2. W. A. Pons, Jr., C. L. Hoffpauir, and T. H. Hopper, J. Agric. Food Chem., 1, 1115 (1953).
3. M. J. Lukefahr, L. W. Noble, and J. E. Houghtalin, J. Econ. Entomol., 59, 817 (1966).
4. A. A. Bell, Phytopath., 57, 759 (1967).
5. R. Adams, T. A. Geissman, and J. D. Edwards, Chem. Rev., 60, 555 (1960).
6. F. E. Carruth, J. Am. Chem. Soc., 40, 647 (1918).

7. E. P. Clark, J. Biol. Chem., 76, 229 (1928).
8. F. H. Smith, Ind. Eng. Chem. Anal. Ed., 18, 43 (1946).
9. W. A. Pons, Jr. and J. D. Guthrie, J. Am. Oil Chem. Soc., 26, 671 (1948).
10. P. K. Raju and C. M. Carter, J. Am. Oil Chem. Soc., 44, 465 (1967).
11. J. B. Martin, Proceedings of the Conference on Chemical Structure and Reactions of Gossypol and Nongossypol Pigments of Cottonseed, National Cottonseed Products Association, Memphis, Tenn., 1959, p. 71.
12. M. Z. Podol'skaia, J. Appl. Chem. USSR, 17, 657 (1944).
13. A. L. Markman and S. N. Kolesov, J. Appl. Chem. USSR, 29, 242 (1956).
14. Official Methods of Analysis, 12 ed., AOAC, Washington, D.C., 1975, sec. 26.019.
15. R. V. Chalam and H. M. Stahr, J. Assoc. Offic. Anal. Chem., 62 (3), 570 (1979).

CHAPTER 38

Thin Layer Bioautography in Residue Analysis

Alvin L. Donoho, William J. Begue, and Paul R. Handy

INTRODUCTION

The subject of bioautography covers a wide range of techniques and a variety of visualization and measurement systems. For example, bioautography may be used for measurement after separation of biologically active compounds by paper chromatography, thin layer chromatography (TLC), and electrophoresis. Biological systems for measurement may range from bacteria to animal cells. This subject as it pertains to paper and thin layer chromatography has been reviewed by Betina (1). The specific area for discussion in this paper is bioautography on thin layer chromatographic plates, and the primary emphasis will be on measurement of antibiotic residues in animal tissues.

In the period of 1966 to 1967, a thin layer bioautographic assay procedure for measurement of the anticoccidial drug monensin in chicken tissues was developed (2). This was the first application of thin layer bioautography for routine analysis of antibiotic residues in animal tissues. Since that time, thin layer bioautography has been used continuously in the Lilly Research Laboratories as a research tool to study pharmacological, metabolic, and tissue residue parameters of antibiotics for use in animals. Furthermore, thin layer bioautographic assay procedures have been submitted to, and used by, various regulatory bodies to monitor the safe use of animal health products. The purpose of this discussion is to review the thin layer bioautographic technique from the standpoint of advantages and disadvantages, factors affecting performance of the technique, and finally some practical considerations for its use in tissue residue analysis.

OUTLINE OF THE TECHNIQUE

The major steps in the thin layer bioautographic technique for assay of tissue residues are listed in Table I. The three main steps in any residue assay procedure are extraction, purification, and measurement of the drug. An assay procedure is frequently characterized by the measurement system alone, that is, gas liquid chromatography (GLC), high-performance liquid chromatography (HPLC), or microbiological assay. Nevertheless, the extraction and purification steps are equally important and must be compatible with the measurement system.

Bioautographic detection involves first the separation of the sample extracts by standard TLC procedures. Next, a layer of agar that has been inoculated with the test organism (seeded agar) is placed in contact with the chromatogram. Generally, this is done by placing the TLC plate in a plastic holder on a flat surface and flooding the plate with melted seeded agar, which is then allowed to solidify. The TLC plates have been prepared by spraying the surface with a thin layer of unseeded agar using an atomizer-type sprayer as described by Kline and Golab (3) before flooding with seeded agar. This step is to prevent buckling and floating of the silica gel from the TLC plate during the agar pouring step. Other researchers (4, 5) have applied the seeded agar directly without this step.

A necessary step in bioautography is the transfer of the antibiotic to the agar layer, where it can exert its inhibitory action upon the test organism. This may be a separate step in the procedure, or it may involve only the diffusion of the antibiotic into the agar layer during incubation of the bioautogram. A common procedure for bioautography of paper chromatograms is to place the chromatogram in contact with the agar surface for a period of time to effect transfer and then remove the chromatogram prior to incubation. This technique has not been particularly successful with thin layer plates, although it has been used (6). It is probably more effective for analysis of water-soluble than lipid-soluble antibiotics because of a more ready transfer to the agar.

The plate is then incubated, usually overnight, to allow the test organism to grow and the zones of antibiotic inhibition to become apparent. Finally, the bioautogram is evaluated to

TABLE I. OUTLINE OF BIOAUTOGRAPHIC TECHNIQUE

1. Sample extraction
2. Sample purification
3. Bioautographic detection
   a. Thin layer chromatography
   b. Spray-sealing the plate with agar
   c. Application of seeded agar
   d. Diffusion of antibiotic into agar
   e. Incubation of plate
   f. Evaluation of plate

determine the level of antibiotic activity in each TLC lane. This may involve only a visual examination to see if there is activity on the plate, or it may involve a careful measurement of zone area for quantitative analysis. Various techniques can be used to improve the contrast on the plate for better measurement or documentation (7).

ADVANTAGES OF TLC BIOAUTOGRAPHY

There are several advantages for the use of TLC bioautography for antibiotics. Among these are sensitivity, selectivity, economy, and capability for screening large numbers of samples.

Sensitivity

Several classes of antibiotics have characteristics that make them unsuitable for measurement by standard high-sensitivity techniques. For example, many antibiotics are relatively large-molecular-weight compounds or highly polar, which makes them nonvolatile and unsuitable for analysis by GLC. Many do not have useful spectra suitable for low-level measurement by ultraviolet (UV), fluorescence, or other spectrophotometric detection. The polyether inophore antibiotic monensin is a good example. It is nonvolatile, has no useful UV spectrum, and is not readily derivatized for measurement by HPLC. Yet the TLC bioautographic measurement can routinely detect as little as 50 ng of monensin applied to the plate. Similar experiences have been reported by

Brodasky et al. (6) for clindamycin and McDonald et al. (4) for lasalocid.

Selectivity

The high resolving power of TLC is well-known and need not be discussed here. When the further specificity of measurement by antimicrobial activity is added, the TLC bioautographic system becomes highly selective.

Economy

Compared to most instrumental analysis systems for residue assay, the cost of TLC bioautography is quite low.

Screening Capability

There are at least three requirements to be met by residue assay methods for use in development and support of animal drugs. These are (1) a method for development of research data for product clearance; (2) a method for monitoring animal tissues for the presence of violative residues; and (3) a method for adequate confirmation and quantitative evaluation once a potential violative sample has been identified. These may be different methods or different applications of the same basic method. TLC bioautography is especially useful as a screening tool for monitoring the presenc or absence of residues. Its utility as a quantitative tool is discussed later.

DISADVANTAGES OF TLC BIOAUTOGRAPHY

The primary disadvantage of TLC bioautography as a measurement technique is that materials and expertise for microbiological assays are not always available for routine use in analytical laboratories.

In the early days of TLC bioautography, it was regarded as a qualitative or semiquantitative technique. This apparent disadvantage can be surmounted by careful sample preparation, application of improved quantitative TLC procedures, and more sophisticated measurement of zones of antimicrobial activity.

FACTORS AFFECTING THE BIOAUTOGRAPHY PROCEDURE

A list of several factors to be considered in TLC bioautography is presented in Table II.

TABLE II. FACTORS AFFECTING BIOAUTOGRAPHIC PROCEDURE

1. Selection of microorganism
2. Microbiological parameters
3. TLC parameters
   a. Selection of plates
   b. TLC developing systems
4. Transfer of antibiotic into agar layer
5. Sample purification

Selection of the Microorganism

Selection of the proper microorganism for the assay is an important consideration. Primary factors in this selection are the sensitivity of the microorganism to the antibiotic, the ease of handling, and safety of handling the microorganism in the laboratory. For routine assay, the microorganism preparation becomes essentially an analytical reagent. Therefore, it must be easily grown and maintained in the laboratory and must be safe for handling by the analysts. Examples of suitable microorganisms include *Bacillus subtilis*, *Staphylococcus aureus*, *Streptococcus fecalis*, *Sarcina lutea*, and *Micrococcus flavus*. A number of others have also been used.

Microbiological Parameters

There are nearly unlimited combinations of microbiological assay variables that could be examined for TLC bioautography. A detailed discussion of these variables is beyond the scope of this report. A few are mentioned for illustration. Incubation temperature, pH of the medium, composition of the nutrient agar, and quantity of inoculum all affect the performance and sensitivity of the bioautography measurement. In general, one should select conditions that give consistent growth of the microorganism but that stress the microorganism sufficiently to give a good response to the antibiotic. Conditions that strongly favor growth of the microorganism, such as optimum pH, rich nutrient agar, and heavy inoculation, tend to give poor sensitivity of detection. Stressing the microorganism too much results in poor growth. This may increase the detection sensitivity, but the contrast on the plate may be poor, resulting in problems with measurement and documentation. These factors need to be evaluated to obtain optimum assay conditions.

Thin Layer Chromatography Parameters

Selection of TLC Plates

Even after the advent of commercially available TLC plates, it was common to use laboratory-poured plates for bioautography. The primary problems with commercial plates were nonuniform adsorbent layers and flaking of the adsorbent during preparation of the bioautogram. In the last few years suitable plates have become available commercially. A "soft" plate, such as the Analtech Woelm silica gel G plate, is suitable for routine use. We have also successfully used Whatman LK5 plates, which have a cellulose preadsorbent layer for application of the sample. No doubt a variety of plates from different manufacturers can be used successfully. Glass-backed plates are more suitable than those with flexible backing (such as aluminum) because of problems with floating or buckling of the latter. Flexible plates are useful for some specific applications (6).

Some commercial plates may give problems with low sensitivity or lack of growth of the microorganism due to various causes. These problems may result from poor desorption of the antibiotic from the adsorbent or from inhibition of growth due to improper pH or inhibiting substances on the plate. Prewashing of the plate or pH buffering may reduce or eliminate the problems. A systematic evaluation of these factors is needed for successful development of TLC bioautography procedures.

TLC Developing Systems

A wide variety of mobile phases may be used for TLC and subsequent bioautographic measurement. These include the use of appropriate acids and bases for chromatography of polar compounds. The primary limitation is the possibility of leaving residual materials on the plate which may interfere with the growth of the microorganism. Usually, this limitation can be overcome by use of purified solvents, careful removal of residuals on the plate by volatilization, and/or neutralization of acids or bases. Gentle heating of plates in a vacuum oven is often an effective means of removing volatile residuals.

Efficient separation and detection of closely related antibiotics can be achieved by careful selection of chromatographic conditions. Pauncz and Harsanyi (5) have reported on separation and detection of aminoglycoside antibiotics. Figure 1 shows the separation of several polyether inophores conducted by one of the authors.

Transfer of the Antibiotic into the Agar Layer

Desorption from the TLC adsorbent and diffusion of the antibiotic into the agar is an essential step in bioautography. Factors that affect this process have a very direct effect upon assay sensitivity. Higher incubation temperatures increase the diffusion rate in the agar. This is used to advantage in the assay for the antibiotic hygromycin. After the seeded agar layer is poured, the plate is incubated for a few hours at 37°C, a temperature higher than the optimum for the assay organism (*Pseudomonas syringae*). During this period the antibiotic diffuses at an accelerated rate while the growth of the organism is minimal. Subsequently, the plate is incubated overnight at 30°C. This gives a more sensitive assay than a direct incubation without the diffusion step. The pH of the medium can also affect the diffusion rate.

Thickness of the agar overlay also has an effect on sensitivity because of the distance the antibiotic has to move. A thicker layer generally gives lower sensitivity. Because of this factor it is imperative to have a uniform thickness of agar over the plate to give a uniform antibiotic response.

Sample Purification

One of the key steps in any residue assay procedure is sample purification. Any assay procedure in which sample purification is inconsistent or marginal is doomed to a high failure rate.

Sample impurities have the potential for either increasing or decreasing sensitivity, depending upon their effect upon the

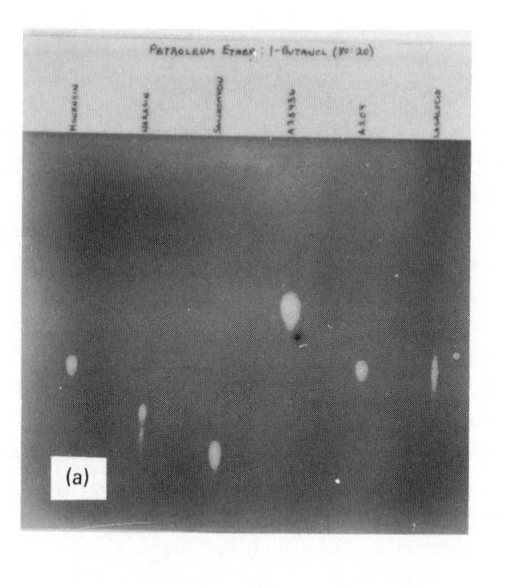

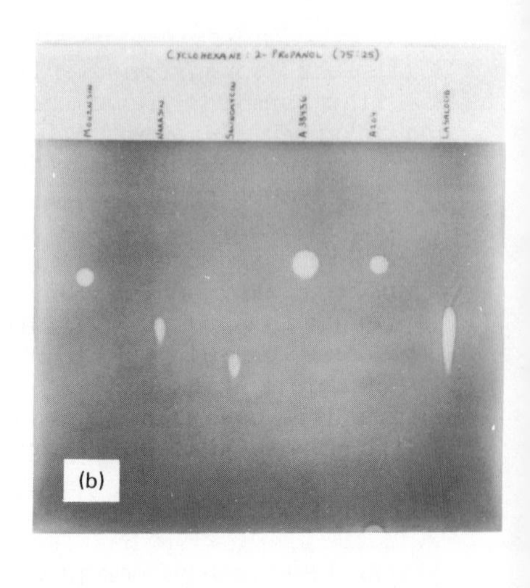

Figure 1. Bioautograms showing separation of monensin, narasin, salinomycin, A38436, A204, and lasalocid (left to right) on two separate Analtech Woelm silica gel plates. Developing solvents were (a) petroleum ether/n-butanol (80:20) and (b) cyclohexane/2-propanol (75:25).

antibiotic, the chromatography, and the microorganism. Sample impurities may affect the desorption or diffusion of the antibiotic, thereby influencing sensitivity. Overloading of the TLC plate with a "dirty" sample may cause distortion of the zone of inhibition, or in extreme cases may change the R_f of the sample. Furthermore, sample impurities may have a direct effect upon the microorganism to potentiate or suppress the antibiotic response. Direct interaction of the sample impurities with the antibiotic could also increase or decrease the antimicrobial response.

Degree of sample purification has a very direct effect upon the limits of detection for the assay procedure. An extensive sample purification allows the application of more grams equivalent of tissue extract to the plate and, thereby, gives a lower limit of detection.

QUANTITATIVE ASSAY BY BIOAUTOGRAPHY

Much greater demands are made upon tissue residue assay procedures of today than those of 10 to 15 years ago. The bioautographic assay for monensin that was developed in 1967 was designed as a simple procedure to detect the presence or absence of monensin at a sensitive detection level (25 to 50 ppb). Semiquantitative measurements were made by bracketing test samples with standard recoveries, but no provision was made for careful quantitation. It should be emphasized that the semiquantitative nature of bioautographic procedures of that era was due to the mode in which they were employed. It was not because the bioautographic technique was incapable of quantitative measurement.

With more stringent requirements for sensitivity, specificity, and quantitation, bioautographic procedures for residue analysis have been refined to meet these requirements. Okada et al. (8) have refined the monensin procedure of Donoho and Kline (2) to give a quantitative procedure with a lower limit of detection in the range of 10 ppb. Zones of inhibition were measured with a ruler. McDonald et al. (4) have reported on quantitative TLC bioautographic procedures for assay of lasalocid in chicken tissues. Measurement of zones was done by employing an electronic planimeter. Quantitative TLC bioautography was employed for studying clindamycin in the rat (6) and for aminoglycoside antibiotics in fermentation broths (5). The utility of TLC bioautography for quantitative residue analysis should increase with the application and further refinements in measuring zones of inhibition. The use of sophisticated image analyzers (9) promises to give an improvement in quantitative measurement of antibiotic response.

To illustrate the quantitative nature of the bioautographic response, graded levels of monensin were assayed in triplicate on two different days. The samples were applied quantitatively to Analtech Woelm silica gel G plates using a Kontes sample applicator. The plates were developed twice in ethyl acetate, and the bioautograms were prepared. Zones of inhibition were measured manually, horizontally, and vertically to the nearest 0.5 mm, and a relative area was calculated by multiplying the two. Results are presented in Table III and Figure 2. These data show that the response of monensin on the bioautogram is sufficiently linear for quantitative measurement even when a relatively crude measurement system is used. More sophisticated measurements of area should reduce the replicates needed for quantitation.

Another illustration is given in Table IV, which lists results of the bioautographic determination of the polyether

TABLE III. TLC-BIOAUTOGRAPHIC RESPONSE FROM VARIOUS MONENSIN LEVELS

Monensin Level (ng)	Antibiotic Response[a]						Mean ± SD
	Day 1 Plate Number			Day 2 Plate Number			
	1	2	3	1	2	3	
30	0	0	0	+[b]	+	+	
60	36	36	27.5	36	27.5	27.5	31.8 ± 4.66
90	49	49	45.5	64	60	45.5	52.2 ± 7.88
120	76.5	64	64	72	72	60	68.1 ± 6.33
150	72	81	81	90	90	72	81 ± 6.05

[a] Length × width of zone in millimeters.

[b] Trace of activity too small to measure.

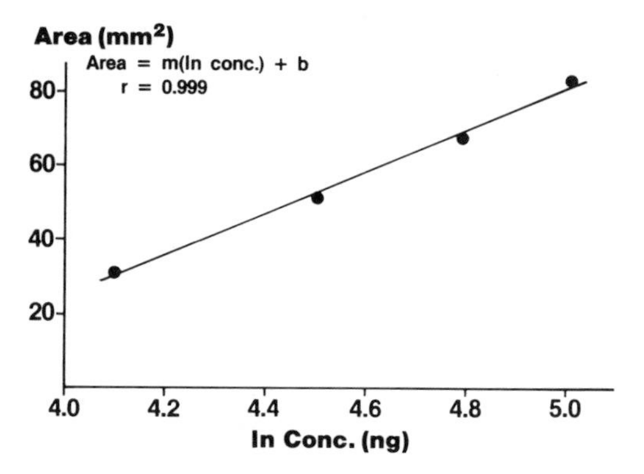

Figure 2. Monensin standard curve.

inophore narasin recovered from chicken fat tissue. Samples of fat from untreated birds were fortified with 5, 10, 20, and 40 ppb narasin and then extracted and purified by silica gel and alumina column chromatography. The purified extracts were applied to each of five Analtech Woelm silica gel G TLC plates using 20-μl glass disposable pipettes. Subsequent to development, the narasin activity was detected using <u>B. subtilis</u> as the bioautographic test organism. Area measurements on the resulting zones of inhibition were made using a manual planimeter.

The results indicate that the bioautographic test responses are suitable for quantitation down into the parts-per-billion range. The zone diameters vary from plate to plate, which suggests that for quantitative measurements the test sample and recoveries or standards must be applied to the same TLC plate. Even the plate-to-plate variability was better than 10% relative standard deviation at recovery levels of 10 ppb or above.

There continues to be a need for semiquantitative screening procedures, especially for use in monitoring the safe use of drugs in food-producing animals. For such use a large number of samples need to be assayed, and the bulk of these samples will be negative or below the assigned tolerance. It is an unreasonable investment to attempt quantitative analysis on all such samples, since this would involve at least duplicate analysis.

Instead, the screening procedure can be used to identify potentially violative samples, and then a separate assay can be done with sufficient replicates on the suspect samples to derive

TABLE IV. RELATIVE ZONE AREAS FROM ASSAY OF CHICKEN FAT SAMPLES FORTIFIED WITH VARIOUS LEVELS OF NARASIN[a]

Plate Number	Fortification Level (ppb) 5	10	20	40	Slope	Correlation Coefficient
1	15.25	44.6	72.75	117.7	48.4	0.993
2	18.7	48.0	75.0	122.5	48.8	0.991
3	17.0	49.7	76.5	139.0	56.7	0.981
4	21.5	55.5	89.0	145.5	58.5	0.991
5	14.0	56.5	80.0	138.0	53.9	0.989
Mean	17.3	50.8	78.6	132.6		
SD	2.9	5.0	6.4	11.8		

[a]Area = m (ln concentration) + b.

a quantitative answer. TLC bioautography is a simple and elegant technique that can be applied in either mode to meet both of these assay needs. With proper attention to sample purification, quantitative TLC principles, and measurement of antibiotic response, TLC bioautography can give quantitative results comparable to other methods of residue assay.

ACKNOWLEDGMENTS

The authors wish to thank J. W. Moran, J. E. Rea, and D. E. Ruggles for technical assistance.

REFERENCES

1. V. Betina, J. Chromatogr., 78, 41 (1973).
2. A. L. Donoho and R. M. Kline, Antimicrobial Agents and Chemotherapy, 763 (1967).
3. R. M. Kline and T. Golab, J. Chromatogr., 18, 409 (1965).

4. A. McDonald, G. Chen, P. Duke, A. Popick, R. Saperstein, M. Kakaty, C. Crowley, H. Hutchinson, and J. Westheimer, Densitometry in Thin-Layer Chromatography, J. C. Touchstone and J. Sherma (Eds.), Wiley-Interscience Publishers, 1979, p. 201.
5. J. K. Pauncz and I. Harsanyi, J. Chromatogr., 195, 251 (1980).
6. T. F. Brodasky, C. Lewis, and T. E. Eble, J. Chromatogr., 123, 33 (1976).
7. W. J. Begue and R. M. Kline, J. Chromatogr., 64, 182 (1972).
8. J. Okada, I. Higuchi, and S. Kondo, J. Food Hygienic Soc. Japan, 21, 177 (1980).
9. A. H. Thomas and J. M. Thomas, J. Chromatogr., 195, 297 (1980).

CHAPTER 39

Quantitative TLC of Cholinesterase-Inhibiting Pesticides Residues by Reflective Scanning

Joseph Sherma and Melvin E. Getz

INTRODUCTION

Direct quantitation is being used to an increasing degree as a primary method of analysis for residues of pesticides and their metabolites (1-4). Excellent accuracy and precision can be obtained using procedures that have been described previously (1, 5, 6). Most detection methods are capable only of detecting 100 to 500 ng of pesticide as a lower limit. An exception is the measurement of zones that exhibit natural fluorescence or can be made to fluoresce by reaction with a fluorogenic reagent, but in many cases a rigorous cleanup procedure is necessary in order to use this approach. One way to selectively enhance the sensitivity limits of thin layer chromatography (TLC) of pesticides is to use enzymatic methods for detection. Enzymes act as biological amplifiers and allow very small amounts to be detected.

Many cholinesterase-inhibiting pesticides have replaced the persistent organochlorine compounds. The inhibiting properties of these pesticides have been used in the past for their quantitative detection by colorimetry (7) and qualitative determination by paper and thin layer chromatography (8). One method that produced blue spots on a yellow background on paper (9) could not be used with inorganic thin layer adsorbents.

Indoxyl acetate (10) and some of its derivatives have been widely applied as a substrate for detecting pesticide residues on silica gel TLC plates. The inhibitors are visualized as light or white spots against a darker blue background. One drawback of this substrate is that it is not specific for cholinesterase, being hydrolyzed by other enzymes such as carboxyesterase and alliesterases. In addition, densitometry based on the scanning of white spots has not been advantageous.

It was preferred, therefore, to produce a colored spot on a white or light background because of improved results when scanning such spots by densitometry. A substrate such as acetylcholine or acetylthiocholine with serum as the source of enzyme was expected to be less susceptible to the formation of artifact inhibition spots and uncontrollable hydrolysis of the substrate. After preliminary studies, it was decided to use acetylthiocholine as the substrate and horse serum as the source of cholinesterase. Horse serum is an ideal source for this purpose since it can be obtained relatively inexpensively in sealed ampoules that have a long storage life under refrigeration.

Since thiocholine is one of the hydrolysis products of acetylthiocholine and is a good reducing agent, its use in conjunction with a redox dye afforded good possibilities for color reversal. As it turned out, a dye commonly used for the colorimetric determination of ascorbic acid proved to be satisfactory. The dye is 2,6-dichloroindophenol, which is reduced to its colorless form by the thiocholine. With this detection system, the zones of inhibition are represented by the unreduced dye, which is blue or bluish purple.

The reagents were used by two modes of application, spraying all the reagents onto the TLC plate or dipping the plate into reagent solutions. Comparisons were made between the use of indoxyl acetate and acetylthiocholine as the substrates. The investigation showed that low nanogram amounts of cholinesterase-inhibiting residues can be quantitated by the combination of the enzyme-inhibition technique and reflectance scanning of the visualized spots.

MATERIALS

Apparatus

Rectangular glass development tanks for 20 × 20 cm plates.
Reagent spray flasks, for example, Kontes K-422500.
Dip tank, Thomas Mitchell, available from Arthur H. Thomas.
Automatic spotter, Chromaflex (Kontes K-416330).
Reflectance scanner, Chromaflex (Kontes K-495000).

Silica gel precoated plates, Whatman K-5 or K-6.

Reagents

Pesticide standards, obtained from government, industrial, and commercial sources.
Horse serum, Diffco 10-ml vials.
Acetylcholinesterase, purified, stabilized, from bovine erythrocytes, Winthrop Laboratories.
Acetylthiocholine iodide, Eastman No. 10587.
Indoxyl acetate, Allied Chemical Company, No. 718.
2,6-Dichloroindophenol sodium salt, Fisher No. S-286.
Bromine, reagent grade.
Tris[2-amino-2(hydroxymethyl)-1,3-propanediol], Eastman No. P-4833.
Pentane, hexane, benzene, ethanol, methanol, acetone, chloroform, pesticide quality.
Ethanol, USP absolute.
Equivalent equipment and reagents can be used in place of any of the above.

METHODS

TLC Procedure

Extracts of food products and other substrates can be cleaned up by any suitable procedure (11). The extracts analyzed in this investigation were purified by column adsorption chromatography utilizing a macroporous silica gel (12).

Thin layer plates were used as received with no pretreatment. Standards and samples were applied onto the layer with the Kontes automatic spotter (13). The same solvent and volume were used for samples and standards; up to 2 ml can be applied. All volumes were transferred to the spotter with volumetric pipettes.

The metabolites of the organophosphate pesticide phorate (Table I) were used in determining standard curves and for fortification of green bean extracts. Carbofuran and its metabolites (Table I) were used to illustrate the inhibition properties of the carbamate family of pesticides. Two mobile phases were used for the resolution of the phorate metabolites.

Benzene/Ethanol (85:15 v/v)

When this system was used, the tank was completely lined with filter paper and presaturated for 15 to 30 minutes with

TABLE I. COMMON NAMES AND CHEMICAL NAMES OF PESTICIDES

Pesticide	Chemical Name
Phorate sulfoxide	0,0-diethyl S-[(ethylsulfinyl)methyl] phosphorodithioate
Phorate sulfone	0,0-diethyl S-[(ethylsulfonyl)methyl] phosphorodithioate
Phoratoxon sulfoxide (phorate thiol sulfoxide)	0,0-diethyl S-[(ethylsulfinyl)methyl] phosphorothioate
Phoratoxon sulfone (phorate thiol sulfone)	0,0-diethyl S-[(ethylsulfonyl)methyl] phosphorothioate
Azinphosmethyl (Guthion)	0,0-dimethyl S-[4-oxo-1,2,3-benzo-triazin-3(4H-yl)methyl] phosphorodithioate
Carbofuran	2,3-dihydro-2,2-dimethylbenzofuranyl-7-N-methylcarbamate
3-Keto-carbofuran	3-keto-2,2-dimethylbenzofuranyl-7-N-methylcarbamate
3-Hydroxy-carbofuran	3-hydroxy-2,2-dimethylbenzofuranyl-7-N-methylcarbamate
Carbaryl	1-naphthyl methylcarbamate
Aldicarb	2- ethyl-2-(methylthio)proprionaldehyde O-(methylcarbamoyl) oxime

the developing solvent before the TLC plate was placed in the tank for development.

Pentane/Acetone/Methanol/Chloroform (70:15:10:5 v/v)

Twenty percent methanol will give a higher R_f value for phorate thiol sulfoxide, which is better for scanning. When this system was used, a thick sheet of 20 × 20 cm chromatographic or filter paper was immersed into the developing solution along with the plate, but on the opposite side of the tank facing the adsorbent surface. No preequilibration period was employed; the paper and plate were introduced immediately after the solvent. Hexane/acetone/chloroform

(70:25:5 v/v) was used for chromatography of carbofuran and its metabolites in an unsaturated tank.

The chromatograms were developed for a distance of at least 10 cm from the origin, which was located 2.5 cm up from the bottom of the plate. After development, the plates were dried in a forced draft oven at 50°C for 20 minutes.

Application of Detection Reagents

Before organophosphate compounds containing P=S groups will react satisfactorily with the enzyme, they must be converted to P=O. This can be done by reacting the compounds in an atmosphere of bromine vapor. Compounds already containing P=O react directly. Place 0.2 ml of liquid bromine in a beaker on the bottom of a glass tank, cover the tank, and allow to react for 30 seconds with the chromatogram. Take out the plate and heat in a forced draft oven at 50°C for 15 minutes. The plate is now ready for treatment with the enzyme.

Spray the plate uniformly with the enzyme solution (20 ml of the horse serum for each 35 ml of H_2O), using the pattern as exhibited by Figure 1. Spraying is continued until the plate is moist but not dripping. Incubate the sprayed plate in a tank containing a water-saturated atmosphere for 30 minutes. When incubation is complete, dip the plate into acetylthiocholine solution (1.0 g of ATC + 0.6 g of Tris in 20 ml of H_2O + 80 ml of acetone) for 5 seconds. Incubate in the water-saturated atmosphere tank for 15 minutes. Serum and ATC solutions are stored in a refrigerator and brought to room temperature before use.

Heat in the oven at 50°C for 15 minutes, and dip into a 0.05% solution of 2,6-dichloroindophenol in acetone/water (9:1 v/v) for 10 seconds. The background should slowly fade, leaving relatively stable dark-blue spots on a white or light-blue background. If the spots should fade too quickly, redip into the dye solution for several more seconds. If the background does not fade completely, the 10-second dip time is too long. The dip time and concentration of the dye should be tested and changed to accomodate local variations in application of the serum and ATC solutions, type of plates used, the batch of serum obtained from the manufacturer, and other conditions. Scan at once and store the plates covered with glass in a refrigerator. This storage preserves the spots for several days.

When indoxyl acetate (0.5% in acetone) is used as the substrate, the plate is dipped into it for 5 seconds after incubation with the enzyme. The background will slowly turn blue, and white

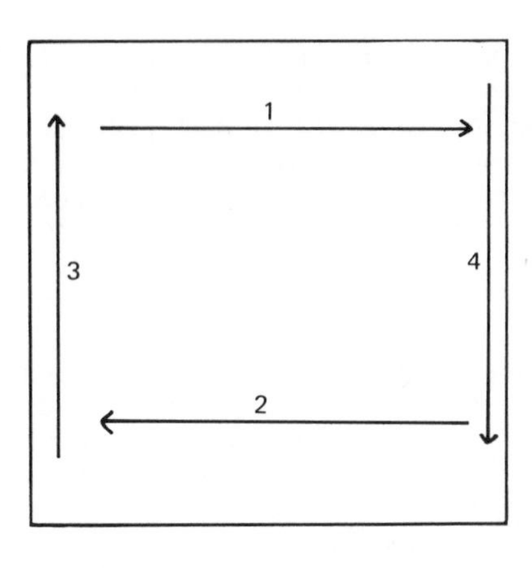

Figure 1. Spray pattern for application of reagents to TLC plates.

spots will form where inhibition has taken place. The chromatogram is scanned after 20 to 30 minutes.

All reagents for both methods can be sprayed by using the pattern exhibited in Figure 1. ATC and indoxyl acetate solutions are sprayed twice in each of the four directions. The indophenol solution for spraying is changed to 0.1% in acetone/water (95:5 v/v), and drying after incubation with ATC is not required. The dye is sprayed evenly until stable blue spots remain on a light background.

Measurement by Scanning

Spots from both methods were scanned in the visible-reflectance mode with the Kontes fiber optics scanner, using the baseline corrector in most cases. A red source filter (Tiffin Photar No. 25) was used to increase selectivity for the indophenol method and a blue-green filter (Roscolene No. 877) for the indoxyl acetate method. The scan speed was 10 cm/second.

Recorder charts were Xeroxed, and peaks were cut and weighed to determine areas. Calibration curves were prepared by plotting weight times attenuation setting versus nanograms spotted. Three bracketing standards were included on the same plate with samples to reestablish the calibration curve, which will vary slightly from day to day. A working range of 4 to 16 ng is convenient owing to the linearity of calibration curves at lower concentrations (Figure 2 through 6).

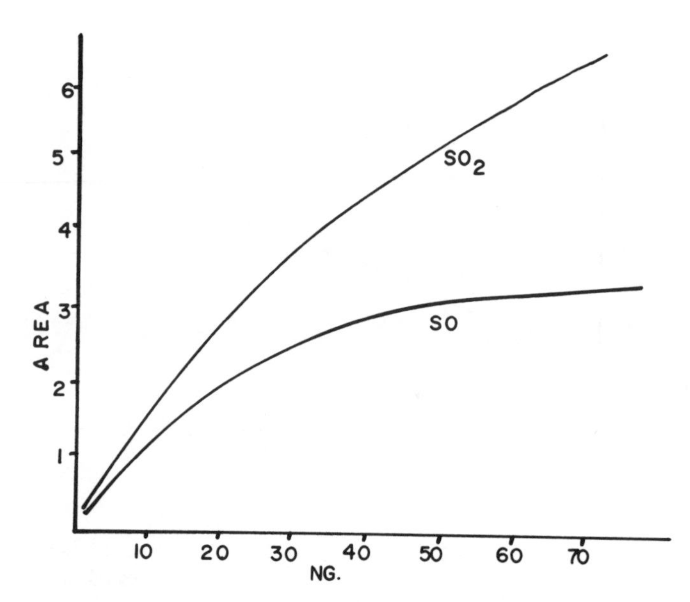

Figure 2. Calibration curves for phorate thiol sulfoxide (SO_2) and sulfone (SO) by the ATC-indophenol method with all reagents sprayed

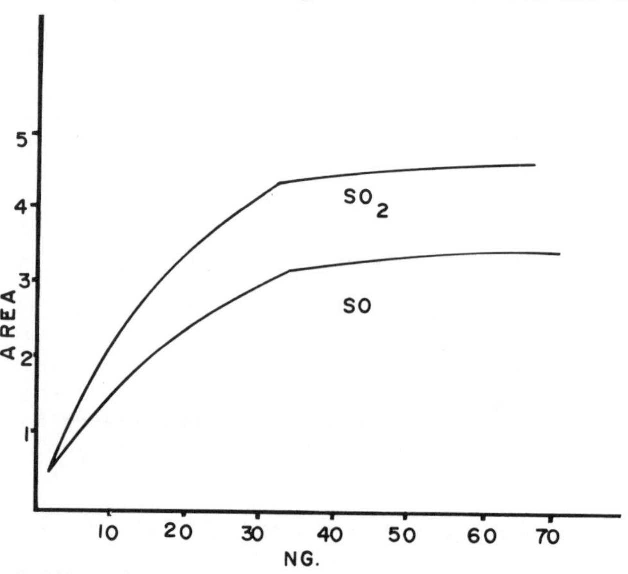

Figure 3. Calibration curves for phorate sulfoxide and sulfone after bromine oxidation.

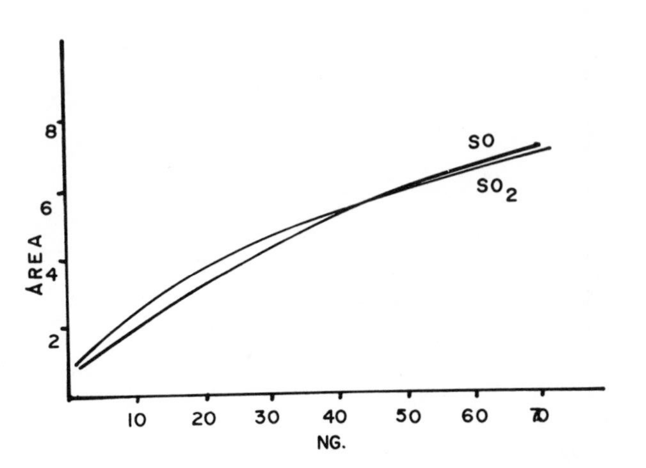

Figure 4. Calibration curves for phorate thiol sulfoxide and sulfone by dipping the plate into ATC and indophenol reagents.

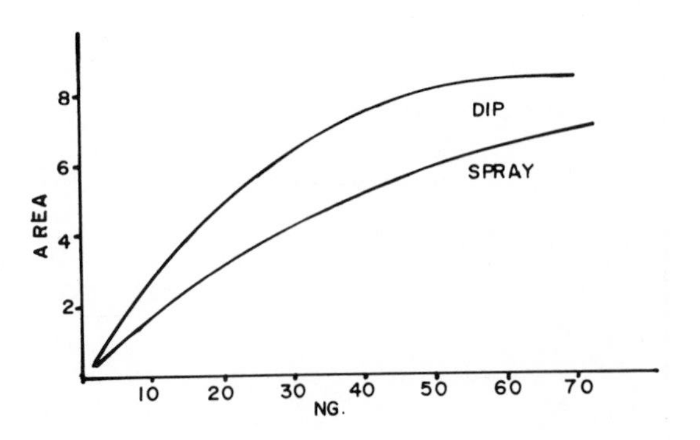

Figure 5. Calibration curves for the indoxyl acetate method by dipping and spraying the subtrate

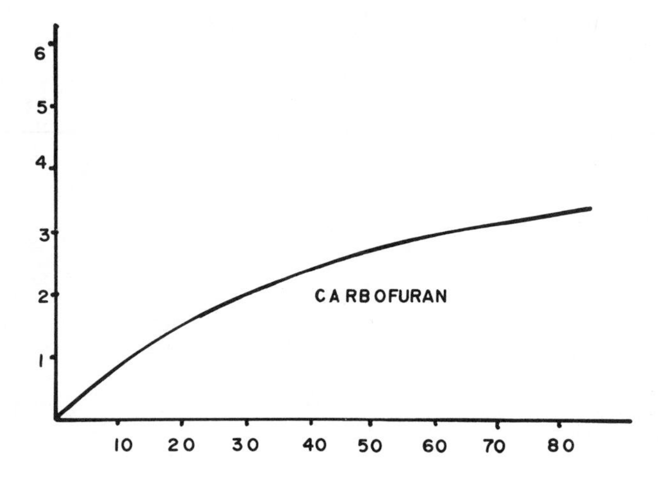

Figure 6. Calibration curve for carbofuran by the ATC-indophenol dip procedure (relative area units versus nanograms spotted).

RESULTS AND DISCUSSION

Figure 2 shows standard curves obtained for phorate thiol sulfoxide and thiol sulfone with the ATC procedure. All reagents were sprayed onto the plate. The concentration range was 2 to 64 ng per spot.

Figure 3 represents standard curves for 2 to 64 ng phorate sulfone and sulfoxide after chromatography and bromine treatment, with ATC as the substrate. No spots were obtained without the oxidation step.

Figure 4 gives standard curves for 2 to 64 ng phorate thiol sulfoxide and sulfone when reagents for the indophenol method were dipped.

Figure 5 compares results for phorate thiol sulfoxide from 2 to 64 ng with the indoxyl acetate procedure when the reagents were sprayed and dipped. The greater sensitivity afforded by dipping is evidenced by the greater slope of the calibration curve.

Figure 6 shows a standard curve for carbofuran using the indophenol method. The concentration range is 5 to 80 ng per spot. The sensitivity of detection for carbofuran is roughly the same as for phorate thiol sulfone, and the shape of the calibration curve is also similar.

Figure 7 is a photograph of a chromatogram containing 0.5 to 16 ng of 3-hydroxy carbofuran. Purified acetylcholinesterase (50 ng/100 ml water) was used in place of serum for the enzyme spray, increasing the sensitivity of the method. The calibration curve was linear over this range of concentrations.

Green bean extracts were fortified with phorate thiol sulfone at the 0.12-ppm level (0.12 ng/mg). One hundred milligrams of extract was cleaned up with the SI-200 silica gel cleanup method (12). The recovery by the indophenol method was 0.14 ng/mg, or 111%. A second extract was fortified and cleaned up, and the recovery was 0.12 ng/mg, or 100%.

Three standard solutions were prepared at 5 ng/ml, 8 ng/ml, and another 8 ng/ml and were given to an analyst as unknowns. The error in the value reported was 10% for the 5-ng, 3.8% for one 8-ng, and 2.5% for the other 8-ng sample.

Six standards of the same concentrations were spotted and developed on the same plate, detected by the indophenol dip method, and scanned. Various concentrations from 4 to 64 ng per spot levels were checked. There was no consistent pattern for the high or low values. The precision (relative standard deviation of the areas for six replicates at different levels varied from 2 to 12%, which is quite satisfactory at such low levels. The average precision of the indophenol spray method with the added step of bromine oxidation for six standards of the same concentration on one plate, repeated for 2-, 4-, and 8-ng levels of the organophosphorous pesticide azinphosmethyl both at the origin and after development, was 10.5%.

Results obtained by spraying or dipping the ATC and indophenol solutions are very similar when both methods are optimized. The advantage of spraying is that application of the dye is more easily controlled in order to produce the best contrast between the spots and background. Because of differences in reagents (e.g., the strength of enzyme in different batches of serum) and conditions such as temperature and humidity in each laboratory, the best dip time and strength of the dye solution should be determined if the dip procedure is preferred.

The indophenol procedure is considered superior to indoxyl acetate for several reasons. Sensitivity of detection was approximately four times as great in terms of peak area per nanogram of pesticide, and colored spots on a white background were more successfully scanned than white spots on a colored background. The indophenol detection method also appeared more selective toward pesticides compared to impurities.

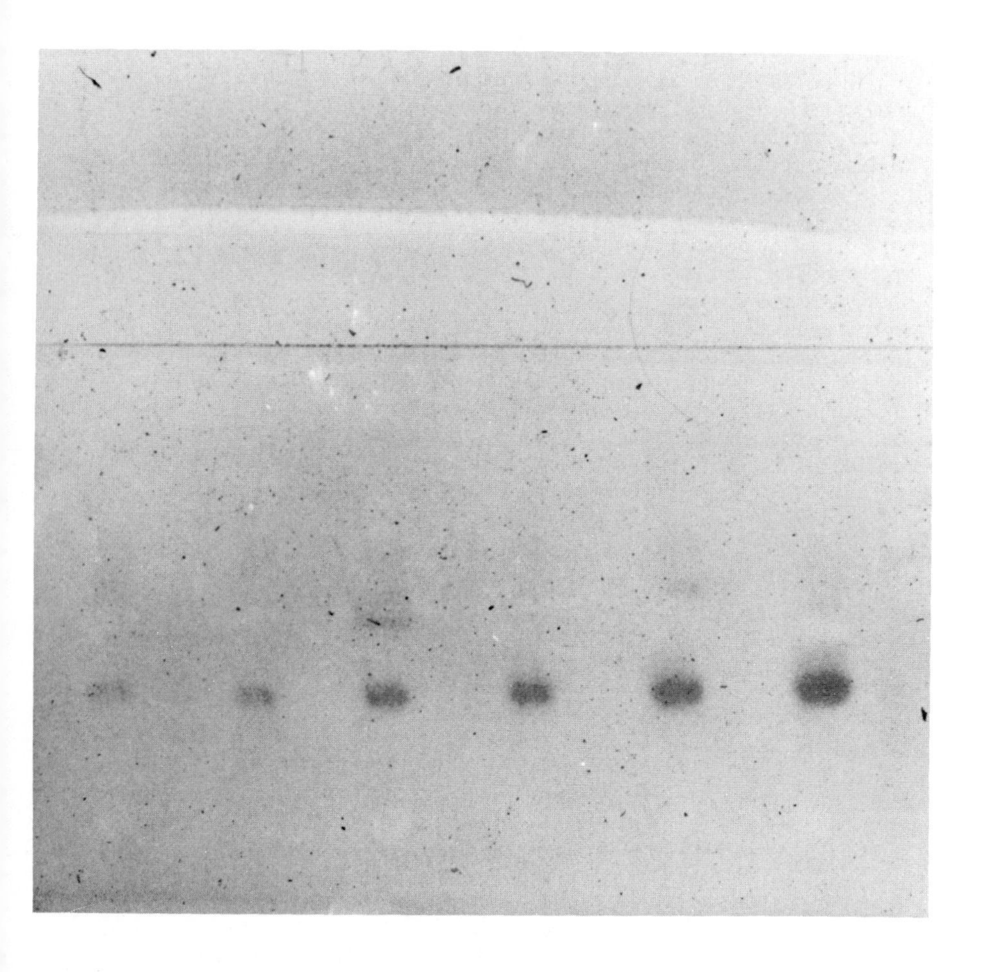

Figure 7. Photograph of chromatogram containing 0.5-, 1-, 2-, 4-, 8-, and 16-ng spots (left to right) of 3-hydroxy carbofuran after development on a Whatman K-6 layer with the hexane/acetone/chloroform mobile phase. Detection was by the ATC-indophenol spray procedure, with acetylcholinesterase in place of horse serum.

SUMMARY AND CONCLUSIONS

Quantitative TLC should be considered along with gas chromatography and HPLC as a primary determinative method for pesticide residue analysis. If analysts carry out all the manipulations for TLC in as precise and quantitative manner as for the other two chromatographic methods, all three should give comparable results.

Since organophosphate and carbamate pesticides are being widely used to replace the more persistent pesticides, it is beneficial to have a quantitative method that would be relatively inexpensive and oriented toward technician use. The results shown are sufficiently sensitive, accurate, and precise for routine residue analyses. Further studies on the application and improvement of the indophenol method have been carried out, and preliminary results indicate that the procedure is applicable to the quantitation of other carbofuran metabolites (e.g., 3-keto) and inhibiting pesticides, such as carbaryl and aldicarb and its metabolites, with sensitivity limits and reproducibilities comparable to those cited in this chapter.

REFERENCES

1. J. Sherma, in Analytical Methods for Pesticides and Plant Growth Regulators, G. Zweig and J. Sherma (Eds.), Academic Press, New York, 1980, Chapter 3, p. 79.
2. J. Harvey, Jr., and G. Zweig (Eds.), Pesticide Analytical Methodology, ACS Symposium Series 136, American Chemical Society, Washington, D.C., 1980, Chapters 8, 9, and 14.
3. G. Zweig and J. Sherma, Anal. Chem., 52, 276R (1980).
4. J. Sherma and G. Zweig, Anal. Chem., 53, 77R (1981).
5. J. C. Touchstone and J. Sherma (Eds.), Densitometry in Thin Layer Chromatography, Wiley-Interscience, New York, 1979.
6. M. E. Getz, in Methods in Residue Analysis, Volume IV, Gordon and Breach, New York, 1971, p. 43.
7. R. L. Metcalf, J. Econ. Entomol., 44, 883 (1951).
8. C. E. Mendoza, Residue Rev., 50, 43 (1974).
9. M. E. Getz and S. J. Friedman, J. Assoc. Off. Agr. Chem., 46, 707 (1963).
10. C. E. Mendoza and W. P. McKinley, Analyst, 93, 34 (1968).
11. J. Sherma, in Thin Layer Chromatography--Quantitative Environmental and Clinical Applications, J. C. Touchstone and D. Rogers (Eds.), Wiley-Interscience, New York, 1980, Chapter 3, p. 17.
12. M. E. Getz, Talanta, 22, 935 (1975).
13. M. E. Getz, J. Assoc. Off. Anal. Chem., 54, 982 (1971).

CHAPTER 40

Analysis of Trichothecenes by TLC

Michael E. Stack and Robert M. Eppley

INTRODUCTION

The trichothecenes are a large group of fungal metabolites produced by various species of Fusaria, Myrothecia, Stachybotrys, Verticimonosporium, Cylindrocarpon, Trichoderma, and Trichothecia (1). A number of trichothecene derivatives have been isolated from the plant, _Baccharis megapotamica_; however, these may also be of fungal origin (2). All of the more than 50 derivatives reported in the literature share the same sesquiterpenoid ring system shown in Figure 1.

The trichothecenes have been implicated as the causative agents in numerous farm animal toxicoses (1). Among those causing the more notable incidents are alimentary toxic aleukia, stachybotryotoxicosis, and dendrodochiotoxicosis in the Soviet Union; red mold disease in Japan; and moldy corn toxicosis in the United States. "Scabby" cereal grains are caused by the presence of Fusarium molds, and in any large-scale occurrence of scabby grains, one or more of the trichothecenes will probably be found to occur in the contaminated cereal.

The observed symptoms in animals vary considerably, but can include feed refusal, vomition, severe dermatitis, bloody diarrhea, decreased body weight, extensive hemorrhaging, and death.

The trichothecenes have been implicated in the two major epidemics affecting large numbers of people and farm animals in the Soviet Union before and during World War II. An epidemic of

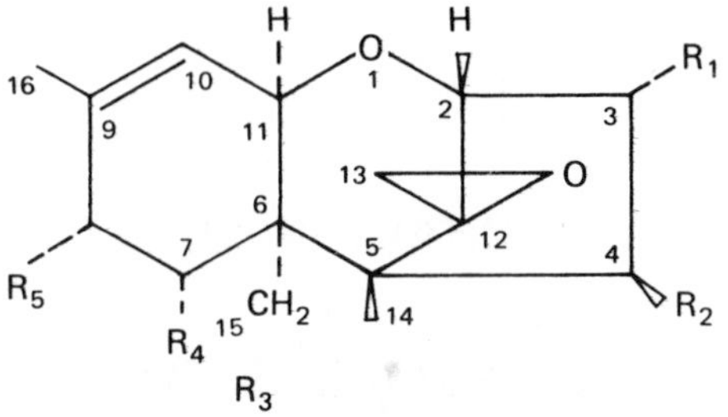

Figure 1. Typical sesquiterpenoid ring system of trichothecenes as exhibited in 12,13-epoxytrichothec-9-ene.

alimentary toxic aleukia (ATA) occurred in the Soviet Union during World War II when cereal grains were left in the field over the winter because of insufficient labor for the fall harvest. The cereals were salvaged in the spring for human food after they had spent the winter in the fields under the snow. Those conditions are now known to be nearly ideal for the production of trichothecenes by some species of Fusarium. The molds isolated from the grains have been shown to produce large quantities of T-2 toxin and related derivatives (Figure 2) under laboratory conditions. Thousands of people were severely affected, and many were reported to have died (1).

The other epidemic, stachybotryotoxicosis, was a severe blow to Soviet farm production in the 1930s; it affected large numbers of work horses and handlers. The Soviet researchers quickly identified the problem as moldy feed and bedding and learned that the mold, Stachybotrys alternans, produced a strong toxin only identified as "stachybotryotoxin." Recently, stachybotryotoxin has been identified as a series of macrocyclic di- and trilactonic derivatives of the trichothecene, verrucarol (Figure 3). Roridin E and verrucarin J had previously been identified as metabolites of Myrothecium species. Satratoxins F, G, and H were new mycotoxins characterized in our laboratory (3-5).

Several of the trichothecenes are highly cytotoxic and inhibit protein and DNA synthesis. Those that have been tested (T-2 toxin, fusarenon-X) were inactive in mutagenicity tests using Bacillus subtilis and Salmonella typhimurium (6). A recent publication (7) has reported the formation of tumors (both benign and malignant) in white rats orally exposed to three to five doses of T-2

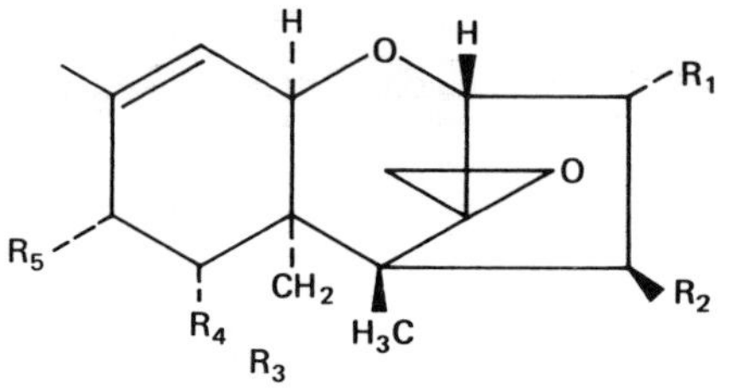

Figure 2. T-2 toxins and derivatives found in grains left under snow all winter: diacetoxyscirpenol: R_1 = OH, R_2 = OAc, R_4 = H, R_5 = H; T-2 toxin: R_1 = OH, R_2 = OAc, R_3 = OAc, R_4 = H, R_5 = OC(O)$CH_2CH(CH_3)_2$; HT-2 toxin: R_1 = OH, R_2 = OH, R_3 = OAc, R_4 = H, R_5 = OC(O)$CH_2CH(CH_3)_2$; neosolaniol: R_1 = OH, R_2 = OAc, R_3 = OAc, R_4 = H, R_5 = OH.

toxin at levels of 0.2 to 4 mg/kg of body weight. Cardiovascular lesions and tumors of the digestive tract and brain were observed in those rats that survived 12 to 27 months. The principal investigator in this study suggested that the Fusarium toxins, especially the trichothecenes and possibly the zearalenones, could contribute to the development of tumors of the digestive tract in humans as well as animals.

Although the molds capable of producing trichothecenes have been recognized as the cause of many toxicoses, few analytical methods have been developed for this large and important group of mold metabolites. Several methods have been published for the detection and analysis of a few of the trichothecenes. To date, most of the analytical interest has been directed toward seven of the trichothecenes: T-2 toxin, HT-2 toxin, neosolaniol, diacetoxyscirpenol (DAS), fusarenon-X, nivalenol, and deoxynivalenol (or vomitoxin). Until recently, T-2 toxin and the associated derivatives--HT-2 toxin, neosolaniol, and T-2 tetraol--were the most frequently detected trichothecenes. Fusarenon-X and nivalenol are reported in the Japanese literature, but have not been reported in other countries. Deoxynivalenol or vomitoxin has been occasionally reported in corn samples from the midwestern United States and South Africa (8); however, vomitoxin has been detected in nearly all the samples of wheat and barley harvested in the fall of 1980 from areas of Canada that had unusually wet conditions during the harvest season (9).

A variety of solvents have been used for the extraction of the trichothecenes from grains. With the large differences in

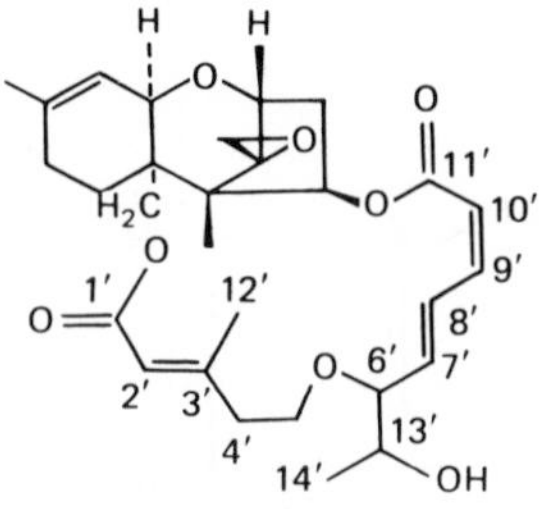

Roridin E
(a)

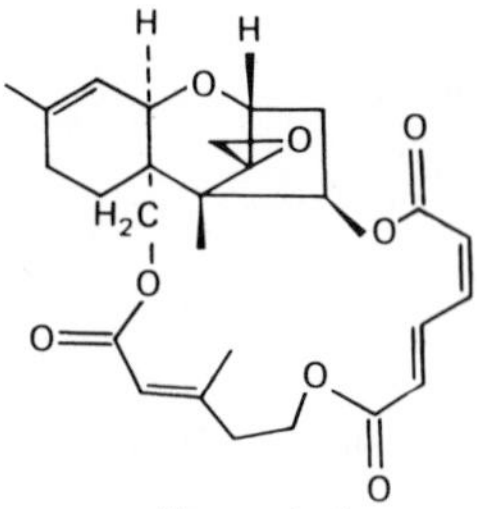

Verrucarin J
(b)

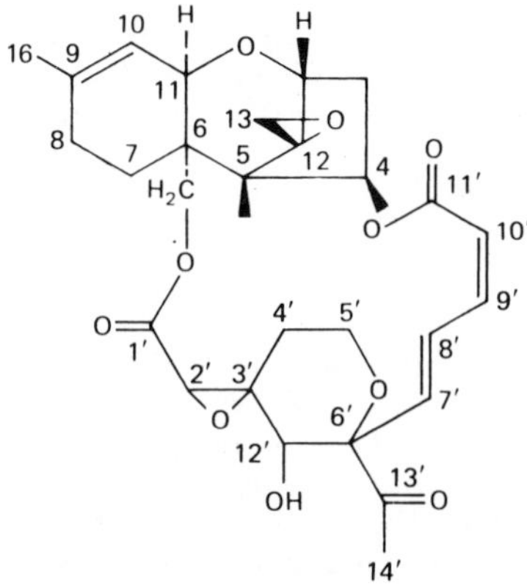

Satratoxin F
(c)

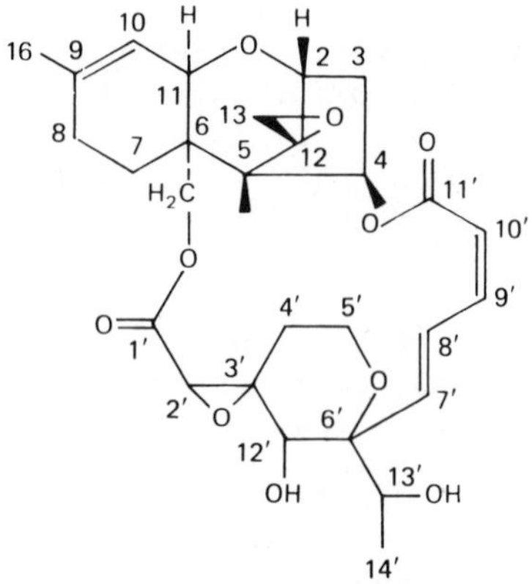

Satratoxin G
(d)

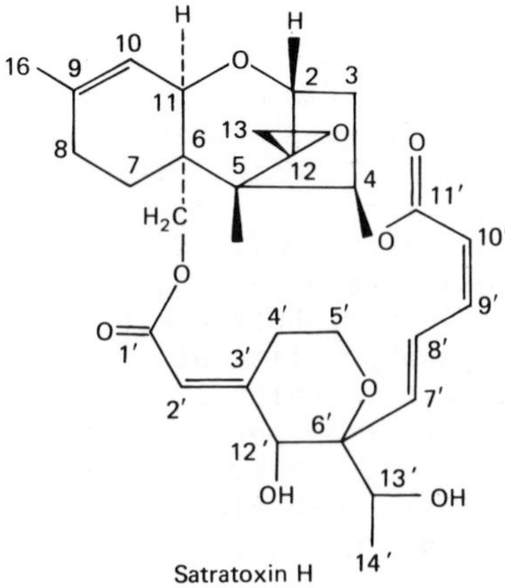

Satratoxin H
(e)

Figure 3. Stachybotryotoxin: a series of macrocyclic di- and trilactonic derivatives of verrucarol.

polarity among these compounds, a mixture of methanol and water has been found to be the most effective (10-14). The ratio of methanol to water has been varied from 50:50 to 95:5. The aqueous methanol extracts are concentrated with vacuum and moderate heat, and the crude extract is then transferred to chloroform or ethyl acetate for additional cleanup. To detect the presence of the trichothecenes extracted from most feed or food samples, some type of additional cleanup is needed. Open-column silica gel chromatography has traditionally been used to clean up mycotoxin extracts; considerable deactivation with water was required, however, to remove the polyhydroxy trichothecene derivatives such as nivalenol, deoxynivalenol, and T-2 tetraol. Unfortunately, deactivation of the silica gel also causes considerable quantities of interferants to elute along with the trichothecenes (10). Recently, several methods have included the use of ion-exchange resins for the initial cleanup of the aqueous methanol extract. Amberlite XAD-2 or XAD-4 eluted with methanol followed by a Florisil column eluted with chloroform-methanol has been used successfully to clean up cereal grain extracts sufficiently for both TLC and gas-liquid chromatographic (GLC) detection of several different trichothecene derivatives (11).

A number of TLC solvent systems have been reported for the separation of the various trichothecenes from each other and from extract components (10, 11, 14-16). Most of these are combinations of a relatively nonpolar solvent (chloroform, benzene, or toluene) mixed with an alcohol or acetone. Typical systems used in our laboratory are acetone/chloroform (5:95) on Adsorbosil 1 (Applied Science) plates, and isopropanol/chloroform (2:98) on SilicAR 7-GF (Mallinckrodt) plates, for the mono- and dihydroxy trichothecenes (T-2 toxin, HT-2 toxin, and DAS). The more polar trichothecene derivatives (nivalenol, deoxynivalenol fusarenon-X, and T-2 tetraol) are separated by use of increased amounts of the more polar solvent, that is, isopropanol/chloroform (5:95).

Most of the TLC detection procedures for the trichothecenes require the use of a spray reagent for visualization and detection. In general, TLC procedures have detection limits ranging from 0.1 μg/spot for T-2 toxin to 1 or 2 μg/spot for fusarenon-X (10), whereas GLC procedures have considerably lower detection limits, that is, 0.1 to 0.2 ng/injection for T-2 toxin (12). A major advantage of TLC is that a relatively large number of samples can be screened by TLC techniques tailored to detect the trichothecenes. As a group, the trichothecenes are particularly difficult to detect by any one analytical technique. The satratoxins, roridins, and verrucarins have a butadiene chromophore that enables their detection on fluorescent background TLC plates. The lack of UV-absorbing or fluorescent chromophores in the other trichothecenes limits the use of TLC spray reagents to those that alter the structure significantly, that is, charring reagents such as H_2SO_4. T-2 toxin and related derivatives form a blue fluorescent spot with H_2SO_4-containing sprays when they are heated at 110 to 120° for 5 to 10 minutes. Generally 20 to 50% H_2SO_4 in MeOH is used, although some workers find that the addition of acetic acid and p-anisaldehyde gives a more intense blue fluorescence (17). DAS gives a very light blue spot with the H_2SO_4 reagents. Ehrlich's reagent (p-dimethylaminobenzaldehyde in concentrated HCl) has been reported to give an intense violet color with DAS but not with T-2 toxin. This test can be used to distinguish between the two toxins, since both have similar R_f values in many of the TLC solvent systems (13).

Those members of the nivalenol group of trichothecenes that contain a carbonyl function in conjugation with the double bond do not form fluorescent products with the H_2SO_4 spray reagents; instead, nivalenol, deoxynivalenol, fusarenon-X, and the various related ester derivatives all give a dark pink to brown spot. A more useful spray reagent for the nivalenol group is 4-(p-nitrobenzyl)-pyridine (NBP), followed by tetraethylenepentamine (TEPA) (14). T-2 toxin, fusarenon-X, deoxynivalenol, and structurally related trichothecenes give a blue spot on a white background.

Two different TLC procedures have been developed for these trichothecenes, one for detection and the other for quantitative analysis. In the detection procedure, the TLC plate is first sprayed with a 1% solution of NBP, air dried, heated for 30 minutes at 150°, cooled, and sprayed with a 10% solution of TEPA. For quantitation, the same procedure is followed except that the initial step is changed to dipping the plate in a 3% solution of NBP. The colors are more stable after the dipping procedure; they last about 1 hr before fading. Quantitative measurements were linear from 0.1 to 10 μg/spot for both the T-2 toxin and the nivalenol groups. The detection limit ranged from 0.02 μg/spot for HT-2 toxin to 0.2 μg/spot for DAS.

Aluminum chloride is another spray reagent that is very useful for the nivalenol group of derivatives (15). When TLC plates are sprayed with a 20 to 40% alcoholic solution of $AlCl_3$ and heated for 10 minutes at 80 to 110°, a blue fluorescent spot is observed. The T-2 toxin group also shows some blue fluorescence with $AlCl_3$ under long-wave UV light; however, the plate should be sprayed a second time with one of the H_2SO_4 reagents to bring out the blue fluorescence of the T-2 toxin group.

A number of other spray reagents have been suggested (1, 10, 16, 17) and used to detect the trichothecenes, but none of them has been found to be as effective as the H_2SO_4, $AlCl_3$, and 4-(p-nitrobenzyl)-pyridine systems.

It is recommended that the developed TLC plate first be sprayed with the $AlCl_3$ reagent, heated, and observed to detect the nivalenol derivatives, then sprayed with an H_2SO_4 reagent, and heated again to detect the T-2 toxin derivatives. When a sample is suspected of containing one or more of these trichothecenes, a second TLC plate should be developed, preferably with another mobile phase and sprayed with the 4-(p-nitrobenzyl)-pyridine-tetraethylenepentamine reagents. Unfortunately, none of these spray reagents is particularly sensitive for the detection of DAS and related derivatives. As with any TLC procedures, standards of the trichothecenes should be included on all plates.

The multiple development of TLC plates with different solvent systems and the use of different spray reagents does not give satisfactory confirmation of the presence of trichothecenes in a sample. Any suspected trichothecene contamination should be further confirmed either by preparatory TLC followed by mass spectrometry (MS) or by combined GLC-MS. The successful use of preparatory TLC for mass spectrometry requires that a standard be placed on the TLC plate in such a way that the developed sample spots can be covered with a clean glass plate while the spray reagent is used to visualize the standard. Chemical ionization MS usually gives better spectra for confirmation by either direct probe or combined GLC-MS. The underivatized trichothecenes give weak or no molecular ions with electron-impact MS.

In conclusion, there are several adequate TLC procedures and spray reagents for the detection and estimation of a few of the trichothecenes. Considerable research is still needed to improve the isolation and cleanup procedures for the various trichothecene derivatives, particularly in those commodities that can be contaminated with the Fusarium molds. The search for more specific and more sensitive detection reagents for the trichothecenes as a group should be continued.

REFERENCES

1. Y. Ueno, in Mycotoxins in Human and Animal Health, J. V. Rodricks, C. W. Hesseltine, and M. A. Mehlman (Eds.), Pathotox Publishers, Park Forest South, Ill., 1977, p.189.
2. S. M. Kupchan, D. R. Streelman, B. B. Jarvis, R. G. Dailey, Jr., and A. T. Sneden, J. Org. Chem., 42, 4221 (1977).

3. R. M. Eppley and W. J. Bailey, Science, 181, 758 (1973).
4. R. M. Eppley, E. P. Mazzola, R. J. Highet, and W. J. Bailey, J. Org. Chem., 42, 240 (1977).
5. R. M. Eppley, E. P. Mazzola, M. E. Stack, and P. A. Dreifuss, J. Org. Chem., 45 2522 (1980).
6. Y. Ueno and K. Kubota, Cancer Res., 36 445 (1976).
7. R. Schoental, A. Z. Joffe, and B. Yagen, Cancer Res., 39, 2179 (1979).
8. W. F. W. Marasas, S. J. Van Rensburg, and C. J. Mirocha, J. Agric. Food Chem., 27, 1108 (1979).
9. P. Scott, Canadian Health Protection Branch, personal communication, 1980.
10. R. M. Eppley, J. Am. Oil Chem. Soc., 56, 824 (1979).
11. H. Kamimura, M. Nishijima, K. Saito, S. Takahashi, A. Ibe, S. Ochitai, and Y. Naoi, J. Food Hyg. Soc. Jpn., 19, 443 (1978).
12. T. R. Romer, T. M. Boling, and J. L. MacDonald, J. Assoc. Offic. Anal. Chem., 61, 801 (1978).
13. J. R. Bamburg, Clin. Toxicol., 5, 495 (1972).
14. S. Takitani, Y. Asabe, T. Kato, M. Suzuki, and Y. Ueno, J. Chromatogr., 172, 335 (1979).
15. Y. Takeda, E. Isohata, R. Amano, and M. Uchiyama, J. Assoc. Offic. Anal. Chem., 53, 573 (1979).
16. A. Gimeno, J. Assoc. Offic. Anal. Chem., 62, 579 (1979).
17. P. M. Scott, J. W. Lawrence, and W. van Walbeek, Appl. Microbiol., 25, 699 (1973).

CHAPTER 41

Analysis of Ethoxylated Nonionic Surfactants

Gary E. Stolzenberg

INTRODUCTION

Commercial ethoxylated nonionic surfactants are produced by reacting a hydrophobic species with ethylene oxide (EO). They are inherently heterogeneous because varying numbers of EO units are coupled to each reactive hydrophobic species present. The homogeneous ^{14}C-labeled hexa- or nona-ethoxylates of p-(1,1,3,3-tetramethylbutyl) phenol (1), abbreviated tOPh·6EO or tOPh·9EO, and of 2,6,8-trimethyl-4-nonanol (2), abbreviated TMN·9EO, were synthesized as model compounds for studies of those formulating agents applied to plants with pesticides. The surfactants tOPh·6EO and tOPh·9EO were absorbed by leaves of barley plants and metabolized. Several of the methanol-extractable metabolites were identified or characterized (3).

In this study, known compounds were tested in TLC systems to relate chromatographic behavior to structural composition. A TLC procedure was developed to resolve polyethylene glycol (PEG) from various nonionic surfactants. Some of the surfactant-derived fractions were characterized or identified by mass spectrometry. These techniques were applied to some of the metabolites from barley tissues.

EXPERIMENTAL

Standard Materials

The ^{14}C-*t*OPh·6EO, ^{14}C-*t*OPh·9EO, and ^{14}C-TMN·9EO were prepared by Tanaka *et* al. (1, 2). Mixed *t*OPh-ethoxylates with (average) EO contents of 1.5 (Triton X-15), 3, 5, and 9.5 (Triton X-100) mol were purchased from various sources. They were designated *t*OPh-EO$\overline{1.5}$, *t*OPh-EO$\overline{3}$, *t*OPh-EO$\overline{5}$, and *t*OPh-EO$\overline{9.5}$, respectively. Mixed TMN-ethoxylates (Tergitol TMN-types) were provided by Union Carbide.* They were designated as either TMN-EO$\overline{6}$ or TMN-EO$\overline{10}$. The *o* and *p* isomers of the lower *t*OPh·nEO adducts were isolated chromatographically from *t*OPh·EO$\overline{1.5}$ (4). The 2-(4-*tert*-butylphenoxy)ethanol was purchased from New Haven Chemical. 2-Phenoxyethanol, alkylphenols and phenyl-substituted alkanols were purchased from Aldrich Chemical. Polyethylene glycols of various average molecular weights (PEG-300, PEG-600) were obtained from various manufacturers. Samples of Surfynol-440, -465, and -485 were provided by Air Products and Chemicals. These mixed ethoxylates of 2,4,7,9-tetramethyl-5-decyn-4,7-diol were designated as Alkyn-EO$\overline{3.5}$, Alkyn-EO$\overline{10}$, and Alkyn-EO$\overline{30}$, respectively. Special samples from the Alfol surfactant series were provided by Continental Oil. These mixed ethoxylates of purified normal alcohols (decanol through octadecanol) were designated nC_{10}-EO$\overline{7}$, nC_{12}-EO$\overline{6}$, nC_{12}-EO$\overline{8}$, nC_{12}-EO$\overline{17}$, nC_{14}-EO$\overline{12}$, nC_{16}-EO$\overline{8}$, and nC_{18}-EO$\overline{6}$, respectively, to indicate the length of each hydrophobe and the average EO content.

Surfactant Metabolism Experiments

Excised leaves of barley seedlings were treated with about 90-ppm solutions of ^{14}C-*t*OPh·6EO or ^{14}C-*t*OPh·9EO for 5 hr and were then placed into fresh water for 18 to 22 hr (3). The methanolic leaf extracts were concentrated and partitioned as shown in Figure 1. The polar, water-soluble metabolites were adsorbed on SepPak-C18 cartridges (Waters Associates) and were eluted with methanol. They were subjected to acid hydrolysis and analyzed (3). Portions of the polar metabolites were dissolved in saline and extracted with ethyl acetate and chloroform to remove components exhibiting

---

*Mention of a trademark or proprietary product does not constitute a guarantee or warranty of the product by the U.S. Department of Agriculture and does not imply its approval to the exclusion of other products that may also be suitable.

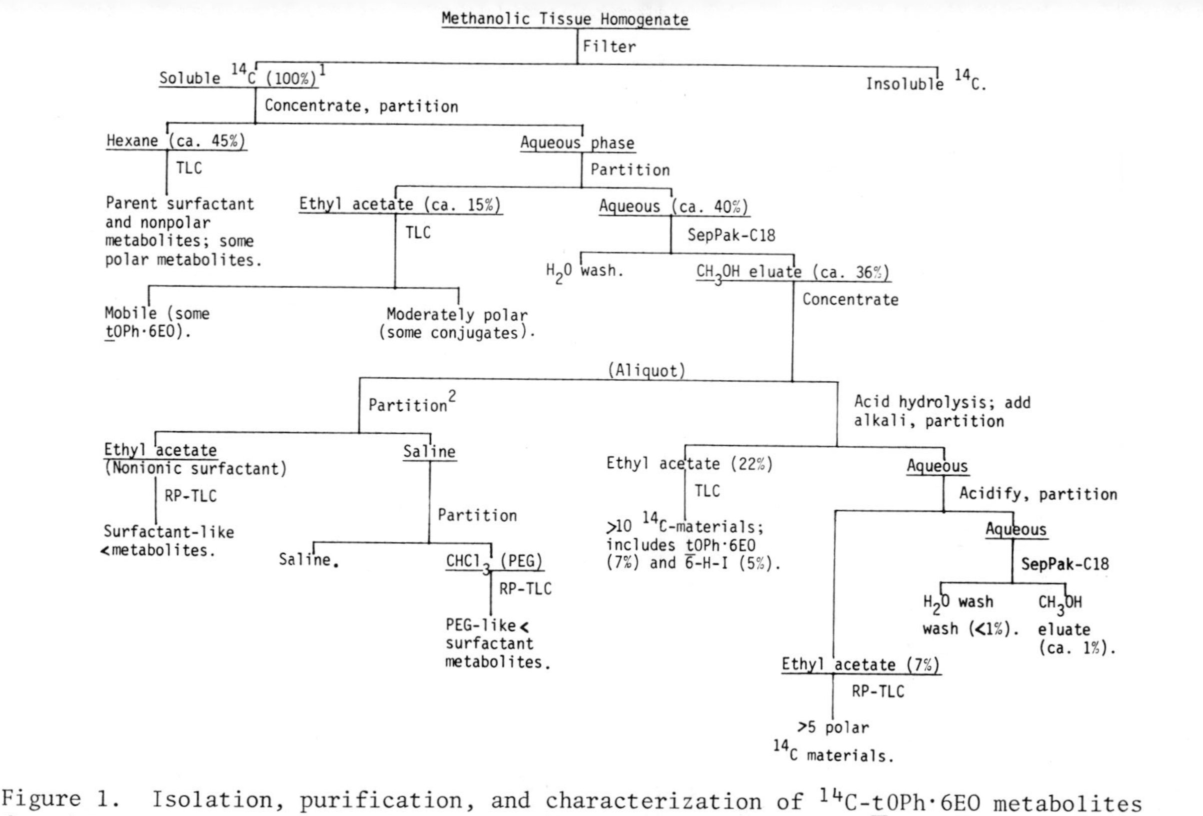

Figure 1. Isolation, purification, and characterization of ^{14}C-tOPh·6EO metabolites from barley tissues (4 lots, total 130 g). Excised leaves were treated in 1 ml/g fresh weight surfactant solutions at 90 μg/ml. The water-soluble polar metabolites were pooled and hydrolyzed. (1) Typically < 5% of the ^{14}C in the excised leaves was methanol-insoluble. (2) Weibull procedure (5) carried out on metabolites from tissues treated with ^{14}C-tOPh·9EO.

nonionic surfactant and PEG properties, according to Weibull (5).

Chromatography

Fluorescent plates with 0.25-mm layers for normal-phase TLC were prepared with silica gel HF (E. Merck); solvent systems used were water-saturated 2-butanone (aqMEK) (6) or toluene/hexane/acetone, 3:6:2 by volume (THA) (4). The availability, limitations, and uses of commercial reverse-phase TLC plates have been described (7). Fluorescent 20 × 20 cm plates had bonded C_{18} (and C_2) groups (Whatman KC18-F, product 4803-800; abbreviated C_{18}-RP-TLC), bonded dimethylsilyl groups (Merck silica gel 60F silanized, product 5747 or 5602; abbreviated C_2-RP-TLC), or were hydrocarbon-impregnated (Analtech RPS-F, product 52011 or 52511, prescored for 5 × 20 cm; abbreviated HC-RP-TLC). Various mixtures of CH_3OH in H_2O were used as the mobile phases. The ^{14}C-surfactants and their metabolites were detected by autoradiography. Radioactive zones were scraped from the TLC plates and quantified by liquid scintillation counting. Standard materials were detected under UV light, with I_2 vapor, or with Dragendorff reagent (6).

Mass Spectrometry

Zones scraped from TLC plates were eluted with methanol and subjected to electron-impact mass spectrometric (MS) analyses at 70 eV using a solid sample probe (3, 8).

RESULTS AND DISCUSSION

The R_f of each substituted phenoxyethanol on the HC-RP-TLC plates decreased as the size of the alkyl substituent increased (Table I). The addition of an EO group to the ethanol moiety caused a small effect on the R_f. These results agreed with reported LC separations of nonionic EO adducts on RP columns (9, 10, 11). On the normal-phase silica TLC plates, the size of the alkyl substituent had a small but significant effect on the R_f (Table I), while an additional EO group caused a large decrease in the R_f (4). Similar results were obtained with the homogeneous ^{14}C-labeled nonionic surfactants (Figure 2). Silica TLC plates consistently resolved <u>t</u>OPh·6EO (R_f 0.51) from <u>t</u>OPh·9EO (R_f 0.32). However, <u>t</u>OPh·9EO and TMN·9EO were resolved poorly, as was reported for other hydrophobes with similar EO contents (6). The results with the ^{14}C-surfactants on RP-TLC plates are given in Table II. The <u>t</u>OPh·6EO,

TABLE I. THIN LAYER CHROMATOGRAPHIC BEHAVIOR OF 2-PHENOXYETHANOLS

R′-C₆H₄—O—CH_2—CH_2—O—R″

R'	R''	R_f[a] Reverse Phase[b]	R_f[a] Normal Phase[c]
None	H	0.77	0.23
_p_-$C(CH_3)_3$	H	0.40	0.27
_p_-$C(CH_3)_2CH_2C(CH_3)_3$	H	0.12	0.34
_o_-$C(CH_3)_2CH_2C(CH_3)_3$	H	0.14	0.43
_o_,_p_-di[$C(CH_3)_2CH_2C(CH_3)_3$]	H	0.01	0.53
_p_-$C(CH_3)_2CH_2C(CH_3)_3$	CH_2CH_2OH[d]	0.14	0.22

[a]Detection under UV light or I_2 vapor.

[b]Hydrocarbon-impregnated silica (Analtech); 58% CH_3OH in H_2O.

[c]Silica gel HF (E. Merck); toluene/hexane/acetone (3:6:2).

[d]Abbreviated as _t_OPh·2EO.

_t_OPh·9EO, and TMN·9EO were difficult to resolve. These RP-TLC plates, however, separated nonionic surfactants as a class from various contaminants or from matrix components in environmental samples.

Aliquots of ^{14}C-_t_OPh·9EO were assayed directly by liquid scintillation counting or by application to silica HF or HC-RP-TLC plates to determine radioisotope recovery. The recovery was >90% when the zones were scraped and counted, and >85% when the scraped zones were extracted with CH_3OH to elute the ^{14}C for counting. Silica HF tended to disperse as a dust, and recoveries of ^{14}C from it were about 5% less than from the HC-RP plates.

Some by-products were separated from various amounts of _t_OPh-EO$\overline{9.5}$ on C_2- and C_{18}-RP-TLC plates (Figure 3). The surfactant with the octylphenyl-moiety and some contaminants could be detected under UV light, while I_2 vapor and Dragendorff reagent also visualized PEG standards (<20 μg) and the PEG (<3%) present with the

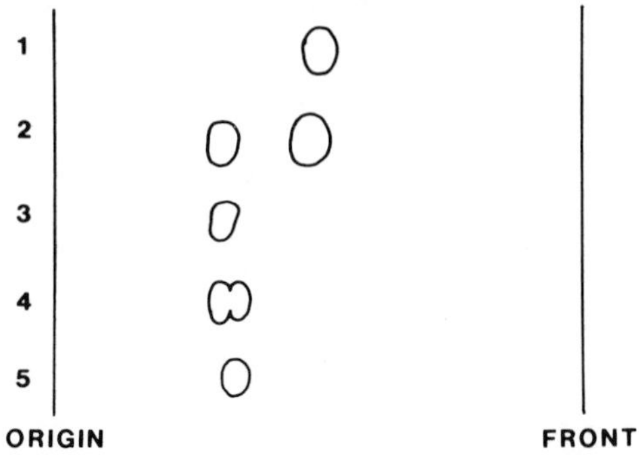

1
2
3
4
5

ORIGIN FRONT

Figure 2. Autoradiogram of a silica HF TLC plate (0.25 mm) with homogeneous ^{14}C-labeled nonionic surfactants. Solvent: water-saturated 2-butanone (aqMEK); samples: 1 = tOPh·6EO, 2 = 1 + 3, 3 = tOPh·9EO, 4 = 3 + 5, 5 = TMN·9EO.

TABLE II. THIN LAYER CHROMATOGRPHIC BEHAVIOR OF HOMOGENEOUS ETHOXYLATED NONIONIC SURFACTANTS IN REVERSE-PHASE SYSTEMS

Surfactant		R_f[a]	
	HC[b]	C_2[b]	C_{18}[c]
^{14}C-tOPh·6EO	0.57	0.48	0.36
^{14}C-tOPh·9EO	0.52	0.48	0.35
^{14}C-TMN·9EO	0.51	0.42	0.33

[a]Detection on X-ray film by autoradiography (≤1 μg surfactant).

[b]Solvent 75% CH_3OH for Analtech (HC) and Merck (C_2) plates.

[c]Solvent 90% CH_3OH in H_2O for the Whatman (C_{18}) plates.

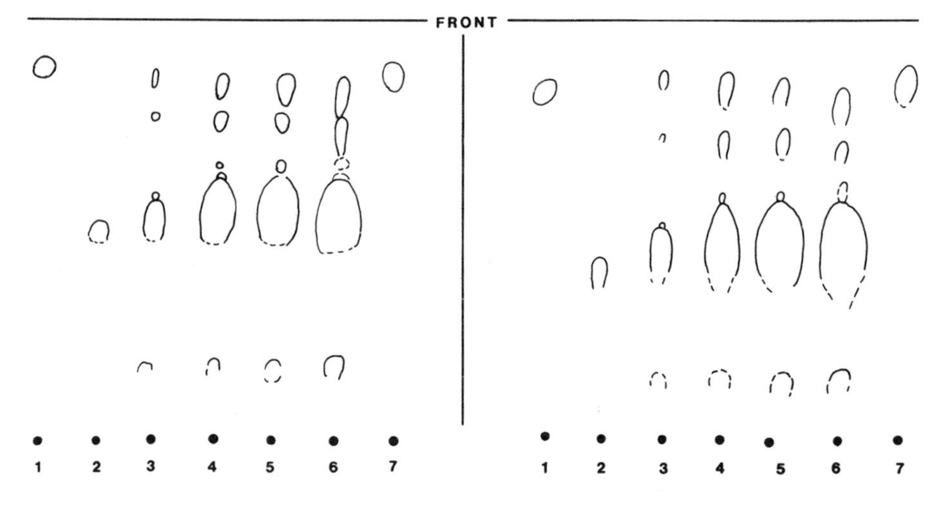

Figure 3. Reverse-phase chromatograms of polyethylene glycols and an ethoxylated alkylphenol surfactant. Left: C2 plate (Merck), 75% CH_3OH in H_2O solvent; right: C18 (Whatman), 90% CH_3OH solvent. Samples: 1--PEG-300 (20 μg), [2 to 6 are Triton X-100 (tOPh·EO9.5)] 2--50 μg, 3--200 μg, 4--650 μg, 5--1.0 mg, 6--1.3 mg, 7--PEG-600 (20 μg); visualized with Dragendorff reagent.

surfactant. Plates sprayed with Dragendorff reagent showed reddish spots with PEG and orange spots with the nonionic surfactant on a yellow-colored background. The colors were stable for many days. The HC-RP-TLC plates developed with 58% CH_3OH also resolved the components in PEG-300 or PEG-600 as a series of spots (R_f 0.5 to 0.7 or 0.3 to 0.6, respectively) from the major surfactant component in tOPh·EO9.5 (R_f ∿0.05). Resolution was lost, however, when about 1 mg of surfactant was spotted onto HC-RP-TLC plates to determine the quantity of PEG present in the sample of tOPh·EO9.5. The resolution attained by RP-TLC agreed qualitatively with reported separations of PEG from various ethoxylated nonionic surfactants using LC columns with bonded phases (9, 10).

A 2-mg sample of tOPh·EO9.5 was streaked onto a C_2-RP-TLC plate and chromatographed with 70% CH_3OH in H_2O (compare with Figure 3, left). The major zone (R_f ∿0.4) and two minor zones (R_f ∿0.1 and 0.6) were visualized under UV, scraped, and eluted with CH_3OH. Each extract was then chromatographed on silica HF with aqMEK to yield a series of EO adducts (UV visualization).

The more abundant components were scraped, eluted, and subjected to MS analyses. The components isolated from the major band on C_2-RP-TLC all gave spectra typical of the *t*OPh-moiety bearing 6 to 14 EO units with ions at m/e 45, 89, 133, and 177 from EO chain fragments (8) and at m/e 135, 161, and 205 from octylphenol ethoxylates (3, 8). The species with 10EO gave the characteristic molecular ion at m/e 646 and an intense tropylium ion at 575; the other EO adducts all gave homologous ions (3, 8).

The more mobile of the two minor zones from the C_2-RP-TLC plate was rechromatographed with aqMEK. The components were examined by MS, but it was not possible to identify them. The less mobile minor zone from *t*OPh·EO9.5 was also rechromatographed, and its components were examined by MS. Although numerous other ions were present in the spectra, those consistent with the di- through pentaethoxylates of a di-*tert*-octylphenol (m/e 335, 379, 423, and 467, respectively) were observed. A reference compound, the monoethoxylate of *o*,*p*-di-*tert*-octylphenol (Table I) (4), gave a weak molecular ion (m/e 362) and an intense tropylium ion at m/e 291 to support these assignments.

These results showed that C_2-RP-TLC was more specific for the size of the hydrophobic moiety than for the EO content of the *t*OPh-derived nonionic surfactants. Similar results were obtained by C_{18}-RP-TLC with 90% CH_3OH for a variety of nonionic ethoxylates (Table III). As in Figure 3, the R_f of the surfactant increased with the amount of sample applied (50 to 600 μg). When larger amounts of samples with relatively high average EO content were applied, PEG ($R_f > 0.75$) was consistently detected as a contaminant (6). Many of the surfactants tested had been prepared from purified hydrophobes, and only one major Dragendorff-positive surfactant component ($R_f < 0.60$) was found. When < 60 μg each of nC_{12}-EO6 and nC_{14}-EO12 were applied together, these EO adducts of dodecanol and tetradecanol were resolved (R_f 0.31 and 0.24, respectively). Standard surfactants can be used in these RP-TLC systems to characterize the hydrophobic moieties of other ethoxylated nonionic samples, and their average EO content can be estimated on silica with aqMEK. Fatty acid ester derivatives of PEG and other polyols can present special analytical problems (11).

Since PEG is a normal contaminant in most ethoxylated nonionic surfactants, the formation of only traces of PEG in plant tissue as a metabolite from *t*OPh-ethoxylates may not be significant. The RP-TLC method was coupled with a solvent-partitioning procedure (5) to separate ethoxylated nonionic surfactants from PEG. Samples of PEG-300 (200 μg) dissolved in about 50 ml of 5*M* NaCl were extracted at room temperature with 2 × 25 ml ethyl acetate and

TABLE III. ANALYSIS OF VARIOUS EO ADDUCTS ON C_{18}-RP-TLC PLATES WITH 90% CH_3OH/10% H_2O[a]

Sample	R_f
Polyethylene glycols	
PEG-300	0.83
PEG-600	0.82
Alkylphenols	
*t*OPh-EO$\overline{1.5}$	0.41
*t*OPh-EO$\overline{3}$	0.42
*t*OPh-EO$\overline{5}$	0.40
*t*OPh-EO$\overline{9.5}$	0.46
Branched alcohols	
TMN-EO$\overline{6}$	0.45
TMN-EO$\overline{10}$	0.46
Linear Alcohols	
nC_{10}-EO$\overline{7}$	0.45
nC_{12}-EO$\overline{6}$	0.35
nC_{12}-EO$\overline{8}$	0.32
nC_{12}-EO$\overline{17}$	0.34
nC_{14}-EO$\overline{12}$	0.27
nC_{16}-EO$\overline{8}$	0.22
nC_{18}-EO$\overline{6}$	0.17
Acetylenic diols	
Alkyn-EO$\overline{3.5}$	0.57
Alkyn-EO$\overline{10}$	0.56
Alkyn-EO$\overline{30}$	0.54

[a] Major component visualized with Dragendorff reagent.

with 2 × 25 ml $CHCl_3$ (5). The organic extracts were concentrated and chromatographed on HC-RP-TLC using 58% CH_3OH. The $CHCl_3$ extracts and standard samples of PEG-300 yielded Dragendorff-positive zones (R_f ∿0.6), while none was found for the ethyl acetate extracts.

Any oligomeric ^{14}C-PEG formed in the treated tissues by removal of the entire EO chain from ^{14}C-tOPh·6EO (labeled in the innermost EO unit) (1) would be in the aqueous phase after the partitioning steps described in Figure 1. Various conjugates of the surfactant and of its polar metabolites probably would remain in the aqueous phase that contained about 40% of the methanol-extractable ^{14}C. The ^{14}C in the aqueous phase was adsorbed extensively by a SepPak-C_{18} cartridge and was eluted with CH_3OH; most of the ^{14}C materials recovered in CH_3OH were less mobile in normal-phase TLC systems that the tOPh·9EO or tOPh·6EO surfactants (Figure 2). These materials had R_f 0.0 to 0.3 (in aqMEK) and could include PEG-300 (R_f ~0.08) or PEG-600 (R_f ~0.04).

Extracts from tissues treated with ^{14}C-tOPh·9EO were processed as above (Figure 1) and were analyzed occasionally for PEG in more detail after the SepPak-C_{18} separation. An aliquot was partitioned against saline and more than half of the ^{14}C was extracted into ethyl acetate, implying that these metabolites had surfactantlike properties (5). Only about one fourth of the ^{14}C in the saline partitioned into $CHCl_3$, a PEG characteristic (5). The $CHCl_3$ extract, mixed with 200 μg PEG-300, was concentrated and subjected to HC-RP-TLC with 58% CH_3OH; less than one third of the ^{14}C was sufficiently mobile to cochromatograph with the Dragendorff-positive PEG standard. Thus, only about 1% of the ^{14}C-tOPh·9EO given to excised barley leaves was extracted as materials having properties consistent with free oligomeric PEG containing the innermost EO unit of the surfactant.

The ethyl acetate from the partitioning with saline was also subjected to HC-RP-TLC. About one half of the ^{14}C was located (R_f 0.15 to 0.40) between the zones for tOPh-ethoxylates and PEG while one third was near the origin (surfactantlike). Although unmetabolized tOPh·6EO or tOPh·9EO was present in the original methanolic extracts from the barley tissues (3), the parent surfactants were removed in the hexane versus aqueous partitioning (Figure 1). Any traces of surfactant remaining, plus some moderately polar metabolites (silica, aqMEK), were then extracted from the aqueous phase with ethyl acetate. These facts, plus the high polarity (silica, aqMEK) of the water-soluble metabolites recovered from SepPak-C_{18} in CH_3OH, exclude the presence of free tOPh-ethoxylates in this fraction. Thus, the metabolites of ^{14}C-tOPh·9EO extracted by the Weibull partitioning into ethyl acetate or into $CHCl_3$ were assumed to have relatively large hydrophobic moieties because of their low R_f in HC-RP-TLC. This suggested that a complex mixture of metabolites was present.

The polar, water-soluble metabolites from ^{14}C-*t*OPh·6EO were subjected to acid hydrolysis in an attempt to obtain less complex materials for analysis. The products were made alkaline and extracted with ethyl acetate (Figure 1). Because most of the ^{14}C partitioned into the organic phase, PEG was excluded as a major hydrolysis product (5). The extracted ^{14}C was considerably more mobile on normal-phase silica TLC (aqMEK) than the original. About one third of the ^{14}C in this organic phase was identified as the parent surfactant, *t*OPh·6EO, by TLC and MS (m/e 470, molecular ion m/e 399, tropylium ion). When excised barley leaves were treated with ^{14}C-*t*OPh·9EO, the parent surfactant was also the major product recovered after acid hydrolysis of the polar, water-soluble metabolites.

When the methanolic extracts from treated, excised barley leaves were partitioned (Figure 1), the hexane phase contained most of the unmetabolized ^{14}C-*t*OPh·6EO or ^{14}C-*t*OPh·9EO. This phase was highly pigmented and also contained metabolites less polar (silica, aqMEK, Figure 2) than the parent surfactant. Repeated TLC on silica (aqMEK, THA) separated the pigments and the parent surfactant from these metabolites. This fraction represented about 13% of the methanol-extractable ^{14}C, and its more abundant components were not affected by acid hydrolysis (3). Chromatography in several systems (3, 4, 6) with reference materials showed that the major nonpolar metabolites from either ^{14}C-surfactant were partially deethoxylated compounds, including *t*OPh·1EO through *t*OPh·5EO. The most abundant usually was *t*OPh·2EO (Table I, R_f 0.22, THA), comprising <5% of the methanol-extractable ^{14}C.

Similar analyses of the products extracted into ethyl acetate after acid hydrolysis of the polar, water-soluble metabolites failed to detect significant quantities of such deethoxylated materials (Figure 1). However, the ethyl acetate extract contained several other ^{14}C materials in addition to the parent surfactant. The next most abundant compound, designated 6-H-I, represented about 5% of the ^{14}C extracted from barley tissues treated with ^{14}C-*t*OPh·6EO. This compound was more polar than the parent surfactant on silica (aqMEK; R_f 0.45 and 0.51, respectively) and was considerably more polar on HC-RP-TLC with 58% CH_3OH (R_f ∿0.2 for 6-H-I and <0.1 for *t*OPh·6EO). The MS of 6-H-I had a molecular ion at m/e 486 and an intense m/e 399 (tropylium ion). This suggested that in 6-H-I an oxygen had been added in the neopentyl moiety of the surfactant.

Acid hydrolysis of the polar, water-soluble metabolites of ^{14}C-*t*OPh·6EO also released materials that were not extracted by the first partitioning with ethyl acetate (Figure 1). These represented about 10% of the ^{14}C extracted from treated barley tissues. They were very polar in HC-RP-TLC (58% CH_3OH; $R_f > 0.3$), and none was identified.

The *t*OPh-ethoxylates have several potential sites for oxygenation in their hydrophobic moiety, multiple potential EO losses, and other potential EO metabolites. The parent surfactant and any of its metabolites could also exist as a variety of conjugates in the methanolic extract from plant tissues. The conjugation of an organic compound with a carbohydrate or peptide moiety greatly increases its polarity on silica (12). Relatively large hydroxylated organic compounds and their *O*-glycosyl derivates can have similar mobilities in RP-HPLC, but the introduction of an additional hydroxyl function into the aglycone moiety can result in increased mobilities (12). Examples of this have already been observed from among the many metabolites of ^{14}C-*t*OPh·6EO (3).

Oxygenated analogs of phenols with large alkyl substituents were not available, but some with smaller alkyl groups, some phenyl-substituted alkanols, and some EO adducts were examined by TLC. The R_f values given in Table IV show that the length of the hydrocarbon chain and oxygenation in a hydrocarbon chain are detected readily by RP-TLC methods. Changes in the EO content and some oxygenations are detected more readily by TLC on silica. It is possible, however, that oxygenated analogs of some nonionic ethoxylates would be difficult to distinguish from PEG solely by TLC techniques. Other methods for surfactant analysis that are available include solvent partitionings (5), MS analyses (8), nuclear magnetic resonance spectra (4), or selective ^{14}C-labeling.

ACKNOWLEDGMENTS

The skilled technical assistance of Prudence A. Olson was invaluable in this work. The MS data were obtained by Richard G. Zaylskie, Eugene R. Mansager, Mark Carlson, and CaroleJean H. Lamoureux. Air Products and Chemicals, Inc., Continental Oil Company, and Union Carbide Corporation provided ethoxylated nonionic surfactants

REFERENCES

1. F. S. Tanaka and R. G. Wien, J. Labelled Cmpd. Radiopharm., 12, 97 (1976).
2. F. S. Tanaka, R. G. Wien, and G. E. Stolzenberg, J. Labelled Cmpd. Radiopharm., 12, 107 (1976).

TABLE IV. THIN LAYER CHROMATOGRAPHIC PROPERTIES OF PHENYL-SUBSTITUTED ALKANOLS, ALKYLPHENOLS, AND SOME EO ADDUCTS

Compound	Hydrocarbon Chain Length	R_f[a] Reverse Phase[b]	R_f[a] Normal Phase[c]
1-Phenyl-2-propanol	C_3	0.65	0.39
2-Phenyl-1-propanol	C_3	0.64	0.41
3-Phenyl-1-propanol	C_3	0.65	0.33
1-Phenyl-2-butanol	C_4	0.56	0.48
4-Phenyl-2-butanol	C_4	0.54	0.39
3-Phenyl-1-butanol	C_4	0.57	0.36
4-Phenyl-1-butanol	C_4	0.58	0.35
2-Methyl-4-phenyl-2-butanol	C_5	0.52	0.43
5-Phenyl-1-pentanol	C_5	0.43	0.37
4-Methyl-4-phenyl-2-pentanol	C_6	0.45	0.47
3-Methyl-2-phenyl-1-pentanol	C_6	0.39	0.49
4-(4-hydroxyphenyl)-2-methyl-2-butanol[d]	C_5	0.76	0.14
4-*tert*-Butylphenol	C_4	0.56	0.48
p-(1,1-Dimethylpropyl)phenol	C_5	0.42	0.48
2-Phenoxyethanol	--	0.74	0.26
2-(4-*tert*-Butylphenoxy)ethanol	C_4	0.40	0.31
Polyethylene glycol-300	--	0.64	0.01

[a]Detection with I_2 vapor (PEG-300) or under UV light.

[b]Hydrocarbon-impregnated silica, 0.25 mm (RPS-F, precoated; Analtech); 58% CH_3OH in H_2O by volume.

[c]Silica gel HF (E. Merck), 0.25 mm layer; toluene/hexane/acetone (3:6:2).

[d]Purified by TLC; MS: m/e 180 (molecular ion), 162, 147, 107 (base peak, tropylium ion).

3. G. E. Stolzenberg, P.A. Olson, R. G. Zaylskie, and E. R. Mansager, J. Agric. Food Chem., in press.
4. G. E. Stolzenberg, R. G. Zaylskie, and P. A. Olson, Anal. Chem., 43, 908 (1971).
5. B. Weibull, Proc. 3rd Intern. Congr. Surface Activity, 3, 121 (1960); International Standard, ISO 2268 (1972).
6. K. Bürger, Z. Anal. Chem., 196, 259 (1963).
7. U. A. Th. Brinkman and G. de Vries, J. Chromatogr., 192, 331 (1980).
8. E. Julia-Danes and A. M. Casanovas, Tenside Deterg., 16, 317 (1979).
9. L. P. Turner, D. McCullough, and A. Jackewitz, J. Am. Oil Chem. Soc., 53, 691 (1976).
10. H. Henke, Tenside Deterg., 15, 193 (1978).
11. N. Parris and J. K. Weil, J. Am. Oil Chem. Soc., 56, 775 (1979).
12. F. Erni and R. W. Frei, J. Chromatogr., 130, 169 (1977).

Index

A

B

C

D

E

N

O

P

Q

R